Creative Sequencing Techniques for Music Production

Creative Sequencing Techniques for Music Production

A practical guide for Logic, Digital Performer, Cubase and Pro Tools

Dr Andrea Pejrolo

AMSTERDAM • BOSTON • HEIDELBERG • LONDON • NEW YORK • OXFORD
PARIS • SAN DIEGO • SAN FRANCISCO • SINGAPORE • SYDNEY • TOKYO
Focal Press is an imprint of Elsevier

Focal Press
An imprint of Elsevier
Linacre House, Jordan Hill, Oxford OX2 8DP
30 Corporate Drive, Burlington MA 01803

First published 2005

British Library Cataloguing in Publication Data
A catalogue record for this book is available from the British Library

Library of Congress Cataloguing in Publication Data
A catalogue record for this book is available from the Library of Congress

ISBN 0 240 51960 4

Typeset by Charon Tec Pvt. Ltd, Chennai, India
www.charontec.com
Printed and bound in Great Britain

Contents

Acknowledgments

This book is the result of years of training and passionate sessions spent in front of a computer and a sequencer. I want to thank all the people that in one way or another have had an impact on my education, musical growth, and musical life over the years. First of all my dear wife, Irache, a never-ending resource for energy, assistance, and enthusiasm. I gratefully acknowledge the immense support and help of my family, relatives, and friends. In particular, I would like to thank my parents, Rosalba and Gianni, and my dear friend Nella. A special thanks goes to my brother Luca, who is responsible for exposing me, at a very early age, to MIDI and to the immense power of music, and to my brother Marco, who keeps reminding me every day through several exceptional collaborations how lucky I am to be able to share my music with other people.

A very special vote of appreciation goes to my dear friends and mentors from Manhattan School of Music, Richard De Rosa, Garry Dial, and Richard Sussman. Through their help, knowledge, and encouragement I have been able to grow not only musically but also personally.

I would like to thank the technical reviewer, Kurt J. Biederwolf from Berklee College of Music, for his extremely helpful contribution in fine-tuning some of the technical details of this book. A special thanks goes to Beth Howard, Emma Baxter, Margaret Denley and Georgia Kennedy from Elsevier U.K. for the interest they showed in the idea for this book and for their precious support.

And finally, a big thanks to my friends and colleagues: Terre Roche, Dion and Livia Driver, Jim Oakar, John Wineglass, Gal Ziv, Lisa Nardi, Victor Girgenti, Mike Richmond, Robert Rowe, Ron Sadoff, The New England Institute of Art in Boston, Berklee College of Music, the Institute of Audio Research in New York, The Audio Recording Technology Institute in Long Island, Francesco Avato, Jonathan Scott and Martin Kiszko in Bristol, U.K., and many others.

Introduction

This book covers the four main sequencers used in professional production environments—ProTools, Digital Performer, Cubase SX, and Logic Pro—and how to get the most out of them, by explaining and revealing advanced techniques such as groove quantizing, sounds layering, tap tempo, creative meter and tempo changes, advanced use of plug-ins automation, and synchronization of linear and nonlinear machines, just to mention a few. The subjects are approached from both the technical and the musical points of view in order to provide the modern composer/producer with tools and inside views on how to treat MIDI and DAWs as the orchestra and musicians of the 21st century.

The main thing that inspired me to write this book was the incredible potential concealed in the modern production tools and in the existing software applications available to contemporary composers and producers. Unfortunately, most of the time these tools are used in very mechanical and nonmusical ways, reducing and limiting not only the potential of the technology involved but also (and mainly) the potential of the composers that utilize these incredibly powerful tools for their productions. In this book I bridge the two "worlds," trying to bring the term *music technology* back to its original connotation: a way to produce music through the help of technology. I want to stress the "help" factor that technology offers to the music production process, since this is what technology is, a tool to help expand and improve the creation process on which the composer relies. Technology applied to music is not the goal of the production process but can be seen as the thread that guides and joins the inspirational process with the final product. In this book you will learn sequencing techniques that always relate to practical aspects of music production, and they are explained as much as possible in a simple yet thorough way. Thus I will refer often to the MIDI/audio computer workstation as "the orchestra of the 21st century," since modern composers find themselves more and more treating the MIDI and audio setup as the virtual musicians for whom they are writing. The MIDI standard plus a professional sequencing program represent the modern score paper, and they provide an extremely flexible medium both for sketching ideas and for full-scale productions. I don't see this approach as limiting in terms of flexibility and sonorities; in fact I believe the opposite. The use of new sounds, techniques, and tools can only expand and improve the palette of contemporary composers, orchestrators, and producers. The reason I decided to take on the four main sequencers at the same time is my belief that these days both professional and beginner musicians need to be able to master and program all of these sequencers in order to have an edge on the competition. It is not enough anymore to be familiar with only one or two applications. It is crucial to be comfortable with all of them, not only to expand one's opportunities, but also to be able to take advantage of specific features that are available only in certain applications. This approach will help you enhance considerably your palette and tools when it comes to sequencing and music production. Each technique explained in this book is first presented on a general level and then further developed with examples and practical applications for each sequencer.

This book was written with four main categories of readers in mind: the professional acoustic composer, the professional MIDI composer, the college educator along with his/her students, and the beginner. The professional acoustic composer who has been "afraid" to approach the digital MIDI and audio workstation or who has been using only basic sequencing techniques will be able to greatly improve her/his skills and will find a familiar environment, since references are made with a musical approach to sequencing in mind. Seasoned MIDI programmers and producers can take great advantage of the multiple environments on which this book is based. With the help of the provided examples and techniques for each of the four main sequencers used in the industry, they can quickly learn for all the other applications the same tools with which they are already familiar on a certain sequencer, giving them an advantage over their competition. College educators and students can use this manual not only for introductory to intermediate MIDI and sequencing classes but also for more advanced MIDI orchestration and production courses. The summary and exercise sections at the end of each chapter were specially designed for educational applications. Beginner readers will be amazed at the improvement in their sequencing skills from reading just a few chapters and using the included exercises to further improve their techniques.

In the first chapter I cover the needs and solutions for a problem-free project studio, in order to enhance the creative flow involved in a production session. In Chapters 2, 3, and 4 I guide readers through, respectively, basic, intermediate, and advanced sequencing techniques, targeted at improving the overall quality of their productions. These chapters will help readers to reach a professional level in terms of MIDI orchestration and programming using the leading and most advanced digital audio sequencers available on the market.

Chapter 5 is dedicated to MIDI orchestration. Here you will learn how to orchestrate for the MIDI ensemble and how to get the most out of your gear. This chapter covers not only acoustic instruments but also synthesizers and some of the most common synthesis techniques available at the moment. Chapter 6 focuses on the final mix and on the premastering process. Here you will learn mixing techniques that take advantage of the plug-in technology. How to maximize the use of effects such as reverb, compressor, limiter, and equalizer, among many others, is crucial to bring your productions to the next level.

At the end of each chapter you will find a comprehensive summary of the concepts and techniques explained in it and a series of exercises oriented to provide practical applications and to further develop the notions learned. These two sections are helpful for both the professional and the student. They provide the former with a quick reference for several techniques and ideas, while the latter can take advantage of their concise layout to become further familiar with the concepts just learned.

The book includes examples of arrangements and sequencing techniques on the bundled CD. They serve the purpose of better demonstrating how to avoid common mistakes as well as how to fix them. Here you can find loops, templates, and comprehensive audio examples to use as a starting point for your productions.

Learn the technology in every detail, but always let the creative flow guide your music. Now let's begin.

1 Studio Setup

1.1 Basic studio information

Here we go. You are now ready to take the next step in sequencing techniques in order to improve the quality of your productions. Remember that the final quality of your music depends on many variables, including your skills with and knowledge of sequencing techniques, the equipment you use, the software, and the environment (meaning essentially the studio) in which you work. In fact the studio is one of the most important elements involved in the creative process of composing your music. I am not talking just in terms of equipment and machines (which I will discuss in detail in a moment), but also in terms of comfort and ease of use of the working environment, qualities that are essential if you are going to spend many hours composing and sequencing your projects. Your studio should have good illumination, both natural and artificial. If you are going to use electric light as a main source for illumination, try to avoid lights with dimmer switches, since they are known for causing interference with studio recording equipment. Acoustic isolation and acoustic treatment of the room are also important elements that will help avoid external noises and create well-balanced mixes.

Even though the subject of acoustic isolation and treatment goes beyond the scope of this manual, here are some basic rules to follow when building your studio. First of all try to avoid (if possible) perfectly square or rectangular rooms. These are the most problematic because the parallel walls can create unwanted phasing effects and standing waves. You will soon realize that, unless you build an environment designed specifically to host a studio, most rooms are in fact rectangular. Therefore I recommend the use of absorption panels to reduce excessive reverberation caused by reflective and parallel surfaces, such as flat and smooth walls. Absorption panels (Figure 1.1) help reduce excessive reverberation, their main function being to stop the reflection of high frequencies. As a rule of thumb, try to avoid covering your entire studio with absorption panels since this would make your room a very acoustically dry listening environment, which not only would cause hearing fatigue but also would mislead your ears during your final mixes.

In order to reduce standing waves, you should use *diffusers* (Figure 1.2) on the walls and ceiling of the room. The main purpose of diffusers is to reflect the sound waves at angles that are different (mostly wider) than the original angle of incidence and thereby to limit the audio artifacts caused by parallel walls.

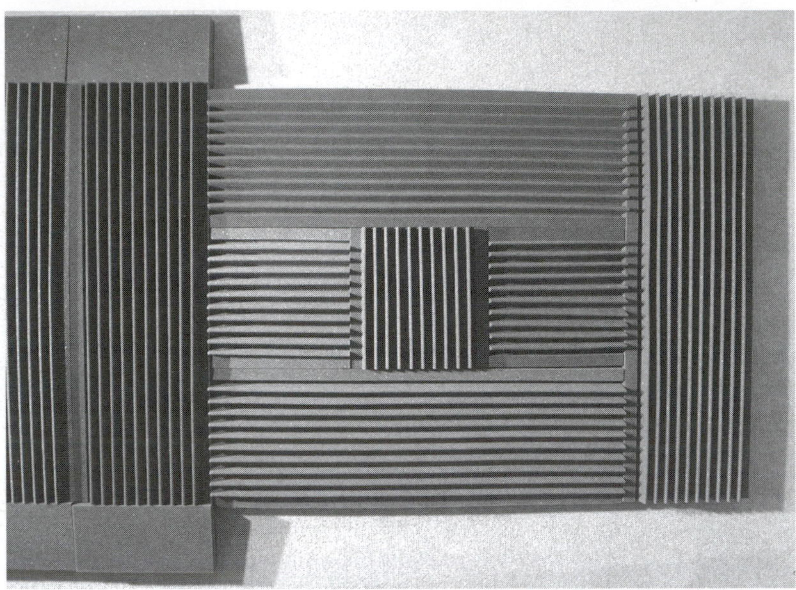

Figure 1.1 Example of absorption panels (Courtesy of Primacoustic).

Figure 1.2 Example of diffusers (Courtesy of Primacoustic).

The use of *bass-traps* will help reduce low-frequency standing waves. By placing them in the top corners of the room you will avoid annoying bass buildup frequencies. In Figure 1.3 you can see a fully acoustically treated studio.

For more detailed information on studio acoustics and studio design I highly recommend *Recording Studio Design*, by Philip Newell and published by Focal Press.

Figure 1.3 Example of an acoustically treated project studio (Courtesy of New England Institute of Art).

1.2 The project studio

A *project studio* originally meant a studio designed and built specifically for a particular project. More recently the term has shifted to indicate a studio that is slightly bigger and better equipped than a *home studio*. A project studio is built around a medium-size control room that serves as the main writing room and that can also be used to track electric instruments, such as electric guitars or basses, if necessary. A small or medium-size iso-booth is often included in order to allow the recording of vocalists, voice-overs, or solo instruments. The size and layout can change drastically, depending on the location and budget, but it is important to understand what the main elements are that are indispensable to create an efficient, powerful, and flexible studio for the modern composer.

The equipment around which your project studio is going to be built can be divided into three main general categories: music equipment, computer equipment, and software. All these elements are indispensable for composers to achieve the best results for their productions. In the modern project studio the music equipment can be further divided into three subgroups: electronic instruments, acoustic instruments, and sound/audio equipment. Remember that every element plays a very important and essential role in the music-making process. Let's take a closer look at each category in order help you make the right decision when building your composing environment.

1.3 The music equipment

The music equipment constitutes the "muscles" of your studio. This category includes everything related to the actual writing/sequencing, playing, and mixing of your production. While the acoustic instruments' setup can vary from studio to studio and from composer to composer, the electronic instrumentation and the audio equipment need to be accurately coordinated and integrated in order to reach the most efficient and trouble-free production environment.

The contemporary project studio is based on two different signal paths: MIDI (musical instrument digital interface) and audio. These two paths interact with one another during the entire production, integrating and complementing each other. The electronic instruments are connected to each other through the MIDI network, while the audio components of your studio are connected through the audio network. At the most basic level the electronic instruments eventually send audio signals to the audio part of your studio. Take a look to Figure 1.4. The MIDI network includes all devices that are connected through MIDI cables. The audio network includes devices connected using audio cables (either balanced or unbalanced connections, depending on the type of device).

1.3.1 The MIDI equipment and MIDI messages

Let's take a look at the MIDI electronic instruments that constitute the "musicians" of your contemporary MIDI orchestra. MIDI was established in 1983 as a protocol to allow different devices to exchange data. In particular, the major manufacturers of electronic musical instruments were interested in adopting a standard that would allow keyboards and synthesizers from different companies to interact with each other. The answer was the MIDI standard. With the MIDI protocol, the general concept of *interfacing* (meaning to establish a connection between two or more components of a system) is applied to electronic musical instruments. As long as two components (synthesizers, sound modules, computers, etc.) have a MIDI interface, they are able to exchange data. In early synthesizers the "data" were mainly notes played on keyboards that could be sent to another synthesizer. This allowed the keyboard players to layer two sounds without having to play simultaneously the same part with both hands on two different synthesizers. Nowadays the specifications of MIDI data have been extended considerably, ranging from notes, to control changes (CCs), from System Exclusive messages to synchronization messages (i.e., MTC, MIDI Clock, etc.).

The MIDI standard is based on 16 independent channels on which MIDI data are sent to and received from the devices. On each channel a device can transmit messages that are independent from the other channels. When sending MIDI data the transmitting device "stamps" on each message the channel on which the information is sent so that the receiving device will assign it to the right receiving channel. One of the aspects of MIDI that is important to understand and remember is that MIDI messages do not contain any information about audio. MIDI and audio signals are always kept separate. Think of MIDI messages as the notes that a composer would write on paper; when you record a melody as MIDI data, for example, you "write" the notes in a *sequencer* but you don't actually record their sound. While the sequencer records the notes, it is up to the synthesizers and

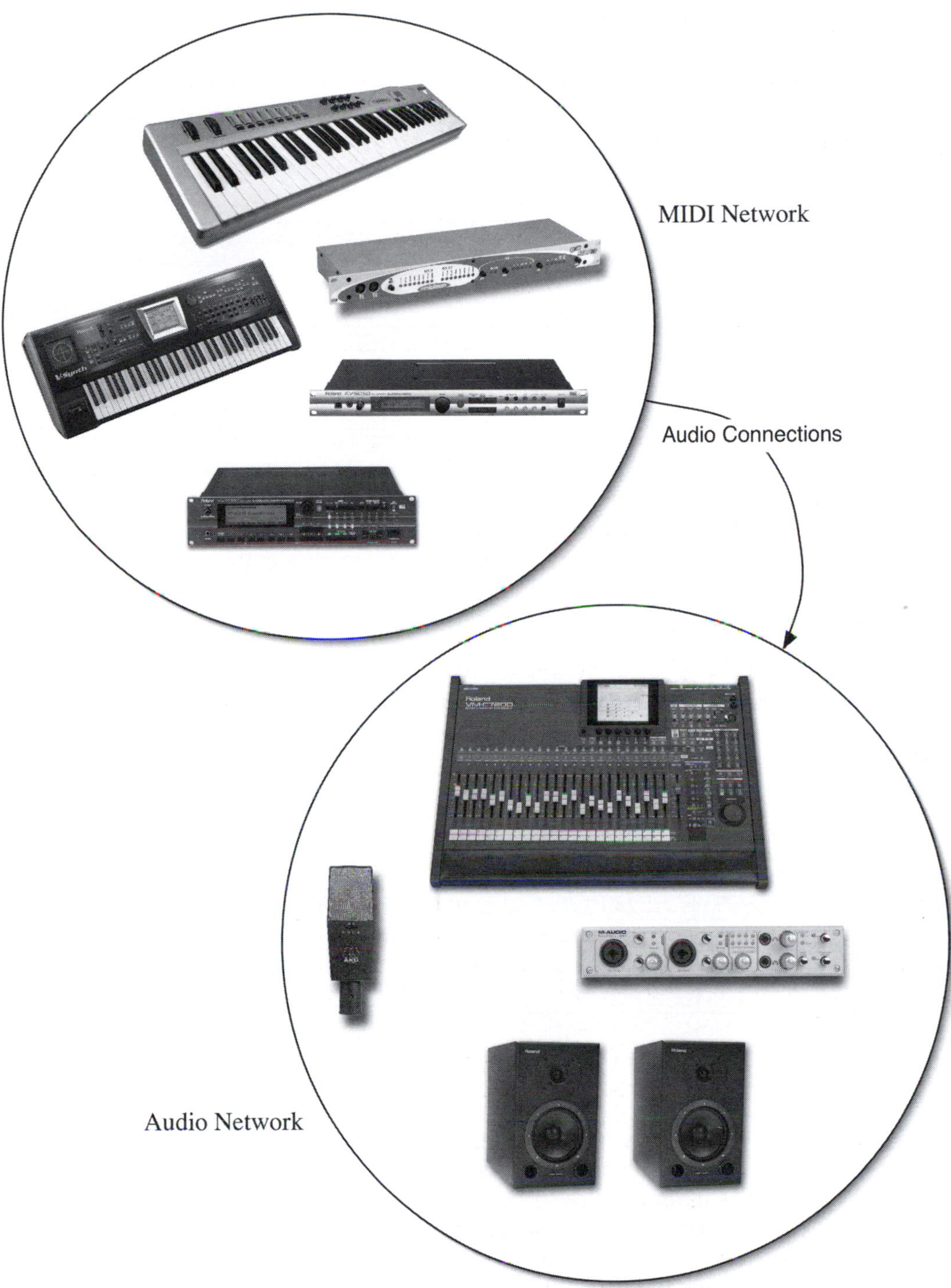

Figure 1.4 MIDI and audio networks (Courtesy of Roland Corporation U.S., AKG, and M-Audio).

sound modules connected to the MIDI system to play back the notes received through their MIDI interfaces. This is the main feature that makes MIDI such an amazing and versatile tool for music production. If you are dealing only with notes and events instead of sound files, the editing power available is much greater, meaning that you are much freer to experiment with your music. Here is a quick example to illustrate this concept. In Figure 1.5 we see a simple melody that was sequenced using a MIDI keyboard controller.

Figure 1.5 MIDI data shown in the notation window in Digital Performer.

The sequencer (in this example, Digital Performer), after recording the notes played on the MIDI keyboard, shows the melody as notation (there are many other ways of looking at MIDI data; we will learn other editing techniques later). Since Digital Performer (DP) is dealing with performance data only, you are free to change any aspect of the music—for example, the pitch of the notes, their position in time, and the tempo of the piece. In Figure 1.6 I have changed the first two notes of the melody, the upbeat of beat 3 in bar 3, and the tempo (from 100 to 120 beats per minute, or BPM).

Figure 1.6 Edited MIDI data.

Every device that needs to be connected to a MIDI studio or system needs to have a MIDI interface. The MIDI standard uses three ports to control the data flow: IN, OUT, and THRU. The connectors for the three ports are all the same: a 5-pin-DIN female port on the device (Figure 1.7) and a corresponding male connector on the cable.

While the OUT port sends out MIDI data generated from a device, the IN port receives the data. The THRU port is used to send out an exact copy of the messages received from the IN port, and it is mainly utilized in a particular setup called *Daisy Chain,* which I will describe in a moment. Of course, as I mentioned earlier, a device in order to be connected to the MIDI network needs to be equipped with a MIDI interface. Nowadays

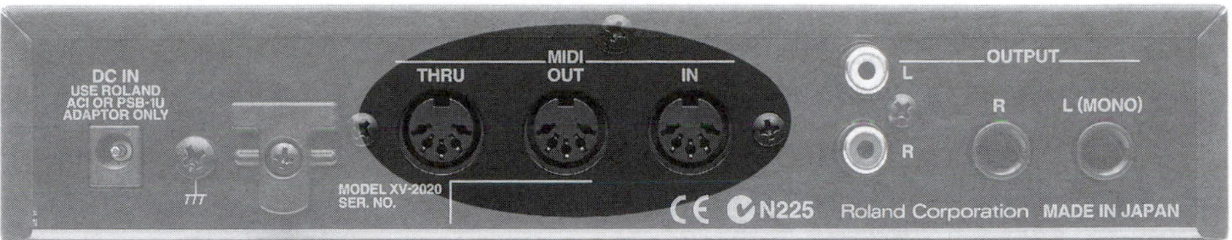

Figure 1.7 MIDI ports (Courtesy of Roland Corporation U.S.).

pretty much all the professional electronic music instruments, such as synthesizers, sound modules, and hardware sequencers, have a built-in MIDI interface. The only exception is the computer, which is usually not equipped with a built-in MIDI interface and therefore needs to be expanded with an internal or external one. The MIDI data can eventually be recorded by a device called a *sequencer*. Such a device records, stores, edits, and plays back MIDI data. In simple words, a sequencer acts as a digital tape recorder for MIDI data; we can record the data on a MIDI track, edit them as we want, and then play them back.

Because the number of messages that constitute the MIDI standard is very high, it is practical to separate them into two main categories: Channel messages and System messages. Channel messages are further subdivided into Channel Voice and Channel Mode messages, while System messages are subdivided into Real-time, Common, and Exclusive. Table 1.1 illustrates how they are organized.

Table 1.1 List of MIDI messages organized by category

Channel messages	System messages
Channel Voice: Note On, Note Off, Monophonic Aftertouch, Polyphonic Aftertouch, Control Changes, Pitch Bend, Program Change *Channel Mode*: All Notes Off, Local On/Off, Poly/Mono, Omni On, Omni Off, All Sound Off, Reset All Controller	*System Real-time*: Timing Clock, Start, Stop, Continue, Active Sensing, System Reset *System Common*: MTC, Song Position Pointer, Song Select, Tune Request, End of Sys. Ex. *System Exclusive*

1.3.2 Channel Voice messages

Channel Voice messages are probably the most used and, in fact, are the most important in terms of sequencing, because they carry the information about the performance, meaning, for example, which notes we played and how hard we pressed the trigger on the controller.

Note On message: This message is sent every time you press a key on a MIDI controller; as soon as you press it, a MIDI message (in the form of binary code) is sent to the MIDI OUT of the transmitting device. The Note On message includes information about the note you pressed (the note number ranges from 0 to 127, or C-1 to G-9), the MIDI channel on which the note was sent (1 to 16), and the velocity-on parameter, which describes how hard you press the key (this ranges from 0 to 127).

Note Off message: This message is sent when you release the key of the controller. Its function is to terminate the note that was triggered with a Note On message. The same result can be achieved by sending a Note On message with its velocity set to 0, a technique that can help to reduce the stream of MIDI data. It contains the velocity-off parameter, which registers how hard you released the key (note that this particular information is not used by most of the MIDI devices at the moment).

Aftertouch: This is a specific MIDI message sent after the Note message. When you press a key of a controller a Note On message is generated and sent to the MIDI OUT port; this is the message that triggers the sound on the receiving device. If you push a little bit harder on the key, after hitting it, an extra message called *Aftertouch* is sent to the MIDI OUT of the controller. The Aftertouch message is usually assigned to control the vibrato effect of a sound. But, depending on the patch that is receiving it, it can also affect other parameters, such as volume, pan and more.

There are two types of Aftertouch: polyphonic and monophonic. Monophonic Aftertouch affects the entire range of the keyboard no matter which key or keys triggered it. This is the most common type of Aftertouch, and it is implemented on most (but not all) controllers and MIDI synthesizers available on the market. The polyphonic Aftertouch, on the other hand, allows you to send an independent message for each key. It is more flexible since only the intended notes will be affected by the effect.

Control Changes (CCs): These messages allow you to control certain parameters of a MIDI channel. There are 128 CCs (0–127); the range of each controller goes from 0 to 127. Some of these controllers are standard and are recognized by all MIDI devices. Among these the most important (mainly because they are used more often in sequencing) are CCs 1, 7, 10, and 64. CC 1 is assigned to Modulation. It is activated by moving the Modulation wheel on a keyboard controller (Figure 1.8). It is usually associated with a slow vibrato effect.

CC 7 controls the volume of a MIDI channel from 0 to 127. CC 10 controls its pan. Value 0 is pan hard left, 127 is hard right, and 64 is centered. Controller number 64 is assigned to the Sustain pedal (the notes played are held until the pedal is released). This controller has only two positions: on (value 127) and off (value 0). Among the 128 controllers available in the MIDI specifications the four mentioned earlier are the ones used the most. Nevertheless there are others that, even though not as common, are recognized by a few MIDI devices, such as CC 2 (Breath), CC 5 (Portamento Time), and CC 11 (Expression).

Pitch Bend: This message is controlled by the Pitch Bend wheel on a keyboard controller (Figure 1.8). It allows you to raise or lower the pitch of the notes being played. It is one of the few pieces of MIDI data that do not have a range of 128 steps. In fact, in order to allow a more detailed and accurate tracking of the transposition, the range of this MIDI message goes from 0 to 16,383. Usually a sequencer would display 0 as the center position (non-transposed), +8191 fully raised, and −8192 fully lowered (Figure 1.9).

Program Change: This message is used to change the patch assigned to a certain MIDI channel. Each synthesizer has a series of *programs* (also called *patches, presets, instruments,*

Figure 1.8 Modulation and Pitch Bend wheels (Courtesy of M-Audio).

or, more generically, *sounds*) stored in its internal memory. For each MIDI channel we need to assign a patch that will play back all the MIDI data sent to that particular channel. This operation can be done by manually changing the patch from the front panel of the synthesizer or by sending a Program Change message from a controller or a sequencer. The range of this message is 0–127. Since modern synthesizers can store much more than 128 sounds, nowadays programs are organized in *banks*, where each bank stores a maximum of 128 patches (Figure 1.10). In order to change a patch through MIDI messages, a

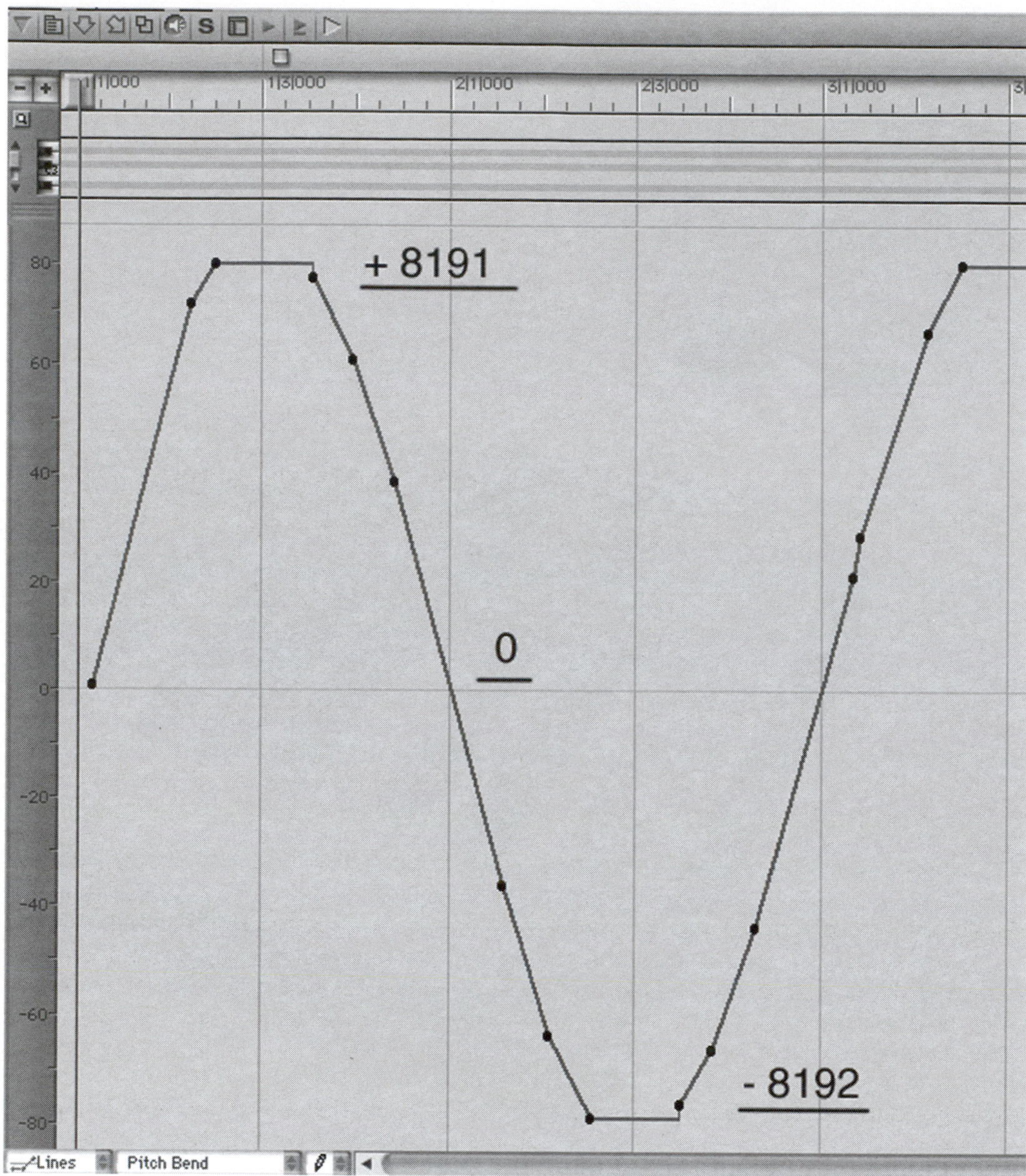

Figure 1.9 Pitch Bend range.

combination of a Bank Change message and a Program Change message is therefore necessary. While the latter is part of the MIDI standard specification, the former depends on the brand and model of MIDI device. Most devices use CC 0 or 32 to change the bank (or sometimes a combination of both), but you should refer to each synthesizer's manual to find out which MIDI message is assigned to Bank Change for a particular model and brand.

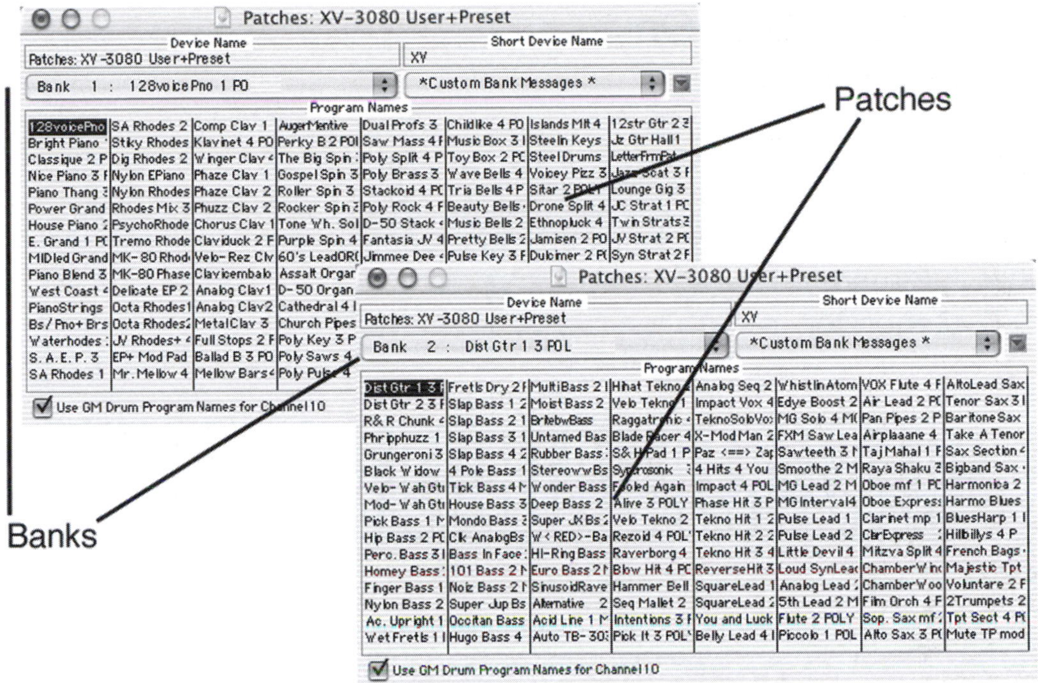

Figure 1.10 Organization of banks and patches.

1.3.3 Channel Mode messages

This category includes messages that affect mainly the MIDI setup of a receiving device.

All Notes Off: This message turns off all the notes that are sounding on a MIDI device. Sometimes it is also called the "panic" function, since it is a remedy against "stuck notes," meaning MIDI notes that were turned on by a Note On message but that for some reason (data dropout, transmission error, etc.) were never turned off by a Note Off message. It can also be activated through CC 123.

Local On/Off: This message is targeted to MIDI synthesizers. These are devices that feature a keyboard, a MIDI interface, and an internal sound generator. The "local" is the internal connection between the keyboard and the sound generator. If the local parameter is On, then the sound generator receives the triggered notes directly from the keyboard and also from the IN port of the MIDI interface (Figure 1.11). This setting is not recommended in a sequencing/studio situation since the sound generator would play the same notes twice, reducing its polyphony (the number of notes that the sound generator can play simultaneously) by half. It is, however, the right setup for a live situation in which the MIDI ports are not used. If the local parameter is switched Off (Figure 1.12), then the sound generator receives the triggered notes only from the MIDI IN port, which makes this setting ideal for the MIDI studio. The local setting usually can also be accessed from the "MIDI" or "General" menu of the device or can be triggered by CC 122 (0–63 is Off, 64–127 is On).

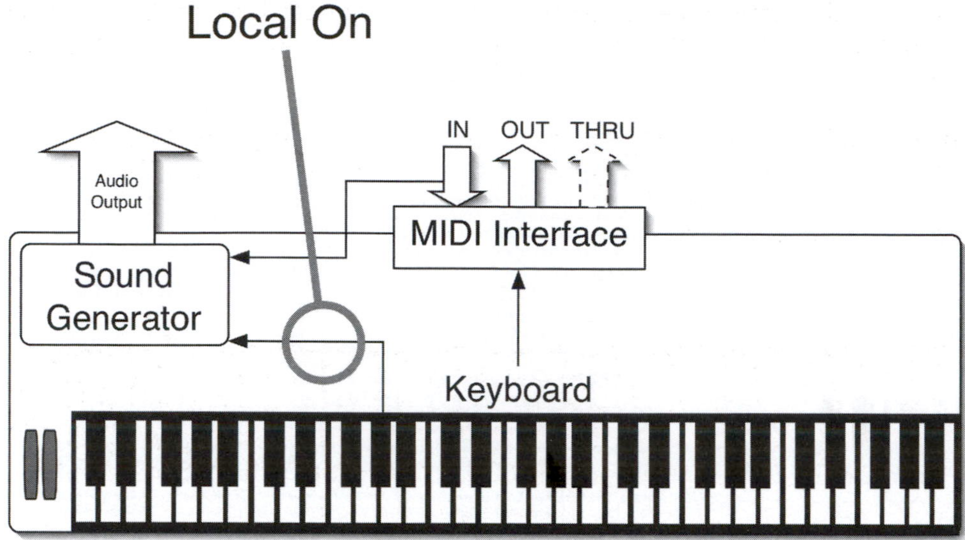

Figure 1.11 Local On.

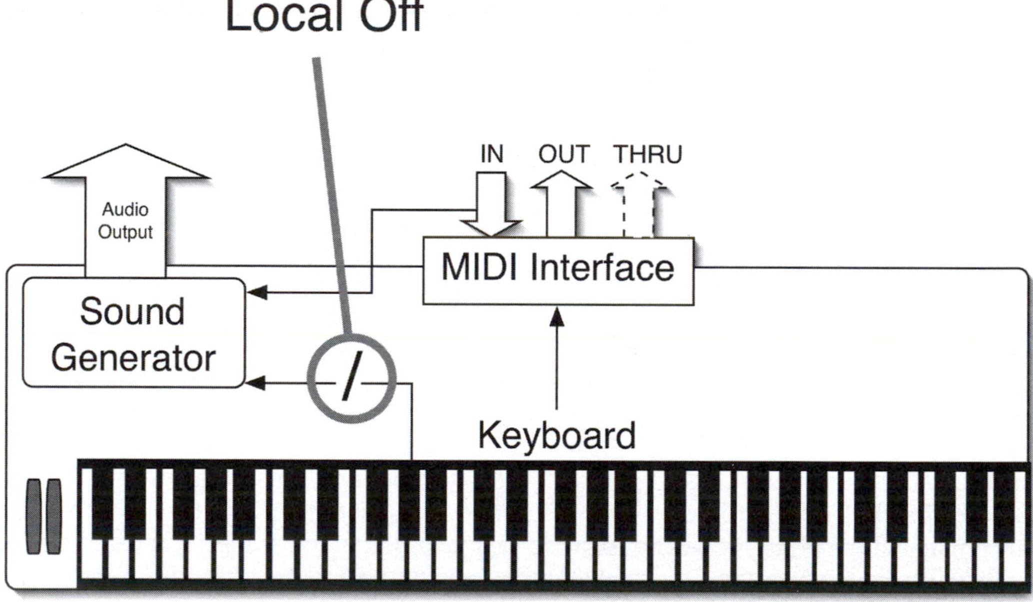

Figure 1.12 Local Off.

Poly/Mono: A MIDI device can be set as polyphonic or monophonic. If set up as Poly, the device will respond as polyphonic, meaning it will be able to play more than one note at the same time. If set up as Mono, the device will respond as monophonic, meaning it will be able to play only one note at a time per MIDI channel (the number of channels can be specified by the user). In the majority of situations we will want a polyphonic device, in order to take advantage of the full potential of the synthesizer. The Poly/Mono parameter is usually found in the "MIDI" or "General" menu of the device, but it can also be selected, respectively, through CC 126 and CC 127.

Omni On/Off: This parameter controls how a MIDI device responds to incoming MIDI messages. If a device is set to Omni On, then it will receive on all 16 MIDI channels but redirect all the incoming MIDI messages to only one MIDI channel (the current one) (Figure 1.13). Omni On can also be selected through CC 125.

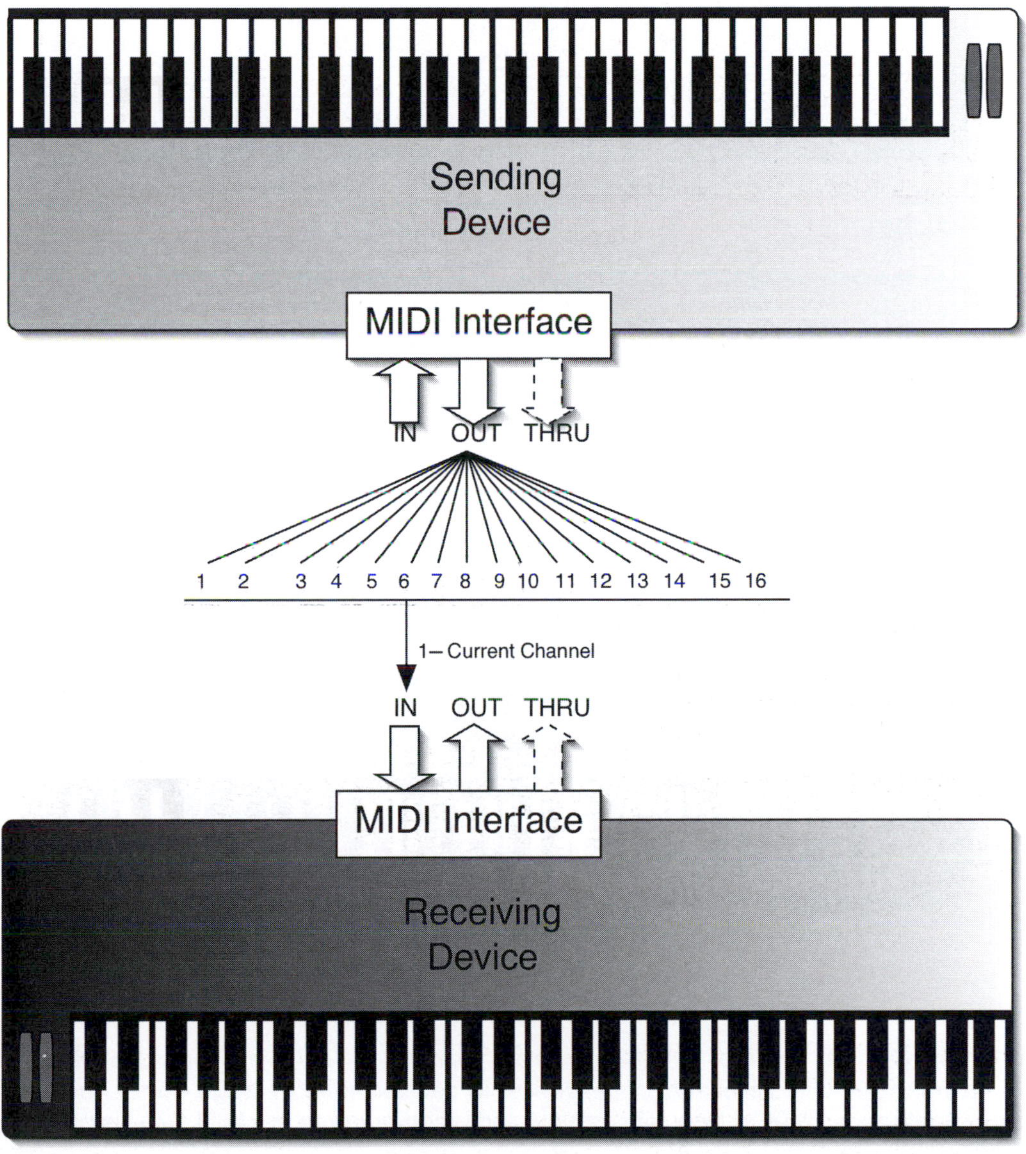

Figure 1.13 Omni On.

If a device is set to Omni Off, then it will receive on all 16 MIDI channels, with each message received on the original MIDI channel on which it was sent (Figure 1.14). This setup is the one most used in sequencing, since it allows you to take full advantage of the 16 MIDI channels on which a device can receive. Omni Off can also be selected through CC 124.

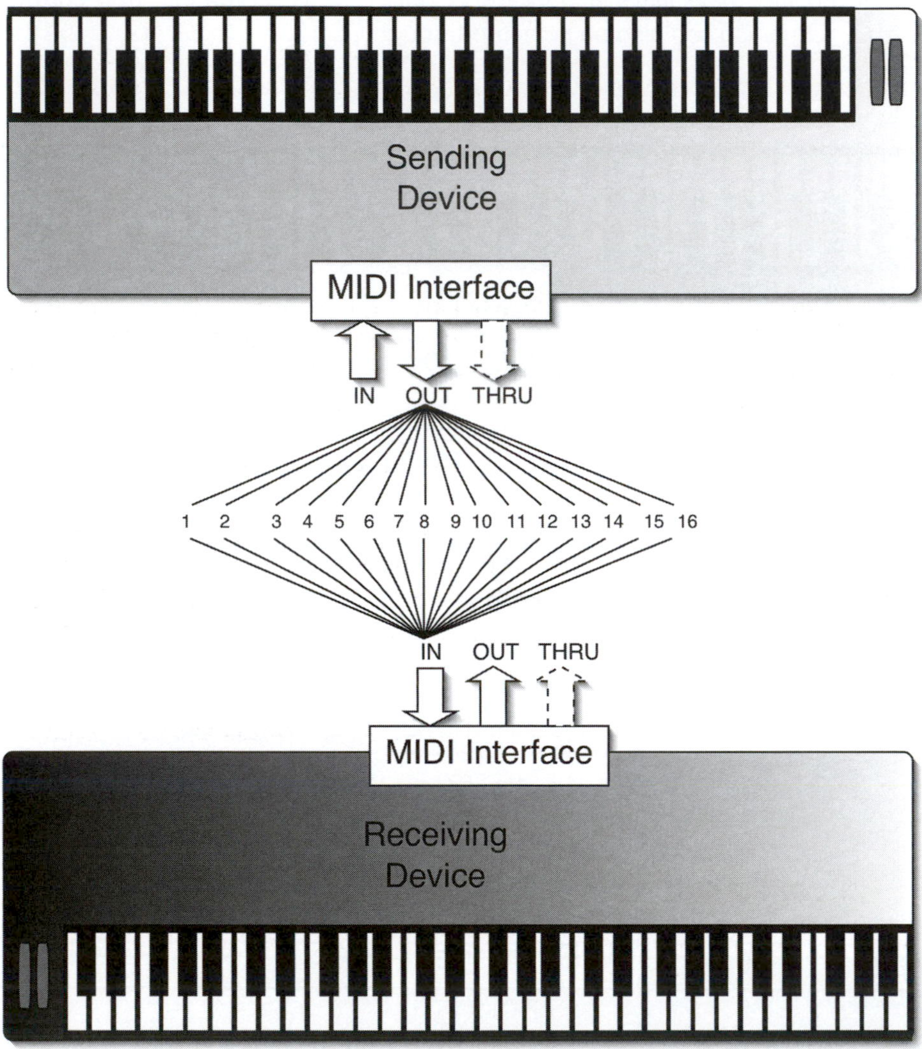

Figure 1.14 Omni Off.

All Sound Off: This is similar to the all notes off message, but it doesn't apply to notes being played from the local keyboard of the device. In addition this message mutes the notes immediately, regardless of their release time and whether or not the hold pedal is pressed.

Reset All Controllers: This message resets all controllers to their default state.

1.3.4 System Real-time messages

The Real-time messages (like all the other system messages) are not sent to a specific channel like the Channel Voice and Channel Mode messages. Instead they are sent globally to the MIDI devices in your studio. These messages are mainly used to synchronize all the MIDI devices in your studio that are clock based, such as sequencers and drum machines.

Timing Clock: This is a message specifically designed to synchronize two or more MIDI devices that need to be locked in to the same tempo. The devices involved in the synchronization process need to be set up in a master–slave configuration, where the master device (sometimes labeled *Internal Clock*) sends out the clock to the slaved devices (*External Clock*). It is sent 24 times per quarter note, and therefore its frequency changes with the tempo of the song (tempo based). It is also referred to as the *MIDI Clock* or sometimes as the *MIDI Beat Clock*.

Start, Continue, Stop: These messages allow the master device to control the status of the slave devices. Start instructs the slaved devices to go to the beginning of the song and start playing at the tempo established by the incoming Timing Clock. Continue is similar to Start, the only difference being that the song will start playing from the current position instead of from the beginning of the song. The Stop message instructs the slaved devices to stop and wait for either a Start or a Continue message to restart.

Active Sensing: This is a utility message that is implemented only on some devices. It is sent every 300 ms or less and is used by the receiving device to detect if the sending device is still connected. If the connection is interrupted for some reason (e.g., the MIDI cable was disconnected), then the receiving device will turn off all its notes to avoid having stuck notes that keep playing.

System Reset: This restores the receiving devices to their original power-up conditions. It is not commonly used.

1.3.5 System Common messages

The System Common messages are not directed to a specific channel and they are common to all receiving devices.

MIDI Time Code (MTC): This is another syncing protocol that is time based (as opposed to MIDI Clock, which is tempo based), and is mainly used to synchronize nonlinear devices (such as sequencers) to linear devices (such as tape-based machines). It is a digital translation of the more traditional SMPTE code used to synchronize nonlinear machines. The format is the same as SMPTE. The position in the song is described in *hours:minutes:seconds:frames* (subdivisions of 1 second). The frame rates vary depending on the format used. If you are dealing with video, then the frame rate is dictated by the video frame rate of your project. If you are using MTC simply to synchronize music devices, then it is advised to use the highest frame rate available. The frame rates are 24, 25, 29.97, 29.97 Drop, 30, and 30 Drop. We discuss the synchronization issues related to sequencing in more detail in Sections 3.9 and 4.6.

Song Position Pointer: This message tells the receiving devices which bar and beat to jump to. It is mainly used in conjunction with the MIDI Clock message in a master–slave MIDI synchronization situation.

Song Select: This message allows you to call up a particular sequence or song from a sequencer that can store more than one project at the same time. Its range goes from 0 to 127, thus allowing a total of 128 songs to be recalled.

Tune Request: This message is used to retune certain digitally controlled analog synthesizers that require adjusting their tuning after hours of use. This function doesn't really apply anymore to modern devices, and it is rarely used.

End of System Exclusive: This message is used to mark the end of a System Exclusive message (see the next section).

1.3.6 System Exclusive messages (SysEx)

System Exclusives are very powerful MIDI messages that allow you to control any parameter of a specific device through the MIDI standard. SysEx messages are specific for each manufacturer, brand, model, and device, and therefore they cannot be listed like the other generic MIDI messages analyzed so far. In the manual of each of your devices is a section in which all the SysEx messages for that particular model are listed and explained. These messages are particularly useful for parameter editing purposes. Programs called *editors/librarians* use the computer to send SysEx messages to connected MIDI devices in order to control and edit their parameters, making the entire patch editing procedure much simpler and faster.

Another important application of SysEx is the MIDI data bulk dump. This feature allows a device to send system messages that describe the internal configuration of that machine and all the parameters associated with it, such as patch/channel assignments and effects setting. These messages can be recorded by a sequencer connected to the MIDI OUT of the device and played back at a later time to restore that particular configuration, making it a flexible archiving system for the MIDI settings of your devices. I discuss the details of this technique in Chapter 4, which is dedicated to advanced sequencing techniques.

1.4 The MIDI devices: controllers, synthesizers, sound modules, and sequencers

It is very important to choose the right MIDI devices and instruments to use in your studio. Remember that they are the "virtual musicians" that will be featured in your music productions, so it is essential to have the right type of equipment, the right variety of instruments, and a very flexible and versatile palette of sonorities to choose from in order to be an all-round composer. MIDI devices can be divided into three main categories: MIDI *keyboard synthesizers* (or MIDI synthesizers), *keyboard controllers*, and MIDI *sound modules* (or sound expanders). The main difference between them is based on the presence or lack of a built-in sound generator and keyboard. Keep in mind that, as underlined earlier in this chapter, all the devices that are going to be part of your MIDI network must be equipped with a MIDI interface. The interface is built into all the professional synthesizers, controllers, and sound modules available on the market. The only exceptions are vintage machine created before 1983.

1.4.1 The MIDI synthesizer

MIDI synthesizers (Figures 1.11 and 1.12) feature a MIDI interface, an internal sound generator, and a keyboard to output MIDI data. If they come equipped with a built-in sequencer, then the term MIDI *workstation* is more appropriate, since they can be used as stand-alone MIDI production studios. The MIDI synthesizer is probably the device you are most familiar with. It is also the most complete, since it allows you to control an external MIDI device through the keyboard and can also produce sounds through the internal sound generator. Notice how the three elements are connected to one another. The keyboard sends signals (according to which note you pressed) to both the MIDI OUT and the internal sound generator, which also receives MIDI messages from the MIDI IN port.

1.4.2 The keyboard controller

A modification of the MIDI synthesizer is the keyboard MIDI controller (Figure 1.15). This device features only a MIDI interface (usually only a MIDI OUT) and a keyboard. There's no internal sound module. In fact it is called "controller" because its only use is to control other MIDI devices attached to its MIDI OUT port.

1.4.3 The sound module

Depending on the equipment, the features of the devices, and the number of MIDI devices involved in your project studios, we can set up the MIDI network in different ways.

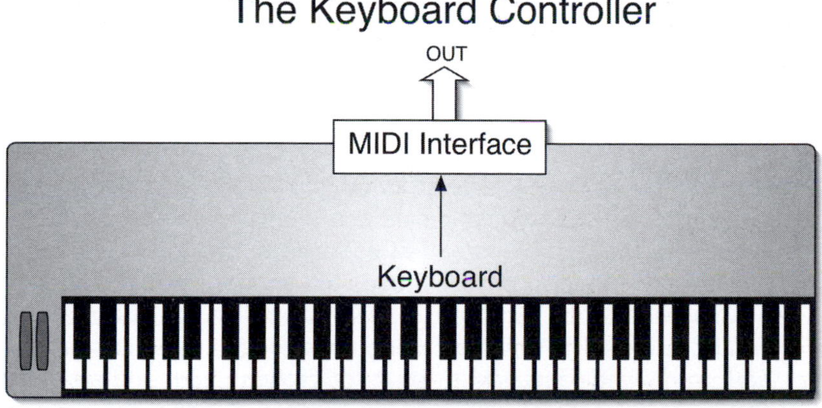

Figure 1.15 MIDI keyboard controller.

In most project studio situations one MIDI controller is sufficient. Through the controller you can output MIDI messages (e.g., Notes, Control Changes) to other MIDI devices and/or to a sequencer. On the other hand a professional MIDI studio will always need new and fresh palettes of sounds. In fact, what characterizes the successful studio and the modern composer is the variety and flexibility of sounds and musical textures available. While in the pre-MIDI era synthesizers were only available with a built-in keyboard, taking therefore a lot of space and costing more money, nowadays we can expand the sound palette of our virtual orchestra by using *sound modules* (or expanders). A sound module (Figure 1.16) features only a MIDI interface and a sound generator but not a keyboard. The advantage of this type of device is that it delivers the same power as a MIDI synthesizer in terms of sounds but in a more compact design and at a lower price.

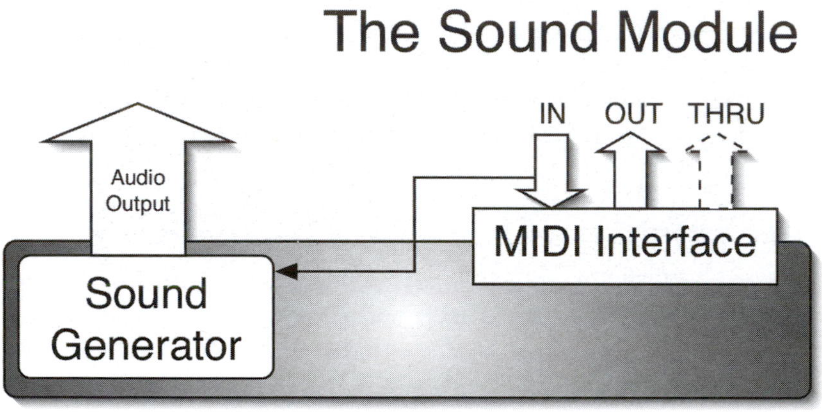

Figure 1.16 MIDI sound module.

Sound modules, along with your controller (or controllers) and MIDI synthesizers, constitute the core of your MIDI network. They also represent your orchestra, with the sound generators of each MIDI device being the "musicians" and the controllers being the "conductor" and the "composer."

1.4.4 The sequencer: an overview

If the MIDI devices and their sound generators symbolize the musicians available to the modern composer, imagine the sequencer as the contemporary score paper that the composer uses to write the music. Instead of scribbling notes, markers, dynamics, and repeat signs on paper, the modern composer uses the sequencer to record, edit, and play back the notes (MIDI messages) sent from a MIDI controller. The advantages of using a sequencer to compose are huge. Infinite editing possibilities, quick comparison of different versions, and immediate feedback in terms of orchestration and arrangements are only a few of the many advantages this working technique offers. The sequencer is the central hub of the MIDI data flow between all the MIDI devices connected in your studio.

There are two types of sequencer: hardware and software (or computer based). In Figure 1.17 you can see an example of a hardware sequencer.

Figure 1.17 Example of hardware sequencer (Courtesy of Roland Corporation U.S.).

These sequencers usually are more limited in terms of editing power and track count than their computer-based counterparts. Their high portability and low price, though, make them an appealing option for the live situation or small home studios. They also have a built-in MIDI interface, which makes them ready to be used right out of the box. Sometimes hardware sequencers are built into keyboard synthesizers in order to offer a complete MIDI production center (usually referred to as a MIDI *workstation*). If you plan to program and record complex sequences with many tempo changes and odd meters and eventually also to record audio tracks (more on this in Section 2.5), I highly recommend investing in a computer-based MIDI/audio sequencer. Nowadays software-based sequencers are the center of every professional studio, mainly because of their versatility, editing power, and multifaceted functionality.

A software sequencer requires three elements in order to integrate seamlessly with the other MIDI devices: a computer on which the software can be installed and run, a sequencer application (meaning the software itself), and a MIDI interface (Figure 1.18).

Whereas a hardware sequencer has a built-in MIDI interface, a computer, in general, doesn't have MIDI connectors directly installed on his motherboard. Long gone are the times of the Yamaha CX5M and Atari ST computers with built-in MIDI interface that were specifically targeted to musicians. Therefore one of the key elements for your studio is to have

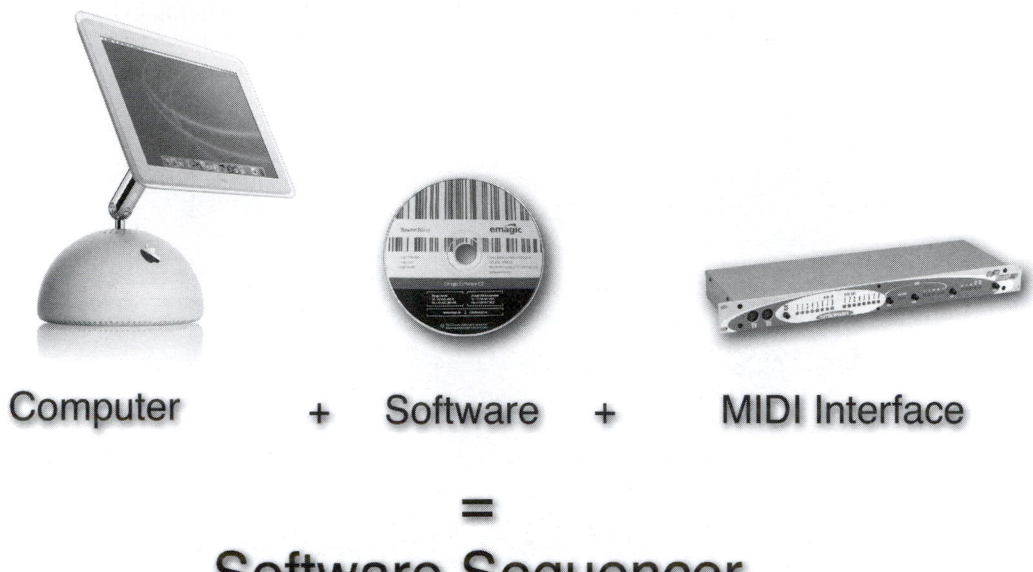

Figure 1.18 Basic elements of a software sequencer (Courtesy of Apple Computer, Apple/Emagic, and M-Audio).

the right computer (more on this subject later in this chapter—Section 1.6.2) and the right MIDI interface.

1.4.5 Which controller?

Depending on how big, complex, and sophisticated your studio will be, you have to choose from among different solutions regarding which type of equipment and devices to use. Every MIDI-based studio needs some type of controller in order to be able to generate and send out MIDI messages to other devices. You have already learned that both the keyboard controller and MIDI synthesizer are MIDI devices capable of transmitting MIDI messages. A controller is indispensable for your studio—no matter what, you need one if you are serious about MIDI. Keep in mind though that keyboards are not the only MIDI controllers available. If you are not a keyboard player there are other options available, and also remember that one does not exclude the others. In fact we will learn in Section 3.6 how to integrate MIDI data sequenced from different types of controllers.

When deciding on a keyboard controller you have to ask a few simple questions: How many keys (extension) does the controller need? How important is it to have weighted-action keys rather than synthesizer-action keys? How important is it to have MIDI controllers such as faders and knobs? How important is it to have a controller with built-in sound capabilities (MIDI synthesizer)? Is portability an important factor? The answers should give you a pretty clear idea about which controller to buy. You can use the chart in Table 1.2 to answer these questions. Let's analyze these factors in more detail.

Table 1.2 Fill-in chart to help choose a keyboard MIDI controller

	Crucial	Marginal	Not important at all
Weighted key action			
MIDI controllers			
Built-in sounds			
Portability			
Aftertouch			
Number of keys or octaves			

Weighted key action: This feature is usually important for professional pianists or musicians that are classically trained. Keyboards with this feature have the same response as (or at least a one very similar to) an acoustic piano. They are usually more expensive and much heavier than keyboards of synthesizers with plastic keys. If you are on a tight budget and the real piano feel is not a major concern, I suggest opting for a MIDI synthesizer (which also has the advantage of the built-in sound generator). If you are planning to have other keyboard players in your sessions and your budget allows it, I would go for a controller with weighted keys. In the long run, like every investment in this business, it will pay to have something a bit more sophisticated. There is also a third option, based on keyboards with a semiweighted action, which feature a slightly heavier action than a regular synthesizer. They are a great solution for musicians who travel a lot and are concerned about the weight of their gear. This type of keyboard is a good compromise between portability and real piano feel (Figure 1.19).

Figure 1.19 Keyboard controller with weighted key action (Courtesy of Roland Corporation U.S.).

If you absolutely cannot do without the feel of a real piano keyboard and while still maintaining control of your MIDI studio, the ultimate solution is to use, as your main MIDI controller, an acoustic piano able to send MIDI data. While there are different brands and models on the market, one of the most successful combinations of acoustic piano and MIDI controller is the Disklavier by Yamaha. Other companies also market devices that can turn any regular acoustic piano into a MIDI controller, such as the PianoBar by the legendary Moog.

MIDI controllers: Some models of controller have one or more faders and knobs that can be assigned to multiple MIDI Control Change messages (Figure 1.20). This means, for example, that you can control the volume and pan of a certain MIDI instrument directly from your keyboard. By moving the controller you send a specific MIDI message to the OUT port. Depending on the message, you can control a specific parameter of another device. You will learn more about Control Change (CC) messages in Section 2.14. This feature can be important to advanced MIDI composers since you can achieve higher control in terms of expression and phrasing of certain instruments, such as strings and pads. It can also help you to create an automated mix without using the mouse, which sometimes can be very useful.

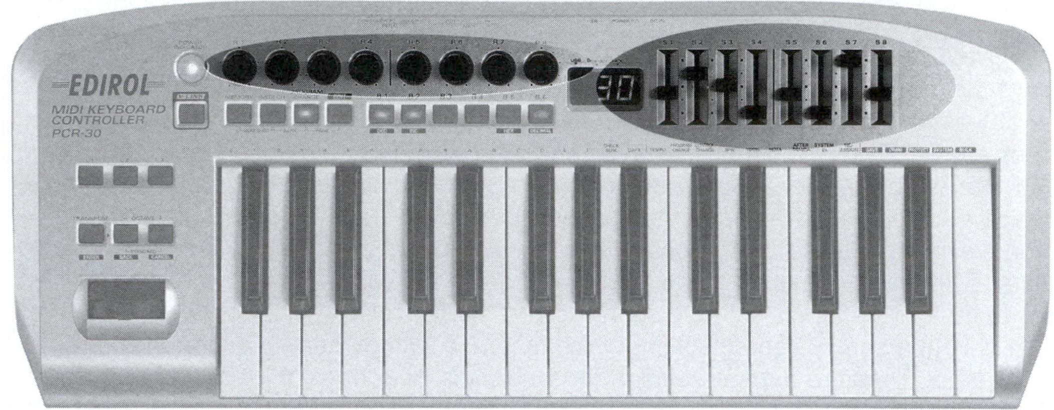

Figure 1.20 Keyboard controller with faders and knobs able to send CC messages (Courtesy of Roland Corporation U.S.).

Built-in sounds: As explained earlier in this chapter, a keyboard controller with a built-in sound generator is called a MIDI synthesizer. This type of device acts as both controller and sound source. It is definitely a good option to use it as a starting point in a MIDI studio because it allows you to sequence right away without also having to buy a sound module. It is usually more expensive than a simple keyboard controller because of the built-in sound generator, and it usually (but not always) comes with unweighted synthesizer-action keys.

Portability: If this is an important issue I recommend avoiding a controller with weighted-action keys. If you use numerous software synthesizers with a portable computer and you plan to do most of your work on location or on the road, then a small portable keyboard is what you need. Portable controllers come in a variety of options. They usually have from 30 to 61 keys (synthesizer action), most of the time with plenty of knobs and sliders assignable to different MIDI CCs, and sometimes with a built-in USB MIDI interface to connect the controller directly to the USB port of your computer (Figure 1.21). If portability is your top priority, these are the best option.

Aftertouch: I highly recommend having a keyboard with Aftertouch, since it gives you much more control over the expressivity of the performance. Devices that respond to polyphonic Aftertouch are harder to find, but the majority are monophonic Aftertouch ready. When you buy your controller, make sure it can send Aftertouch messages.

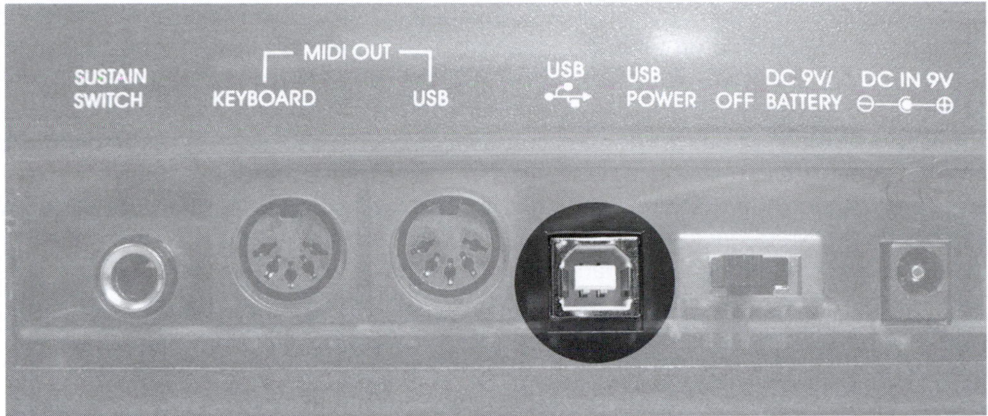

Figure 1.21 Built-in USB MIDI interface on portable keyboard controller.

Number of keys: The number of keys and octaves varies from model to model. The range goes from a small 2-1/2 octaves (30 keys), extra light and portable controller to a more "grown-up" version that features 4 or 5 octaves (49 or 61 keys), to a full 88 keys (7-1/3 octaves). For a serious project studio I highly recommend using a weighted-action keyboard with 88 keys, while for a portable studio configuration the best solution is to use a 49-key controller, which still provides good flexibility in terms of sequencing range.

1.4.6 The sound palette

Along with your MIDI controller another very important aspect of a MIDI studio is the variety of sound sources used to generate the sounds triggered by the MIDI messages. The sound generators found in the MIDI devices determine the sound palette you will be working with. Remember, the sound modules and synthesizers that "populate" your studio are your virtual musicians. The way they are chosen, organized, and arranged has a direct impact on your personal sound and style.

As we learned earlier, you have two main options when considering hardware MIDI devices with sound generators: MIDI synthesizers and sound modules (and software synthesizers to a certain extent). Many manufacturers offer some of their products in both versions: keyboard synthesizer and sound module. After choosing a good and solid controller, in general it is better to opt for the sound module version of your favorite synthesizer, since it is cheaper and takes much less space in your studio without sacrificing any of the features of the sound engine. In fact, keep in mind that in most cases a sound module has the same sound chip as its keyboard version. The more MIDI devices with sound generators you equip the studio with, the better. Imagine the difference between having available for your production only a local band with 8 musicians instead of a full orchestra with 80 musicians. Well, the same can be said for your virtual ensemble: the more devices you have, the higher the sound versatility and control power over your productions you can exercise. I usually like to keep every MIDI device I buy, and I am very reluctant to sell old ones, mainly because I always like at least a few patches from each device, no matter how old it might be—I still hold on to an old Roland U-110 and an "ancient" Yamaha FB-01!

The power and flexibility of each sound source depends mainly on three factors: the number of MIDI channels on which the device can receive simultaneously, the polyphony (the number of notes that can be played at the same time), and the number (and quality) of patches available on the machine. When the MIDI standard was adopted, the major synthesizer manufacturers decided to allow up to 16 different channels to be transmitted on a single MIDI cable. Think of MIDI channels as "musicians" in your orchestra that can play almost any instrument you want. Each channel is independent of the others in terms of notes, velocity, dynamics, volume, pan, etc. The number of simultaneous channels on which each device can receive may vary, depending on the level of sophistication of its sound generator. While the majority of the sound modules and synthesizers available on the market can play 16 different parts at the same time (one for each channel), this is not always the case. Some devices may be able to respond to just two MIDI channels—useful for creating split or layered sounds, such as bass in the left hand and piano in the right hand or piano plus strings layered. Others may respond to 8 or 9 channels. If your synthesizer can respond to more than 1 MIDI channel at the same time, it is called a *multitimbral* device (meaning "many-toned"). An additional 16 parts (for a total of 32) can be added if the device allows for a second set of MIDI IN. Usually, in order to have the maximum flexibility in terms of sounds and instrumentation when sequencing, you will want to have 16- or 32-channel multitimbral devices. Many multitimbral devices can also work in a "single" playback mode such that they play back only one sound on one MIDI channel. If this is the case, you will need to change its mode to work multitimbrally. To make sure the synthesizer is set as a multitimbral device, navigate through its MIDI parameters (usually found in the "MIDI" or "Global Settings" menu) and change the "Mode" parameter to "Multi." Keep in mind that, depending on the manufacturer, the multitimbral setup is sometimes referred to as "Poly," "Multi," or "Mix." The opposite of multitimbral is a device that can receive on only one channel at a time, effectively reducing the number of your MIDI "musicians" to one. This setting, called *Omni*, is mainly used for live settings or to test the MIDI connections rather than for real studio and sequencing situations.

The polyphony (or number of voices) available on a MIDI device represents the number of simultaneous notes that can be played on that device across all 16 channels. This is a very important feature, since it has an impact on the flexibility and power of the synthesizer. A low-end sound engine usually has 32-note polyphony, making it barely acceptable for sequencing. More professional devices feature 64-note polyphony, while top-level synthesizers can take advantage of sound engines with 128-note polyphony. Vintage synthesizers don't usually go higher than 32 voices, but recent models can easily reach 64 and 128. A synthesizer can have mainly two different ways of allotting the available voices among the MIDI channels. Older models usually require you to manually reserve the number of voices for each part (e.g., the Roland JV-880 sound module). On the other hand, more recent devices dynamically assign the voices that are still available to the parts that require them at any given moment. Keep in mind that even though 32- or 64-note polyphony may seem like plenty of voices for a device, the real number of notes you can have a synthesizer play simultaneously also depends on the type of patches you are using and on their level of complexity. Advanced patches are programmed using different layers, each one using a voice. Therefore if you choose a patch that is built on, say, four layers, then every time you press one key on the controller you actually trigger four voices. Keep this in mind when you work on your sequences. If the sound engine runs out of polyphony, then it will "steal" voices from parts that have not been active for a while, and it will assign them to the most recently triggered ones. This may lead to sustained notes that suddenly stop playing due to the lack of voices available.

The number of patches available on a device varies tremendously, depending on the type of device. Nowadays even the most basic synthesizers are equipped with around 1000 waveforms and patches that can be modified to create even more sounds. As a general rule when planning and building your studio, I recommend planning for as many sound sources and sound modules as you can afford and also trying to diversify as much as possible the types of synthesis (analog, FM, sampling, wavetable, etc.), the brands, and the features of your devices. As a modern and versatile composer you need to be able to have almost unlimited sources in order not to run into any technical limitation. Plan to have at least one synthesizer or sound module from each of the leading manufacturers. Usually each brand has a very distinctive sonority that is the result of decades of research. Some textures may suit your style better than others, but do not be afraid to experiment with new sonorities that might not appeal to you right away. As is often the case with writing and composing, you might get inspiration from a new sound, or a new patch may fire up your creativity. Therefore keeping the flow of inspiration running is crucial. Nowadays in the modern studio this flow is often nourished by active listening, programming, and experimentation with new sonorities.

I usually like to have available in my studios some Roland synthesizers (e.g., XV-5080, JV-2080, or similar), which give me a solid palette of sounds to work with, some Yamaha machines, which feature more "edgy" and modern sonorities (such as the Motif series), and some Korg synthesizers, which usually are more trendy, such as the Triton and the Karma. I also recommend keeping one or two hardware samplers handy (such as the AKAI S or Z series or the EMU E series), since they are very reliable and give you fairly good flexibility in terms of sound library availability. For the real analog feel, I usually have some of the Clavia synthesizers, such as the Nord Lead 2 and Nord Lead 3.

These are only general guidelines to help you understand the importance of a wide and diverse sound palette. The best approach to building such a creative and powerful environment is to experiment and listen to as many sound sources as you can and to decide which ones are more suitable for your studio.

Another factor strictly related to the flexibility of the sound palette available on a MIDI synthesizer or sound module is the expendability of the device. Most entry-level machines do not provide expendability of their internal waveforms and sound banks, limiting their flexibility. More advanced models allow you, on the other hand, to increase the number of patches by using internal or external expansion cards with megabytes of new waveforms and patches. Usually each manufacturer uses proprietary expansion boards that are compatible with only certain models of their machines (Figure 1.22).

Use Table 1.3 to outline the most important features of the MIDI synthesizers and sound modules you are planning to buy for your studio.

Figure 1.22 Internal expansion boards (Courtesy of Roland Corporation U.S.).

Table 1.3 Chart for comparing synthesizers and sound modules

Receiving MIDI channels	16	8	Less than 8
Polyphony	128 note	64 note	32 note
Number of patches			
Expendability	Very important	Marginal	Not important at all

1.5 Connecting the MIDI devices: Daisy Chain and Star Network setups

After choosing the right devices for your MIDI studio it is time to connect them together and analyze the MIDI signal flow. There are two main types of MIDI configurations to connect your devices: Daisy Chain (DC) and Star Network (SN). The DC setup is found mainly in the very simple studio or live situations where a computer is (usually) not used, and it involves the use of the THRU port to cascade more than two devices to the chain (Figure 1.23).

1.5.1 Daisy Chain setup

In a Daisy Chain configuration the MIDI data generated by the controller (device A) are sent directly to device B through the OUT port. The same data are then sent to the sound generator of device B and passed to device C using the THRU port of device B, which sends out an exact copy of the MIDI data received from the IN port. The same happens between devices C and D.

A variation of the original Daisy Chain configuration is shown in Figure 1.24, where in addition to the four devices of the previous example, a computer with a software sequencer and a basic MIDI interface (1 IN, 1 OUT) are added.

In this setup the MIDI data are sent to the computer from the MIDI synthesizer (device A), where the sequencer records them and plays them back. The data are sent to the MIDI network through the MIDI OUT of the computer's interface and through the Daisy Chain. This is a basic setup for simple sequencing, where the computer uses a *single-port* (or

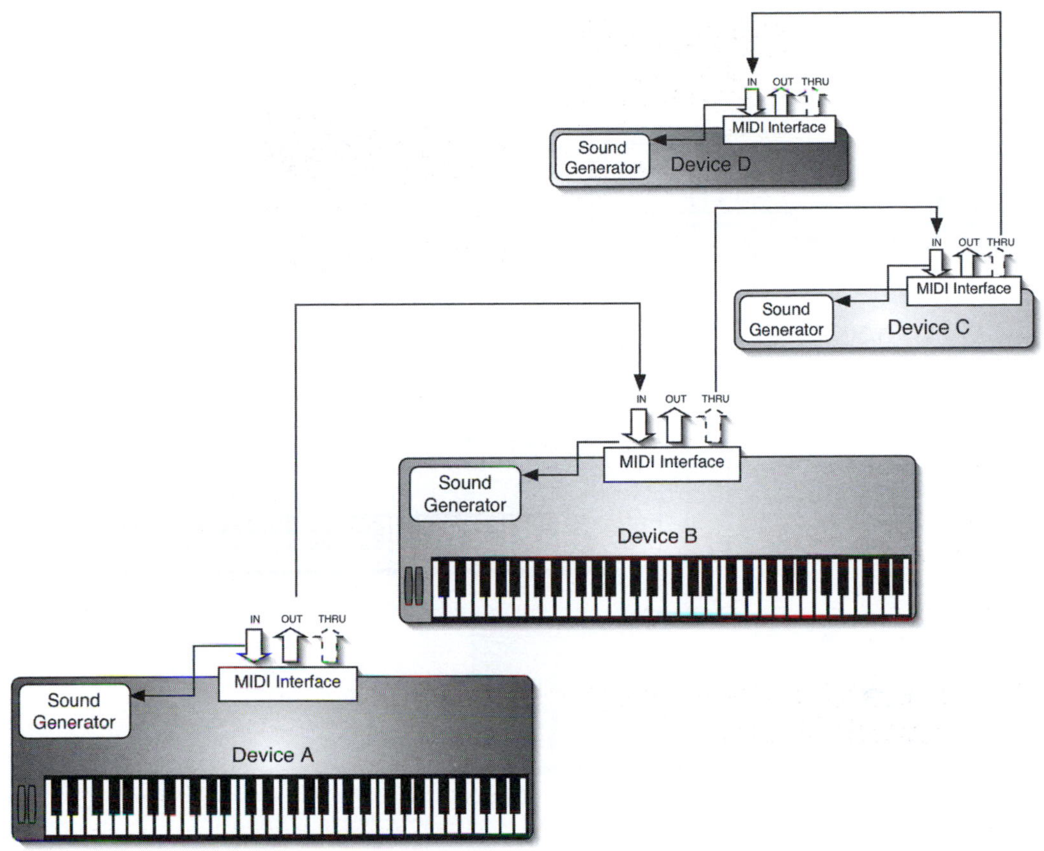

Figure 1.23 Daisy Chain setup 1.

single-cable) MIDI interface (Figure 1.25), meaning an interface with only 1 set of IN and OUT. There are many brands and types of computer MIDI interface, which usually are connected to the computer through a USB (universal serial bus) interface. Sometimes you can install a PCI card that has built-in MIDI connectors instead of using an external USB device.

A Daisy Chain configuration has several limitations, though, which make it a less than perfect solution for a complex MIDI studio. Its two main drawbacks are the delay/error factor introduced by long chains and the fact that you will have to split the 16 MIDI channels available among all the devices connected on the same chain. Usually it is not advised to chain more than three devices using the THRU port, since this may cause delays that can be measured in milliseconds to which we have to add the time it takes for the sound generator, filters, and other components to be triggered before the sound is actually generated. Unless you want to layer sounds by sending the same note to more than one MIDI channel to multiple devices, you will also have to split the 16 MIDI channels available among all the sound generators of your studio. As you can see in Figure 1.26, each device has to be assigned to a specific range of channels in order to avoid having two sound generators receiving on the same channel (which would create a layer of sounds). This will reduce the overall potential of your studio since in theory each multitimbral device is capable of receiving on 16 channels at once.

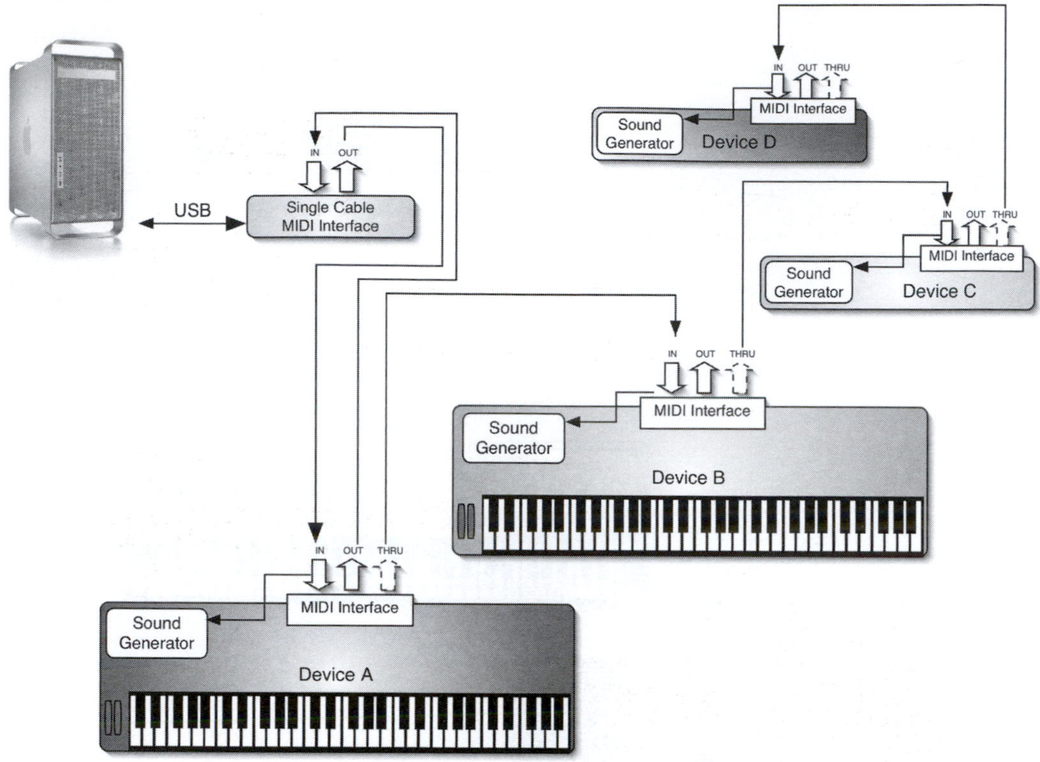

Figure 1.24 Daisy Chain setup 2 (Courtesy of Apple Computer).

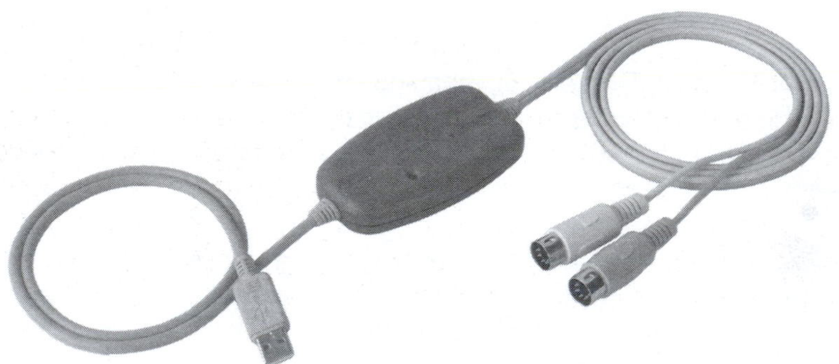

Figure 1.25 Single-cable MIDI interface (Courtesy of Roland Corporation U.S.).

At the moment the majority of MIDI interfaces is on USB, with features that range from a basic 1 × 1 (like the one shown in the previous example) to a top-of-the-line 8 × 8 with SMPTE, ADAT, and Word Clock synchronization options, like the widespread MTP AV by Mark of the Unicorn, the Unitor 8 by Apple/Emagic, and the MIDIsport 8 × 8 by M-Audio (Figure 1.27).

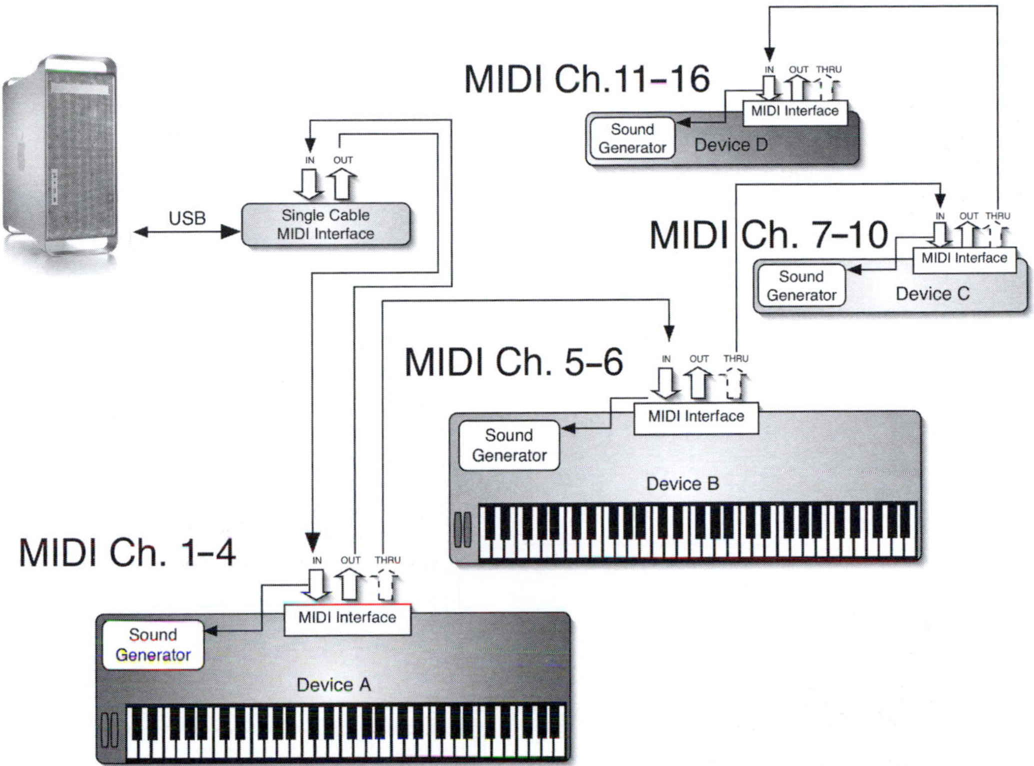

Figure 1.26 Daisy Chain with channel assignment (Courtesy of Apple Computer).

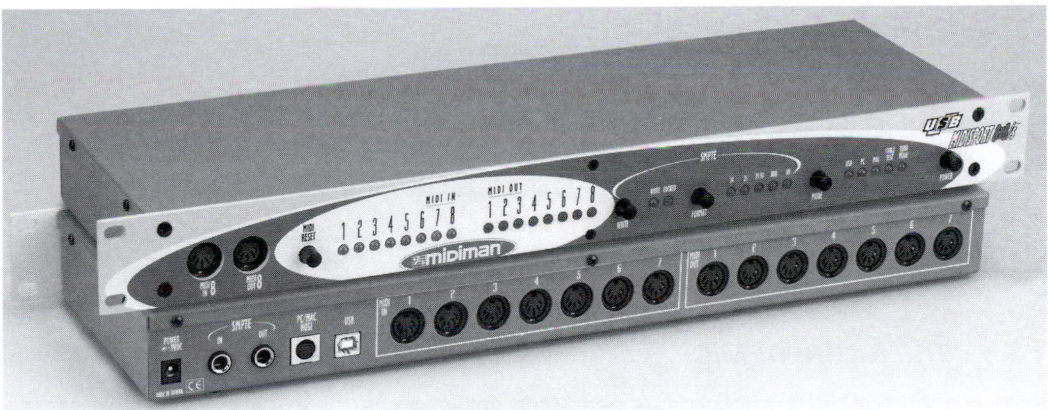

Figure 1.27 Multicable MIDI interface, M-Audio interface MIDIsport 8 × 8 (Courtesy of M-Audio).

1.5.2 *Star Network setup*

A MIDI interface that has more than 1 set of INs and OUTs is called a *multicable* MIDI interface. For an advanced and flexible MIDI studio, a multicable interface is really the best solution, since it allows you to take full advantage of the potential of your MIDI devices and also prevents the delay problems related to a Daisy Chain setup that we saw earlier in this chapter. By using a multicable interface, all the devices connect to the computer in parallel, meaning that the MIDI data won't experience any delay related to the Daisy Chain setup. This configuration involving the use of a multicable MIDI interface is referred to as *Star Network*. One of the big advantages of the Star Network setup is that it allows you to use all 16 MIDI channels available on each device, since the computer is able to redirect the MIDI messages received by the controller to each cable separately, as shown in Figure 1.28.

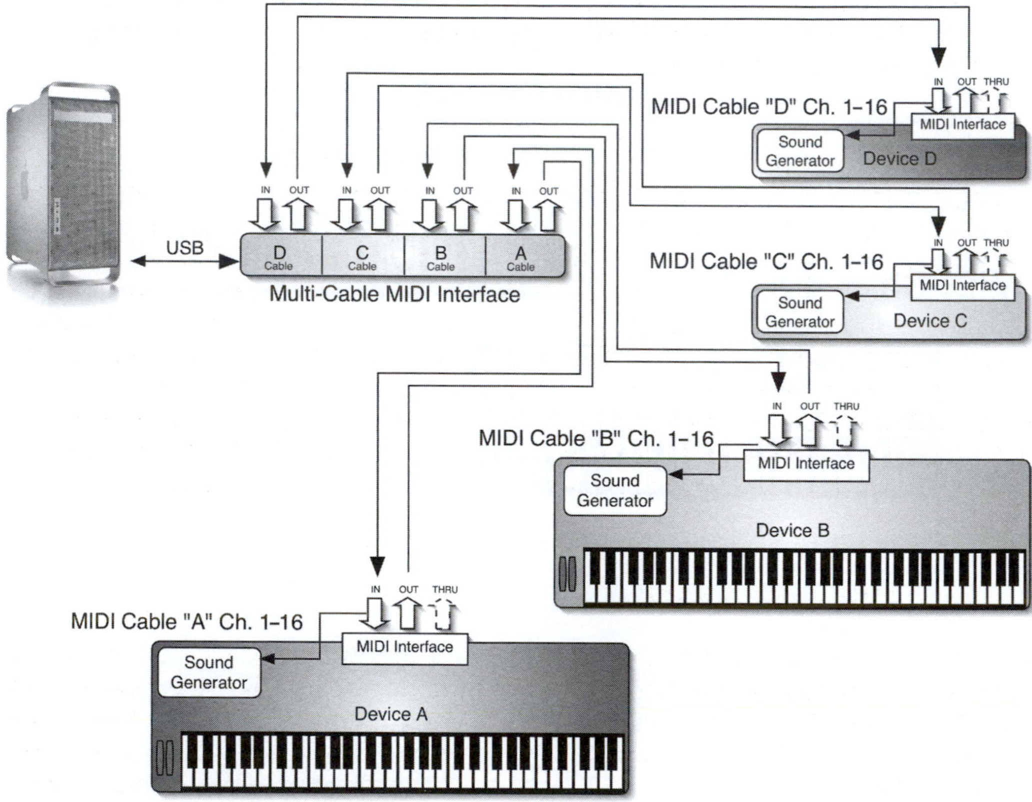

Figure 1.28 Star Network setup (Courtesy of Apple Computer).

1.5.3 The future of MIDI

Even though the MIDI standard has been (and still is) an extremely successful tool for the interaction of several components in a modern studio, its age starts showing when complex MIDI networks need to be set up. The MIDI protocol has kept evolving in order to accommodate some of the newest technologies. To address the constant evolution of complex MIDI networks for large studios, two new protocols have emerged lately to facilitate and increase the potential offered by the MIDI standard: mLAN and MIDI over Ethernet.

mLAN is an acronym for "music local area network." It is a digital networking system developed by Yamaha and based on the FireWire high-speed protocol. It is capable of transmitting both MIDI and audio information at the same time. In addition it is capable of adaptively managing devices and connections without the need of network reconfiguration. This means that the connections between devices are automatically reestablished if a device is disconnected and reconnected at a later time. On a computer the only requirement is to have a FireWire port and an mLAN driver (already available are drivers for Mac OS9 and X and for Windows XP). While the number of audio and MIDI channels that can be carried simultaneously by mLAN can vary depending on the speed of the bus used, at the moment using an FW *a* bus (400 Mbps) you can easily transfer 150 channels of audio (at 24 bit/48 kHz) and literally thousands of MIDI channels. This protocol holds great promises, and it can really simplify the studio connection setup by eliminating most of the cables that lie around a studio. Its success depends on the number of manufacturers that will adopt it.

MIDI over Ethernet also takes advantage of an existing network technology (Ethernet) that is used regularly to transfer binary data. The challenge that Ethernet has to face with MIDI messages is that they need to be transferred in real time and that time accuracy is crucial. The main idea behind this technology is to utilize existing networks to be able to use several computers at the same time to share track count and power in order to construct a more flexible and overall powerful studio. At the moment there are several standards and organizations that are working on developing a reliable and latency-free protocol capable of transferring MIDI messages over Ethernet. Among them the most credited organizations are the DMIDI (Distributed MIDI), which is working on a LAN (local area network) version of MIDI over Ethernet, NetMIDI (by Richmond Sound Design Ltd.), MIDIOverLAN+ (by Music LAB), iMIDI (based on the TCP protocol), and the MWPP (MIDI wire packetization protocol), which is designed to support both LANs and WANs (wide area networks).

1.6 The audio equipment

While the MIDI network is the spinal cord of your studio, the audio network constitutes the necessary link between the MIDI devices and the delivery of a final product in the form of a CD, tape, or other media. As I mentioned earlier in this chapter, on the MIDI cables only a description of MIDI events is carried and there is no audio information whatsoever. The "translation" from MIDI to audio is made by the sound generator built into the MIDI devices that receive the notes to trigger from the MIDI IN port (Figure 1.29).

In order to complete your studio you need a mixing board to "collect" all the audio outputs from your MIDI synthesizers and sound modules. From the mixing board the audio signal is sent to your amplifier and speakers or directly to your active speakers for monitoring. The mixing board will also receive the input from your acoustic instruments, microphones, and other sources you will use for your productions. Figure 1.30 illustrates the two signal paths: MIDI and audio.

1.6.1 The mixing board and the monitors

The mixing board becomes the central hub of the audio signal flow. The type of mixing board you will use in your studio depends on several factors. The main aspect to consider is the number of channels (or inputs) you need. In general the higher the number of your

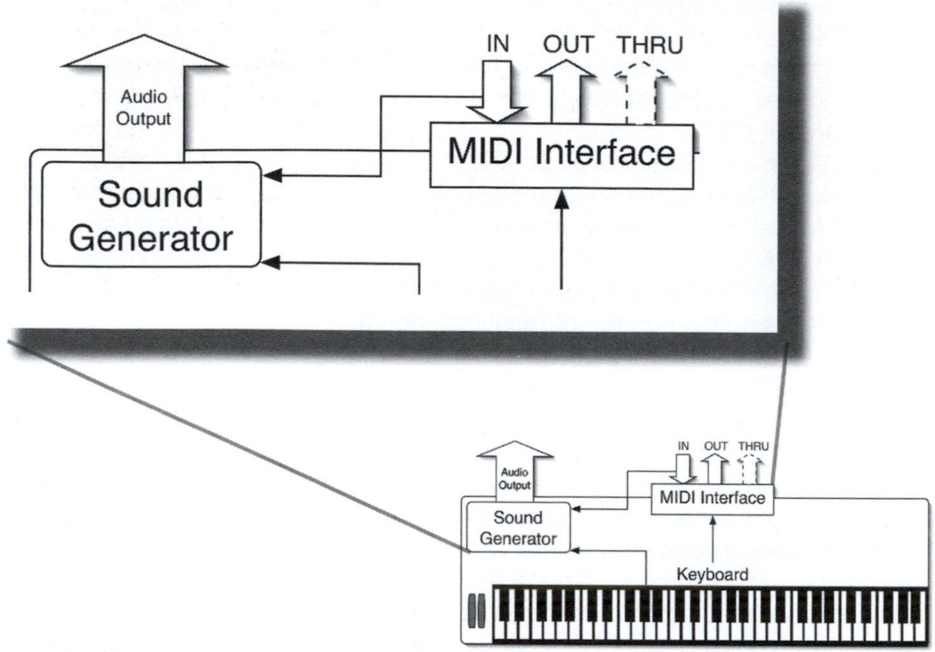

Figure 1.29 The sound generator: from MIDI to audio.

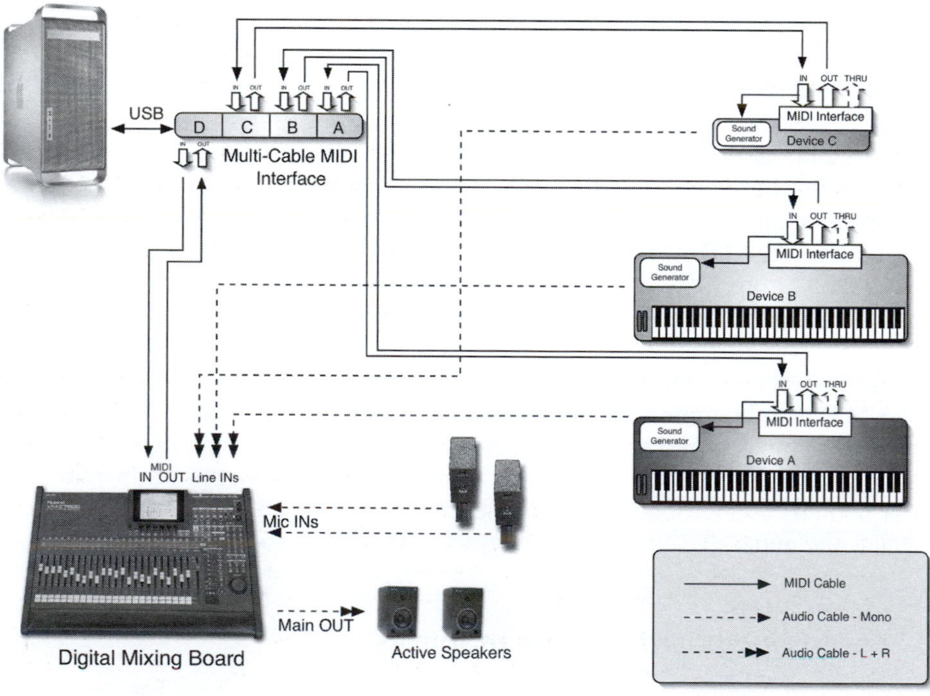

Figure 1.30 MIDI and audio signal paths (Courtesy of Apple Computer, Roland Corporation U.S., and AKG).

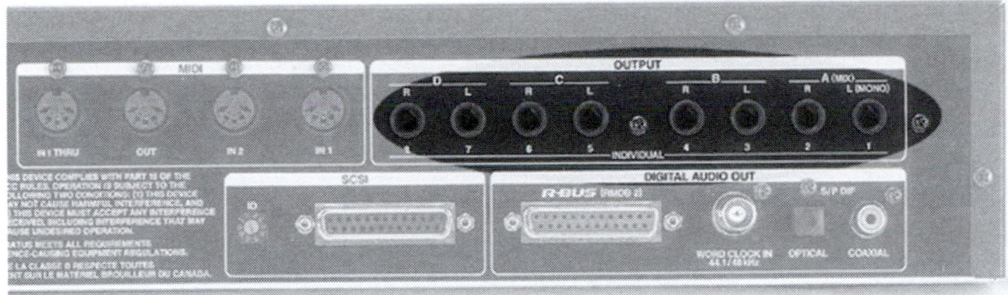

Figure 1.31 Multiple audio outputs on the Roland XV-5080 (Courtesy of Roland Corporation U.S.).

hardware MIDI synthesizers and sound modules, the higher the number of channels you will need on your board. Keep in mind that each device will have at least two audio outputs (left and right channels), but more sophisticated devices come equipped with multiple outputs, sometimes up to eight mono, as in the Roland XV-5080 shown in Figure 1.38, or up to 16, as in the case of the AKAI S6000.

A good starting point for a small to mid-size MIDI studio is a mixing board with 16 channels. If you are planning a bigger studio, consider a 24- or 32-channel board—the output of a sound generator is a "line level out," and it can be either balanced or unbalanced, depending on the model. Remember also that in most cases you want to have "room" for other live/acoustic instruments, such as acoustic/electric guitars and basses (which use a Hi-Z input), and microphones (which use a mic input equipped with mic preamps). In addition to an adequate number of inputs, your mixing board needs to be equipped with a flexible output system. While the *main out* connectors usually feed your monitor speakers and/or your mix-down machine (DAT, reel-to-reel, DAW, etc.), a series of extra outputs called *busses* allows you to send the signal of individual channels to the multitrack recorder inputs. With a 4- or 8-bus board you will be able to send, respectively, four or eight independent channels out from your board to the inputs of you multitrack system. Therefore the higher the number of busses available, the more flexible your working environment will be.

The audio signals collected by the mixing board are sent to the monitor section and the main outputs and from there to the main set of speakers. The audio signal in order to be heard through the monitors needs to be amplified first. You can opt for passive speakers combined with a separate amplifier or for active speakers that have built-in amplifiers. I recommend the second option since usually the amplifier in active speakers is calibrated to give you the best sound for those particular monitors. Depending on the design, use, and distance at which they provide the best performance, speakers can be divided into *near-field, mid-field* and *far-field* types. In a project studio you usually are going to use near-field or mid-field monitors, which are designed for mixing at a short or medium distance, respectively, in order to avoid the disturbances introduced by a faulty design of the room. For detailed information on audio monitors and speakers I recommend the comprehensive text by John Borwick, *Loudspeaker and Headphone Handbook,* published by Focal Press.

One of the big questions that for the past few years have sparked lively discussion among engineers and producers around the world is whether digital boards are better than analog ones. It is not within the scope of this book to settle this dispute, but I would like nevertheless to give some advice and clarify a few points in order to help you make the best decision when planning your studio (and, yes, I have a favorite, and you are going find out soon which one it is!).

First of all let me explain the main differences between an analog and a digital board. In an analog board the audio signal is always kept at its original analog stage (that is, continuous variations in electric voltage) all the way from the input section to the output (Figure 1.32), after passing through the equalizer (which can be more or less sophisticated, depending on the board), pan, and volume sections. In a digital board, on the other hand, the signal is converted after input by the analog-to-digital converter (ADC), as shown in Figure 1.33.

Very simply put, the ADC translates the analog signal coming from the inputs into a digital stream of binary numbers (1s and 0s) by sampling the analog waveform at regular intervals

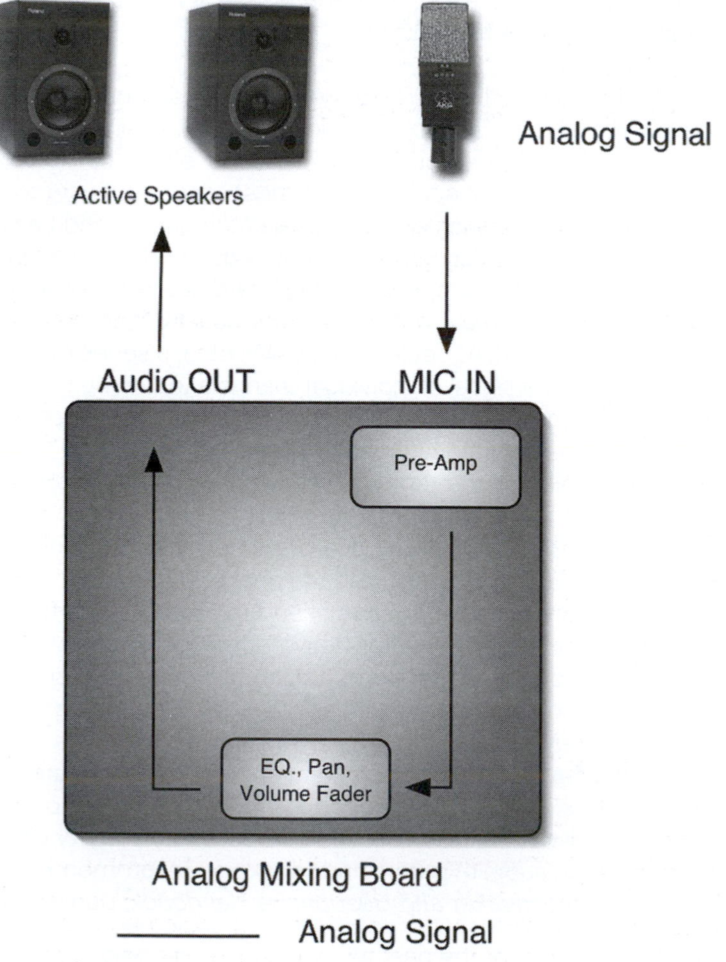

Figure 1.32 Analog board channel strip (Courtesy of Roland Corporation U.S. and AKG).

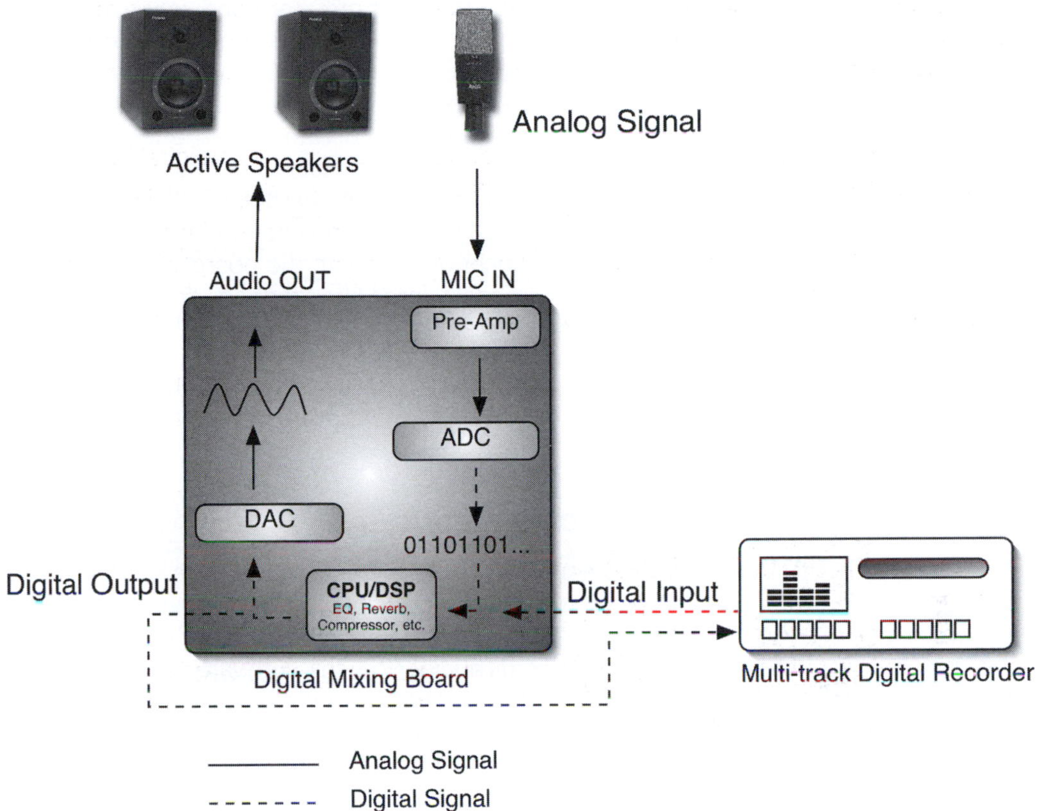

Figure 1.33 Digital board channel strip (Courtesy of Roland Corporation U.S. and AKG).

and representing the amplitude of each sample in binary format. At this point all the internal connections inside the board are digital. The signal will stay digital up to the output stage, where an inverse process, through a device called a digital-to-analog converter (DAC), is applied to the signal, converting it back to analog form. I discuss this process in a little more detail in the next section. It is very common for a digital board to have a built-in digital signal processor (DSP), which, among other functions, allows you to add a series of effects (such as reverb, delay, chorus, EQ, compression) to the digital signal. All digital boards also feature at least one digital input and one digital output that bypass the internal converter and are able to receive and send out digital data directly.

Both analog and digital boards have pros and cons. There is no doubt that a top-of-the-line (and also very expensive) analog board will sound better than anything else (even though less expensive digital mixers are catching up fairly quickly), so if what you are after is the best sound and you can afford it, go with an analog board—you won't be dissatisfied. On the other hand, analog boards have a few drawbacks, one being that the average such board lacks both the ability to recall the settings of a mix (such as equalization, pan, volume) and the option of automating the mix (high-end analog boards are an exception) or serving as control surface for your sequencer. A digital mixer, on the other hand, has many advantages that can make up for some of the acoustic disadvantages. For a project studio, where several assignments can pile up quickly, the ability to store and recall mixes and setups at the touch of a button is very important, and it can save time and money. Just imagine how painful it would be to recall every setting of your board for a 32-channel mix. With a digital board every detail of a mix can be stored in its internal memory and recalled whenever you need. Digital boards usually come equipped with built-in effects (such as reverb, delay, chorus, compressor) that can be very handy for last-minute touches or quick mixes. In addition to the audio inputs and outputs, a digital board has MIDI IN and OUT ports, which allows you to connect it to your MIDI interface and use the faders to send out CC messages to control your software sequencer or automate the board with MIDI messages sent from the sequencer. The price of digital boards has dropped recently, making a digital mixer affordable for many composers and independent producers. Several top-of-the-line manufacturers have digital boards in their catalogs. Among them, Yamaha, Sony, Roland, and Tascam have several digital boards available at affordable prices.

1.6.2 The computer and audio connections

Yes, you read it right! As I mentioned earlier, not only is the computer the central hub of the MIDI network, but in recent years it has become capable of handling digital audio. Over the years since the mid-1990s, computers have become powerful enough to handle the incredible amount of data required by digital audio. Now the software sequencers running in your computer are capable of recording, editing, and playing back not only MIDI data but also audio tracks (meaning audio signal coming from microphones, guitars, bass, acoustic instruments, sound generators of synthesizers, etc.), replacing less flexible and less powerful multitrack recording media (i.e., tape and reel-to-reel machines). Remember that a computer and its software are capable of understanding only binary data, so everything that needs to be sent to a computer must be converted to digital format (binary code) for the computer to be able to process it. Thus in order for a computer and its software sequencer to be able to deal with audio information coming from an external source, we have to translate the analog signal into a digital signal. In fact we just saw something very similar when I described how a digital board works. The ADC (analog-to-digital converter) takes the signal from an analog source (e.g., a microphone) and translates it into binary format by *sampling* the waveform and assigning a definite value to each sample of the waveform. If the number of samples in a certain time frame (also called *sampling frequency,* which is measured in samples per second, or hertz, Hz) is high enough and the resolution of the amplitude of each sample is accurate enough (also called *bit rate*, which is measured by the number of bits allocated by the converters to describe the amplitude value of each sample), then the digital reconstruction will be very similar to the original analog waveform, making the original analog wave and the digital one almost identical (Figure 1.34).

After the original analog signal is converted to binary code it is sent to the computer's central processing unit (CPU), which processes the data and stores them on the hard disk (HD) as audio files. A similar but reverse process is applied in playback, when the digital signal is read from the HD, processed by the CPU, and translated back into an analog signal by the DAC (digital-to-analog converter) (Figure 1.35).

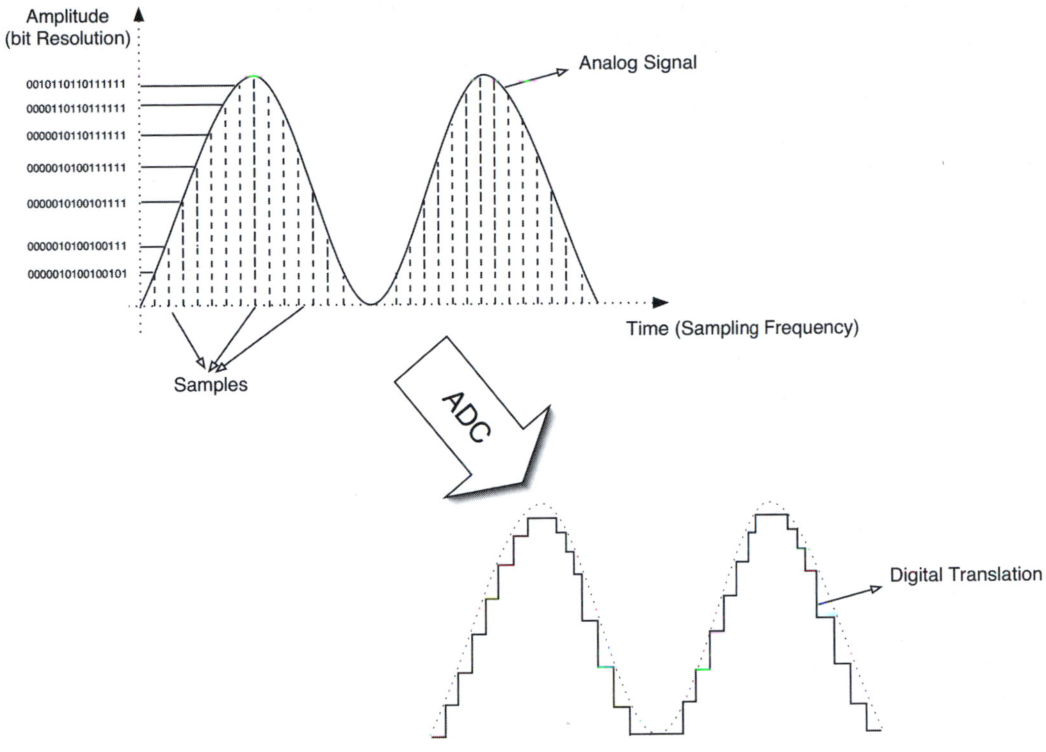

Figuer 1.34 ADC.

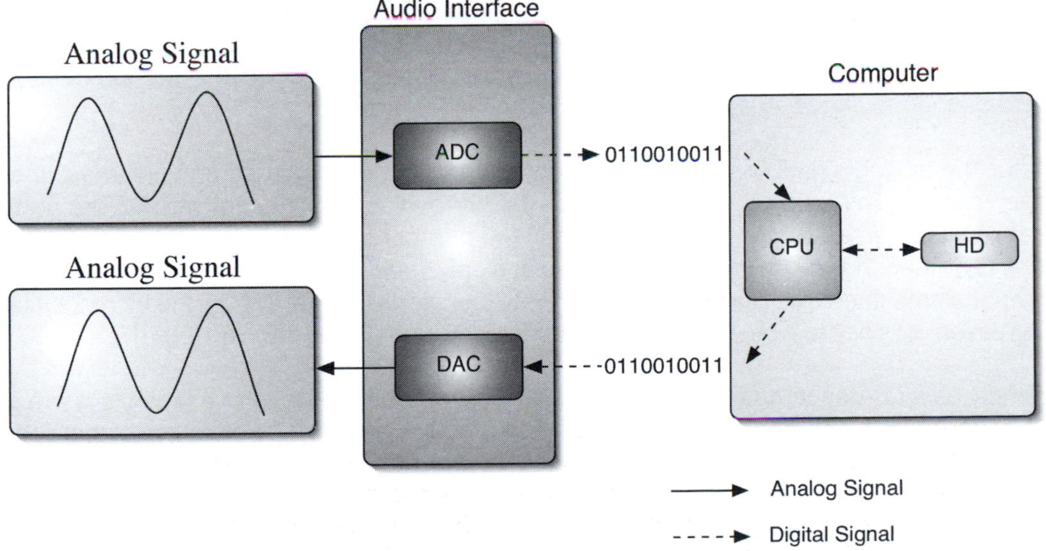

Figure 1.35 ADC–DAC process.

Usually your computer comes equipped with very basic converters that are bundled with the machine. If you have a Macintosh, the converters are built into the motherboard; if you have a PC, most likely you have a preinstalled SoundBlaster (or similar) card. While the internal converters are good enough for basic tasks, their quality is definitely insufficient for professional work. Therefore another very important piece of equipment in your studio is the audio interface, which contains the ADC and DAC. An audio interface varies in price, depending on its specifications and the quality of its converters. The factors to consider when analyzing an audio interface are mainly the following: the maximum sample rate and the bit resolution at which the ADC and DAC work, the number of inputs and outputs available, the type of interface used to connect it to the computer, and the type of software that supports that particular interface.

The sample rate and the bit resolution of your audio interface have a direct impact on the quality of your recordings. While here I don't pretend to cover the sampling theorem in detail, it is important for modern composers and producers to understand the basic concept behind how a converter works, in order to make an educated choice when buying and using an audio interface. As stated earlier, the higher the sample rate and the bit resolution of the converters, the higher the sound quality and the definition of the recording. Let's analyze how these two parameters affect the sound.

The sample rate determines the number of samples per second taken by the converters. At CD quality we have 44,100 samples per second, and we say we use a 44.1-kHz sampling frequency, which means that the analog waveform fed to the converters is "sliced" 44,100 times every second. It might seem like a lot, but in fact we will see that this is just the starting point nowadays for digital audio. Because of the Nyquist sampling theorem, the sampling frequency has a direct impact on the highest frequency we can sample. Thus, if we have a sampling frequency of x, then the highest frequency we can sample without introducing artifacts in the digital domain will be $x/2$. At a sampling frequency of 44,100 Hz, the highest frequency the converters are allowed to sample will be 22,050 Hz. In order to avoid having frequencies higher than the Nyquist point reach the ADC, a low-pass filter is inserted before the ADC, with its cutoff frequency set to half the sampling frequency. If you want more information on this subject, refer to the comprehensive *Introduction to Digital Audio*, by John Watkinson (Focal Press).

After the analog signal is sampled, the converters need to assign an amplitude value to each sample. This value is set according to the bit resolution of the converters. The number of bits of an audio converter determines the number of steps in which the amplitude axis is divided. If the converters can sample at a only very low bit resolution, let's say 2 bit, then the amplitude axis can be divided into only four steps (Figure 1.36); if the bit resolution increases, let's say to 4 bit, then the amplitude axis will be divided into 16 steps (Figure 1.37).

Simply as a reference, the formula to calculate the number of steps of a binary system is 2^n, where n is the number of bits of the system. For example, an audio converter that uses only 8 bits would allow for only 256 steps on the amplitude axis ($2^8 = 256$), resulting in a low definition and grainy digital conversion (keep in mind that a low bit resolution has also a negative impact on other aspects of the digital conversion, such as dynamic range, but here the main concern is the most obvious and overall sound degradation). The main problem of having a low bit resolution is that if the amplitude of a sample doesn't fall exactly

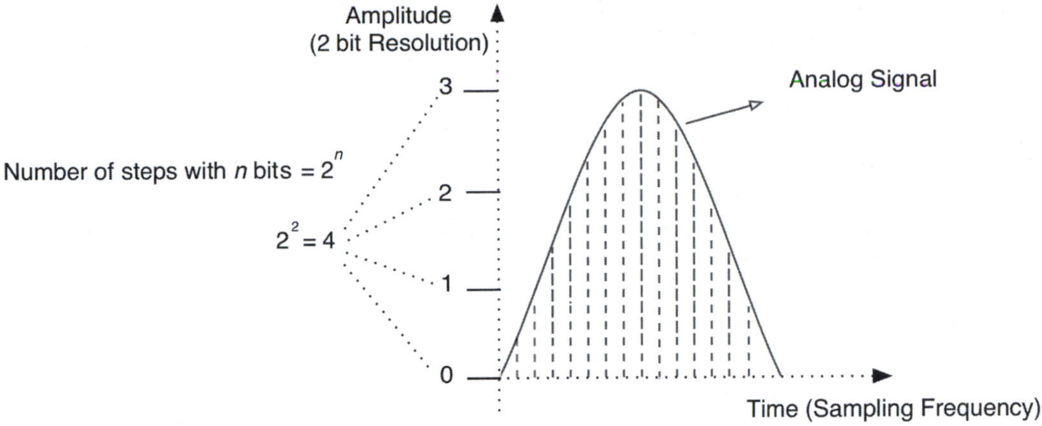

Figure 1.36 2-bit resolution.

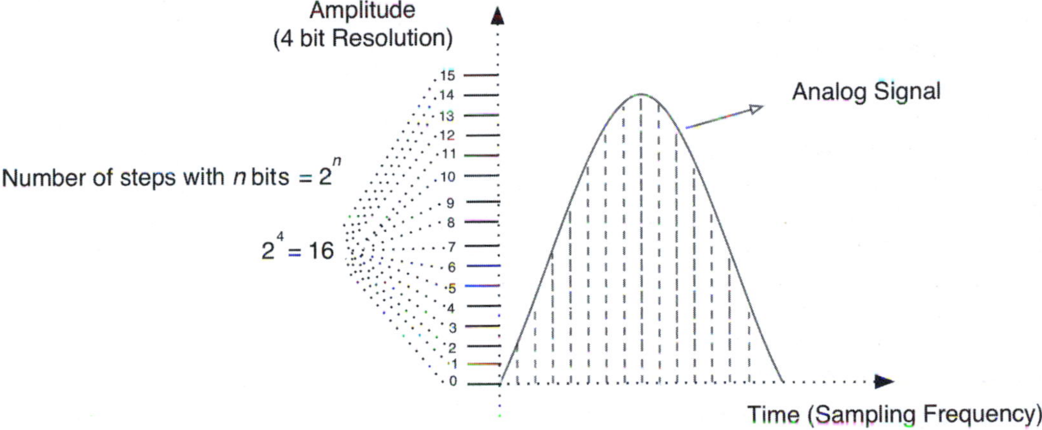

Figure 1.37 4-bit resolution.

on one of the steps into which the amplitude axis is divided, then the converters will have to move it, or, better, to *quantize* it, to the closest available value, introducing a misrepresentation of that sample in the digital domain that will result in what is called *quantization noise* (Figure 1.38).

The bottom line is that a higher bit resolution allows for a much more detailed representation of the amplitude of each sample, resulting in a much more detailed and defined sound. At CD quality we sample with a 16-bit resolution, which means 65,536 steps! But wait, while this was "groundbreaking" technology up to just a few years ago, nowadays the most advanced audio converters are capable of sampling at 24 bits, and most software sequencers are able to handle digital audio up to 32 bits. Do the math to see what an improvement this is over the "old" 16-bit system! Another way to understand and have a visual representation of how the sampling frequency and the bit resolution affect the final sound quality of your recording is to compare the audio signal to a picture taken by a digital

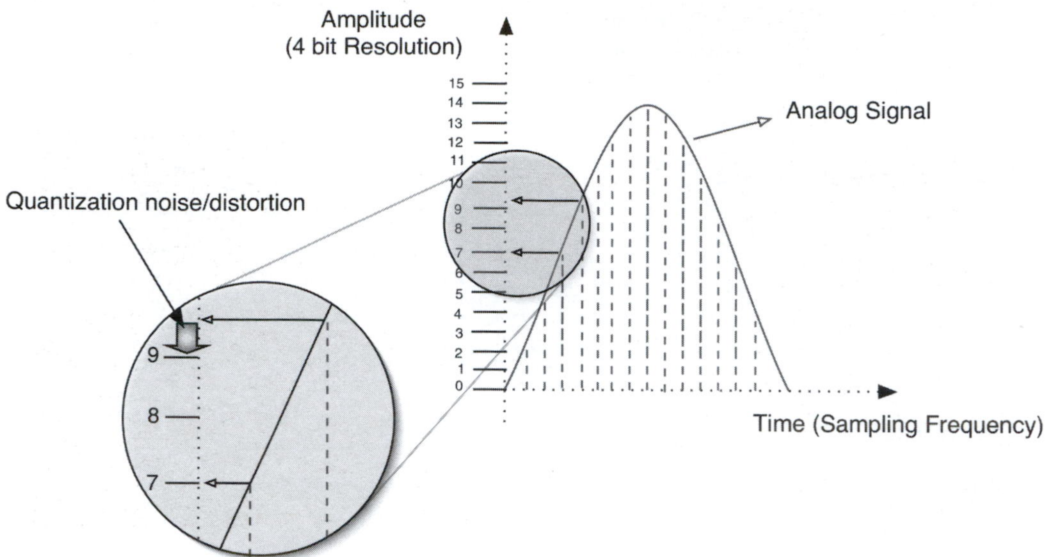

Figure 1.38 Quantization noise.

camera. The process used to digitize, or scan, a picture is very similar to the audio sampling process found in a digital audio interface. If we compare the picture to the audio signal we see that the sampling frequency has a direct impact on how "wide" the spectrum of the sampled signal will be. With higher sampling rates we are able to expand the range of the converters and therefore can capture higher frequencies (Figure 1.39).

The bit resolution has an impact on the clarity and definition of the digitized sound. As shown in Figure 1.40, an analog signal sampled at a low bit resolution has a low definition, and the details of the sound are lost, while at a higher resolution the sound will be crispier and more defined.

Listen to Examples 1.1 through 1.6 on the audio CD and compare the same audio material recorded with a sampling frequency of 44.1, 22.05, and 11 kHz and at a resolution of 16, 10, and 8 bits.

1.6.3 The audio interface inputs and outputs

Another important factor to consider when buying an audio interface is the number of inputs and outputs available. Obviously the higher the number, the more flexible the interface. Usually you should consider an interface with at least eight inputs and eight outputs; this will guarantee enough flexibility to handle a multitrack recording situation in your project studio. When talking about IN and OUT of an audio interface we have to differentiate between two types of connections: analog and digital. Analog connections (line or mic) allow you to connect to the converters of the audio interface microphones, acoustic instruments, sound generators from synthesizers, etc. Digital connectors transfer digital signals without any conversion, and they bypass the internal converters. While analog IN and OUT

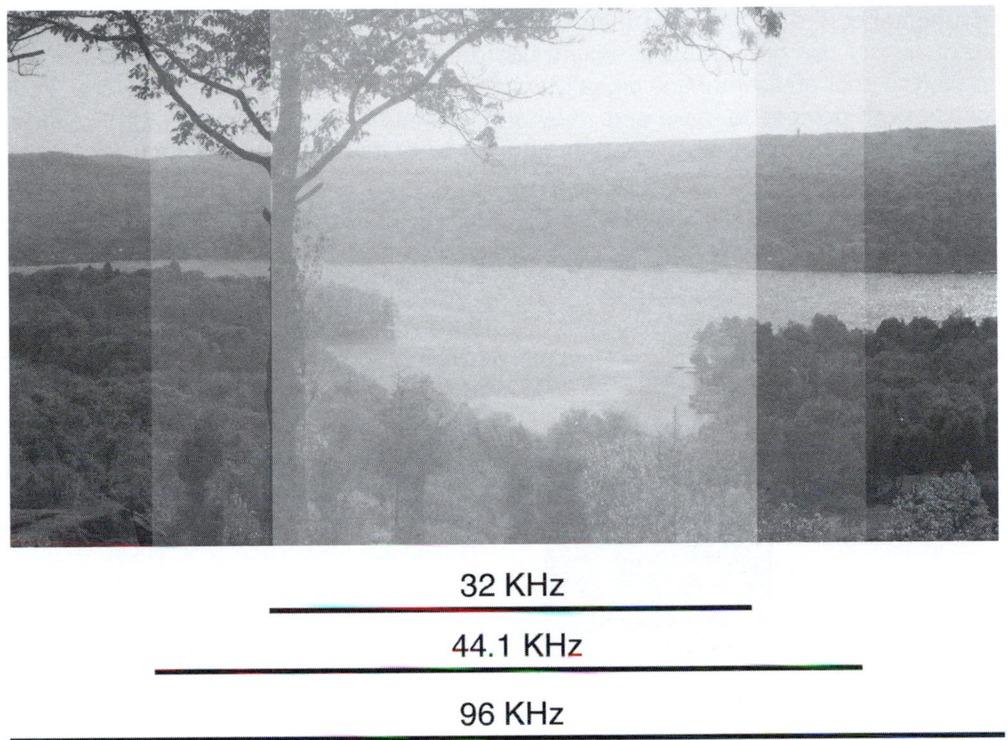

Figure 1.39 Impact of sampling frequency.

Figure 1.40 Impact of bit resolution.

use popular connectors such as XLR, ¼-inch jacks, and RCA, digital IN and OUT can use a variety of connectors and format. The names, pictures (Figures 1.41–1.45), and specifications of the most common digital protocol connectors are presented in Table 1.4.

Usually even the most basic audio interfaces come equipped with S/PDIF or TosLink connectors. If you are planning to have several digital components in your studio, it is important to use digital connectors so that they can all be connected digitally without any loss of audio quality. If you have a digital mixing board it is particularly interesting to be able to connect it to your audio interface digitally, in which case you have to check the connection format of both devices to make sure they are compatible. Keep in mind that it is always possible to buy converters between the different formats mentioned in Table 1.4.

Table 1.4 Digital connectors and specifications

Connector type	Picture	Specifications
S/PDIF Sony/Philips Digital Interface	**Figure 1.41**	Uses 75-Ω coaxial cable and RCA connectors to carry two channels of digital signal of up to 48 kHz and 24 bits The cables can reach up to 10–15 m in length
AES/EBU Audio Engineering Society/European Broadcasting Union	AES/EBU IN AES/EBU OUT **Figure 1.42**	Carries the same digital signal of S/PDIF, with the addition of subcode information Uses XLR connectors on 110-Ω balanced cable that can run for at least 50 m
TosLink	**Figure 1.43**	An optical version of S/PDIF Carries two channels of digital signal of up to 48 kHz and 24 bits Carried on fiber-optic cable
ADAT Lightpipe	**Figure 1.44**	Carries eight channels of digital audio of up to 48 kHz and 24 bits Carried on fiber-optic cable
TDIF Teac Digital Interface Format	TDIF-1 **Figure 1.45**	Carries eight channels of digital audio of up to 48 kHz and 24 bits Carried on 25-pin D-subconnectors

In order to guarantee a perfect transition of the samples in the digital domain from the digital output of one device to the digital input of another, the digital stream needs to be appropriately synchronized so that the bits and bytes are transferred without errors. For this reason one of the key points in connecting two or more digital devices together is to make sure they all follow a "clock" to transfer the digital data. This particular clock is called "*word clock*" since it synchronizes the digital *words* transferred between the devices. The word clock, depending on the type of connection used, can be embedded in the digital stream of data, as in the S/PDIF and AES/EBU protocol, or transmitted separately on BNC connectors. Depending on your configuration and devices' features you can have the word clock simply transmitted from the output of one device (the master device) to the input of the receiving device (the slave). If your studio has many digital devices that need to be connected digitally with one another, then the best option is to buy a word clock generator/distributor that will make sure all the digital devices in your studio are fed with the same word clock. This option will guarantee a smooth and error-free digital network.

A typical audio interface is equipped with eight analog inputs (with at least two inputs with mic preamps), eight analog outputs, eight digital inputs, eight digital outputs (most likely in ADAT Lightpipe format or TDIF), and S/PDIF or AES/EBU stereo IN and OUT. This type of interface would provide a good starting point for your digitally equipped project studio. More expensive devices will provide additional inputs and outputs, a higher sample rate, and a higher-bit resolution.

1.6.4 Audio interface connections

An audio interface can connect to your computer in three main ways: through a PCI (peripheral component interconnect) slot, a USB (universal serial bus) interface, or a FireWire (FW, also called IEEE 1394) interface. The type of connection to your computer doesn't really have a definite impact on the sound quality, but you have the option to choose among these three main types of connection based primarily on the features of your computer. PCI-based audio interfaces are attached directly to the motherboard of the computer, reducing the need for extra cables. Another advantage of this type of interface is that, because of its direct connection, it doesn't tax the CPU (central processing unit) of the computer as much as FW or USB interfaces. The analog and digital connections on a PCI interface can be either built into the back of the card (Figure 1.46a) or sent to a breakout box connected to the PCI card with a proprietary cable (Figure 1.46b).

a

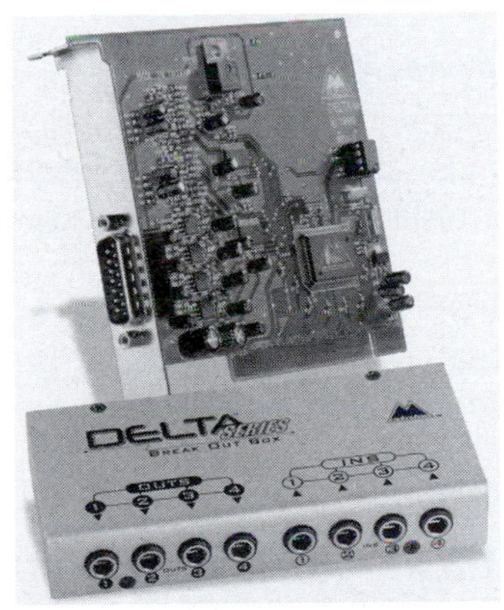

b

Figure 1.46 PCI audio card with (a) built-in connectors and (b) breakout box (Courtesy of M-Audio).

While the design with built-in connectors is more compact, its connections are awkward to reach and sometimes, due to the vicinity of the circuitry of the computer, noisier than the breakout box version. The breakout box design offers instead the best and most flexible design for your project studio. Both models need a free PCI slot on your host computer to accommodate the card. While this is not a problem if you have a desktop machine with at least one PCI slot, it is definitely a restrictive factor if you own a notebook computer or an "all-in-one" computer, such as the iMac or the eMac. There are a few options available if you really need to use a PCI audio card with your notebook computer, but it is pretty expensive and not worth the investment unless portability is mandatory for your working situation. A company called MAGMA Mobility Electronics Inc. manufactures a series of expansion chassis for both desktop and notebook computers that allow you to increase the number of PCI slots available on your machine. In the case of a notebook computer, the expansion chassis is connected through a card-bus expansion that slides into the card-bus slot of your notebook computer. If you have an "all-in-one" machine, then you are out of luck in terms of PCI audio cards. But don't despair, for the answer to your problem is in the next paragraph!

Another way of connecting an audio interface to you computer is through a FireWire or USB connection. These types of interfaces have the advantage of not requiring a PCI slot and therefore are ideal for notebook computers. But this is becoming true of other types of computers as well, for more and more manufacturers are switching to USB and FW interfaces to address the needs of professional studios. These devices are connected to the CPU by a single USB or FW cable, minimizing the setup and installation time. The main difference between the two is, of course, the type of cable but also the performance features. While in terms of audio quality the type of connection has no impact whatsoever, it makes a difference in terms of simultaneous audio channels that can be transferred to and from the audio interface. Since the information traveling from the interface to the computer, and vice versa, is digital, the bandwidth of the connection has a direct impact on the amount of data (meaning the number of channels) that can be transferred at the same time. Table 1.5 pictures the two types of USB and two types of FW interfaces currently available (Figures 1.47–1.50) and lists their performance differences.

At the moment, among the options presented in Table 1.5, FW *a* interfaces and FW *b* are better for professional audio interfaces that require stability, a high number of audio channels, and high sample frequencies (Figure 1.51). USB 1.1 is a good solution for portable interfaces that require a lower number of simultaneous channels in record (Figure 1.52).

USB 2 is fairly new in the audio interface industry but it is a promising standard, since it features a slightly better bandwidth than FW *a* but is cheaper to build. At the moment though, there are not too many commercially available audio interfaces using this protocol.

1.6.5 Software and audio interface compatibility

An important factor in deciding which audio interface to base your studio on is the software you are going to use. There are really two main approaches when dealing with software and interface. The first is a so-called *open* approach, which means that manufacturers are willing to have their software or hardware work with other company's products. The

Table 1.5 USB and FireWire connector specifications

Type of interface	Connector	Burst speed (MB/s)	Audio applications
USB 1.1	**Figure 1.47**	1.5	Basic audio applications Not recommended for more than two/four tracks in record at 44.1 kHz and 24 bits
USB 2	**Figure 1.48**	60	Advanced audio applications Can easily handle at least 16 tracks in record at 96 kHz and 24 bits Fairly new in the audio interface market
FW 400*a*	**Figure 1.49**	50	Advanced audio applications Can easily handle at least 16 tracks in record at 96 kHz and 24 bits Well established
FW 800*b*	**Figure 1.50**	100	Only very few audio interfaces available at the moment Advanced audio applications Theoretically twice the bandwidth of FW 400*a*

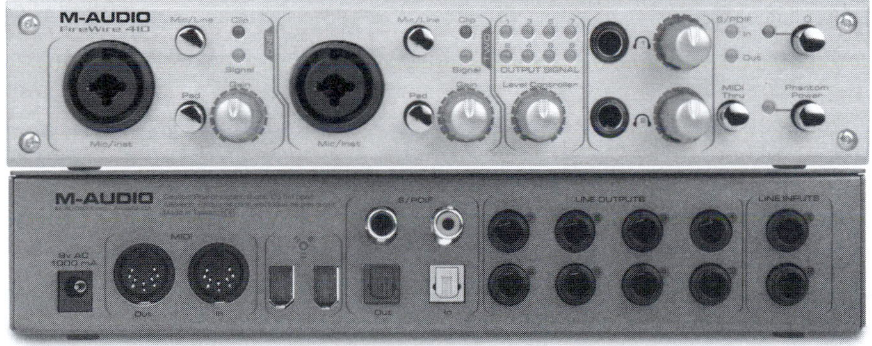

Figure 1.51 FW audio interface (Courtesy of M-Audio).

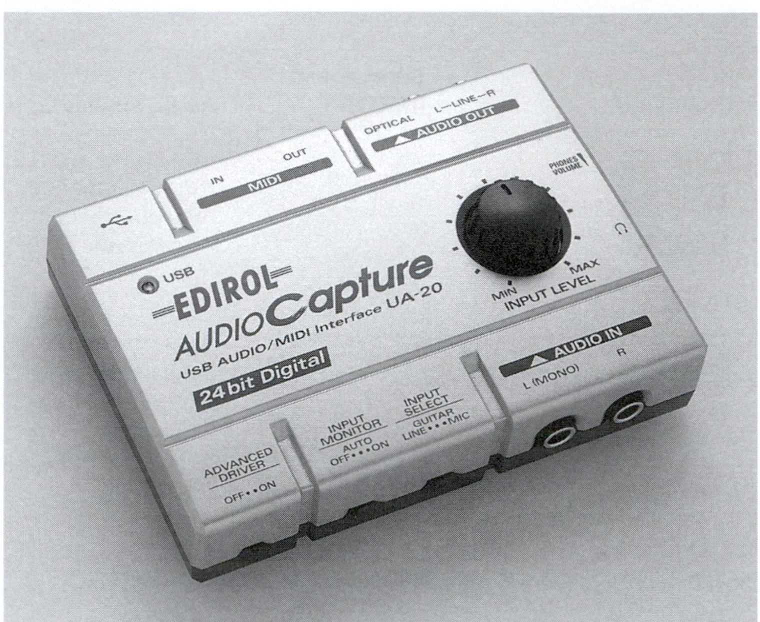

Figure 1.52 USB 1.1 interface (Courtesy of Roland Corporation U.S.).

second option is a *closed* approach, where the manufacturer wants to create a protected market and therefore produces hardware specifically targeted for its software, or vice versa. The main advantage of the open approach is that consumers can choose from a wide variety of sequencers and audio interfaces to select the option that best fits their needs. The software and hardware interact in this type of approach through the use of software drivers that allow the sequencer to exchange data with the audio interface. One of the drawbacks of this choice lies in the reliability, since these drivers sometimes do not guarantee total compatibility with all products.

In the closed approach, the choice is more limited because usually the software manufacturer also delivers the audio interface and uses proprietary drivers to interface the software and the hardware. This mean that if you buy a particular brand of software, then only audio interfaces of that brand will be compatible with that software. The advantage of this approach is that the setup is usually more stable, since there is only one company in charge of both hardware and software. A typical example of this approach is the one used by Digidesign, which makes the software ProTools and has a solid line of audio interfaces, such as Mbox, Digi 002, and HD. While these three interfaces work with other sequencers, ProTools works only with Digidesign audio interfaces, making it a very reliable setup.

Among the audio sequencers that follow the open approach, the most widely used in the professional production environment are Digital Performer (by Mark of the Unicorn), Cubase SX and Nuendo (by Steinberg), and Logic Pro (by Apple/Emagic), with Sonar (by Cakewalk) catching up with new, interesting features. Before buying your audio interface, choose the software you will use in your studio and make sure that it supports the audio interface you are planning to buy. In Table 1.6 I sum up the main audio sequencers and the audio drivers they support.

Table 1.6 Audio drivers supported by the main sequencers

Software	Mac OS X	Windows XP
Digital Performer 4.12	Core Audio Digidesign Audio Engine (DAE)	—
Cubase SX 2.2	Core Audio	Windows MME ASIO
Nuendo 2.1	Core Audio	Windows MME Direct Sound ASIO
Logic Pro 6.4.2	Core Audio Digidesign Audio Engine (DAE)	—
ProTools 6.2.2	Digidesign	Digidesign

1.7 Which software sequencer?

This is another one of the hottest topics among musicians and producers. Just to clarify the matter, keep in mind that there is no such thing as a universally perfect audio sequencer. Each application has strengths and weaknesses that make the product appealing to different categories and levels of users. When considering the software that better fits your needs, analyze the following issues: What are the primary goals you want to achieve with your audio sequencer? What is your level of knowledge in terms of MIDI, audio, and troubleshooting? Which features are most important for your work? Which software are your colleagues mainly using? Do you plan to do a lot of freelance work, or will you be using your studio mainly for you own productions? Which platform (PC or Mac) will you use? By answering these simple questions you will have a much better and clearer idea of the perfect match for your studio. Let's analyze each of these points in detail.

1.7.1 The primary goals you want to achieve with your audio sequencer

The perfect fit for your audio sequencer depends greatly on the type of work you plan to do in your studio. If you will mainly concentrate on tracking live instruments, then you will need a program that guarantees the highest reliability. ProTools has been renowned for its dependable performance in studios all around the world. Other programs, though, such as Nuendo, Logic, and Digital Performer, are quickly catching up in terms of performance and reliability. If you plan to use your studio mainly for multitrack mixing, then pretty much all the main programs are well suited. They all offer the same features in terms of editing power, automation, sound quality, and performance. Keep in mind that for multitrack mixing what can really make a difference is the quality and selection of the plug-ins you use. The graphic interface and the working area layout will also have a pretty big impact for this type of task. If, on the other hand, you plan to do more work composing and producing music and you consider yourself more a composer than an audio engineer, then you should opt for a program oriented toward a musical/creative working approach, such as Digital Performer, Logic, and Cubase. These three audio sequencers were originally created with the composer as the final user in mind. Their MIDI capabilities and editing features are much greater than those of ProTools (which, for example, doesn't have a music notation editor or a MIDI list window), while their audio features are as good as those of ProTools. If notation and score printing are crucial features for you, then Logic or Cubase might be a better choice, since their scoring properties are definitely more complete than the ones offered in Digital Performer.

1.7.2 Ease of use and learning curve

Each program has different levels of interactivity and different learning curves. While a sequencer might appeal to certain users for its clear and straightforward interface, another program might be more appropriate to advanced users in search of a more customizable environment. If you are at your first trial sequencing and you are looking for a simple yet powerful way to start producing your music, I suggest Digital Performer, ProTools, or Cubase SX (Cubase SX and Nuendo's sequencing features are almost identical and therefore I will refer to Cubase SX only from now on). Their graphic interface is simple, clean, and straightforward. Their logical approach to sequencing makes them easy to control and fairly accessible after just a few weeks of use, even to the novice. Though their learning curve is not particularly steep, they can be used at different levels of complexity and therefore appeal to the beginner as much as to the professional. Logic Pro has a much steeper learning curve, and I wouldn't recommend it to the inexperienced MIDI composer. Where Logic Pro really shines is in its infinite ability to be customized in any small detail, making it a great choice for the advanced producer. Keep in mind that all four programs are very similar in their overall features and that they share about 80% of them. Therefore the choice sometimes can really be based on their ease of use or graphic interface.

1.7.3 Which features suit you best?

Even though these programs have very similar characteristics, each has a few distinctive features that might make it more appealing to you and your sequencing style. Identifying these features is crucial in order to make the right choice. As I mentioned before, ProTools (PT) lacks the MIDI editing power of other programs, but its audio features, which have been tested and developed for decades, are excellent. PT has an easy and intuitive interface and a very rich set of keyboard shortcuts to expedite the editing work. It is based on only three main windows (transport, mix, and edit), which makes it a breeze to handle and navigate (Figure 1.53). Therefore if you are looking for ease of use, reliability, and powerful audio editing, PT is perfect for you.

Digital Performer (DP) has strong and very powerful MIDI editing features, making it the perfect choice for the modern composer. Its graphic interface is among the best available, very clean, linear, yet appealing to the eye. The main window (called the track list window)

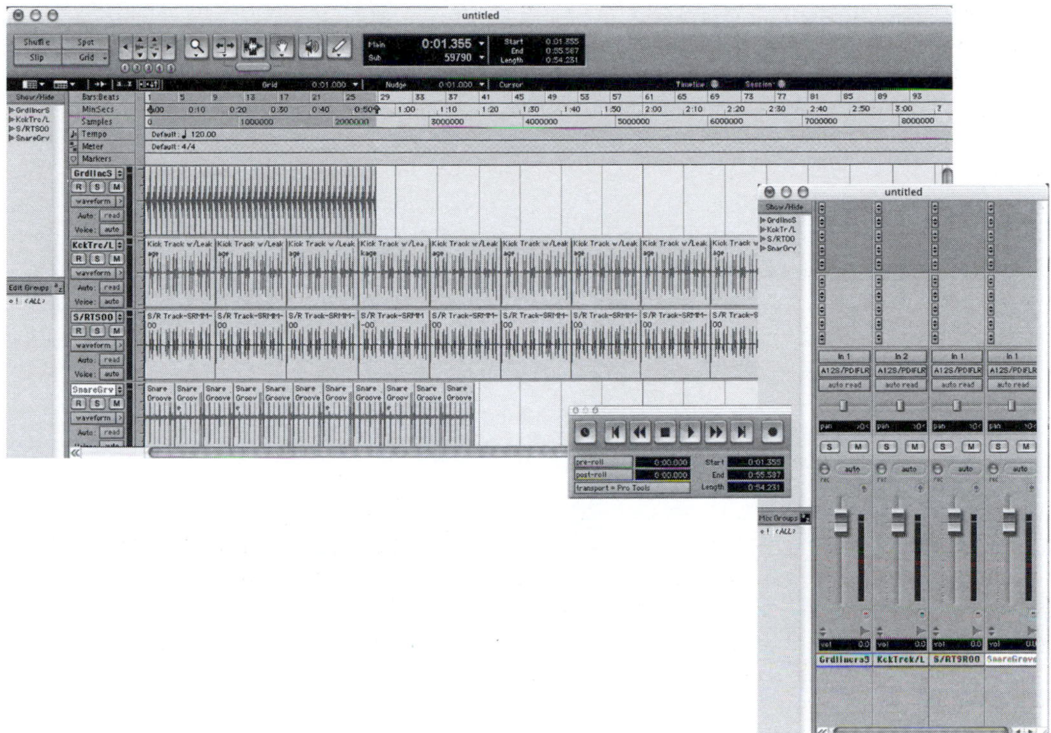

Figure 1.53 ProTools 6.2.2.

shows an overview of the project with MIDI and audio track assignments. DP's learning curve and setup procedures are quick and straightforward; it takes only days to be up and running, even if you are not very experienced in sequencing. While only some basic editing actions can be accomplished from this window, DP offers comprehensive edit modes for both audio and MIDI, including score, list, graphic (or piano-roll style), mix, sequencer (audio and MIDI together), drums, and graphic and list tempo changes (Figure 1.54). The audio features of DP are as advanced as those of PT, with a graphic interface that resembles the one used by Digidesign's software. If you are looking for a program that doesn't get in the way of your creative process, enhances your inspiration, and is a breeze to set up, DP might be what you are looking for.

Cubase SX (CSX) has a very similar approach to DP, with some enhancement in terms of graphic interface and score editing/printing. CSX has an advanced track list window that allows you to control different parameters of a track, such as volume, pan, effects, EQs, graphic editing and automation, right from the main screen (Figure 1.55). This approach is particularly helpful for projects that need to be done quickly. As in DP, more advanced editing can be done using the several edit windows available, including list, graphic, mix, score, drums, and graphic and list tempo changes.

Logic Pro (LP) is probably the most advanced program of the bunch in terms of customizable features and advanced editing options. It can be personalized to fit almost any need. The editing options are very extensive and include the classic graphic/piano roll (in LP called

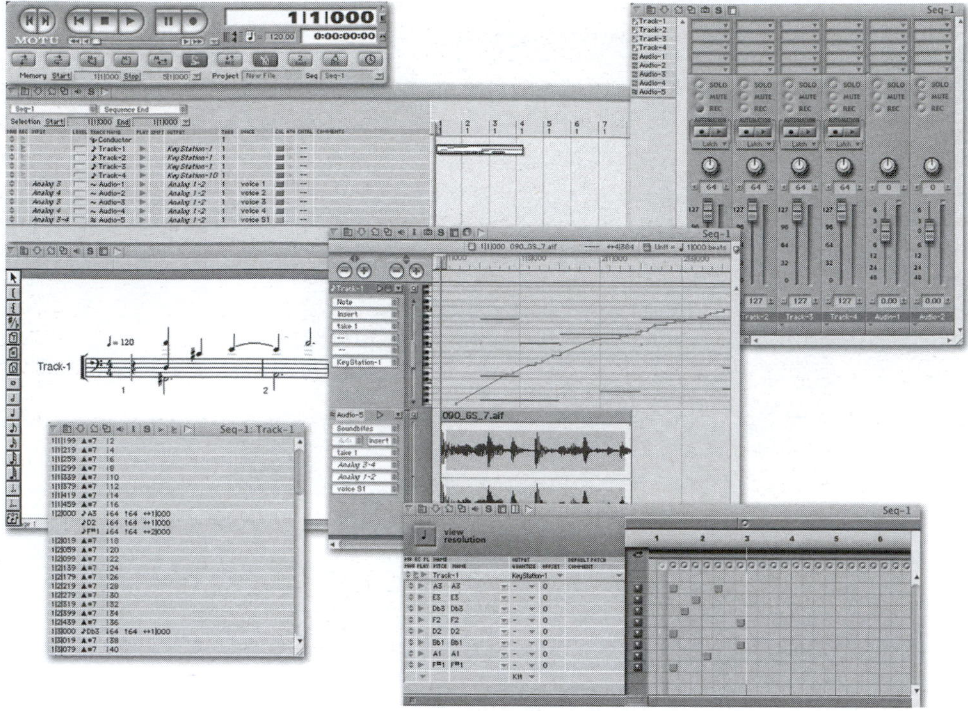

Figure 1.54 Digital Performer 4.12.

Figure 1.55 Cubase SX 2.2.

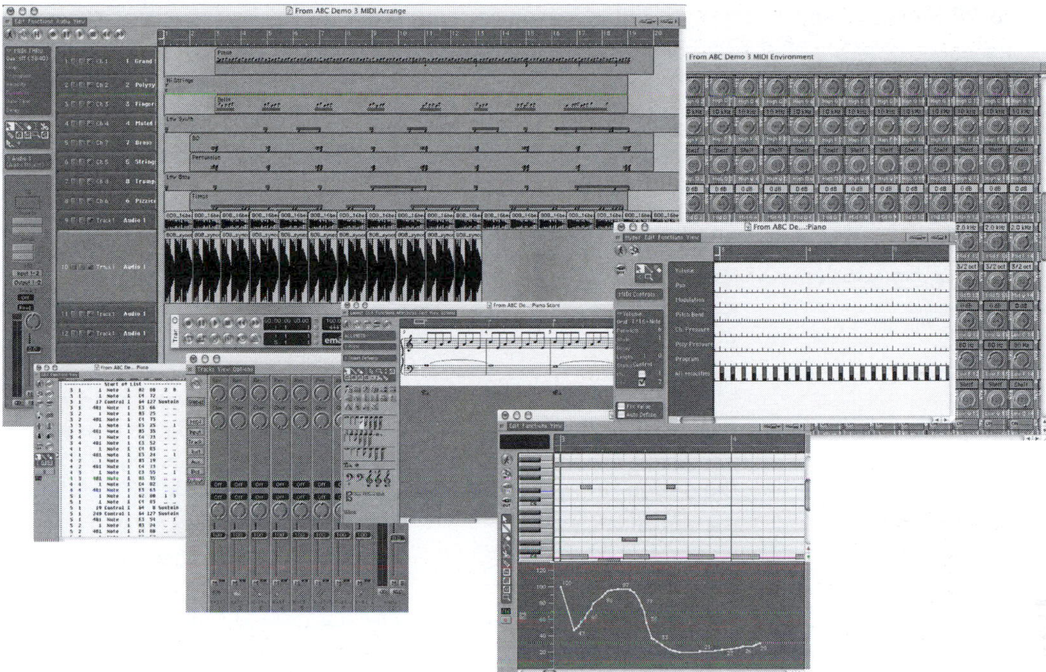

Figure 1.56 Logic Pro 6.4.2.

the Matrix edit window), the list editor, the score editor, the mixer, and the Hyper Editor (used mainly to edit MIDI CC messages) (Figure 1.56). The main track window (called the Arrange window, as in CSX) allows you to control almost every aspect of your production, such as automation, plug-ins, effects, and basic parts editing. LP can be a bit intimidating at the beginning. The fairly complex handling of multiple windows is very powerful but a bit confusing for the inexperienced musician. Every aspect of your sequencer (MIDI and audio) is based on the "environment," which represents your virtual studio. Before using LP for the first time you have to configure the main environment, which, at the most basic level, is a digital description of your studio. The bad part is that the configuration of the environment can be pretty confusing at the beginning. The good part is that its engine is very powerful and allows you to achieve very complex tasks that are not possible with other programs. Included with LP is also a very extensive suite of effects and software synthesizer plug-ins.

1.7.4 Other factors to consider

An important aspect to consider when analyzing which audio sequencer to use is based on which program the majority of your colleagues and clients have in their studios. Even though exchanging projects between programs is possible and becoming easier with interchange formats such as the OMF 2, it is much faster to collaborate with studios and composers that work with the same platform. Do a quick survey among all your colleagues and determine which is their sequencer of choice. Also take into account how much work and how many collaborations you expect to share with them. Do the same with your most important clients, and try to get an idea about which program would make

Table 1.7 Strengths and weaknesses of the four audio sequencers

Program	Strengths	Weaknesses
ProTools 6.2.2 by Digidesign	Very reliable Solid audio editor Solid mixing features Easy learning curve Available for Mac and PC	Basic MIDI editing features Supports Digidesign audio interfaces only
Logic Pro 6.4.2 by Apple/Emagic	Most advanced editing features Advanced score editor Stable Highly customizable Supports a variety of audio interfaces, including Digidesign HD systems	Steep learning curve Cumbersome graphic interface Unconventional nomenclature and shortcuts Graphic interface can sometimes get in the way of the creative process Time consuming to configure Available for Mac only
Cubase SX 2.2 by Steinberg	Extremely intuitive interface Advanced score editor Advanced MIDI editing features Extremely versatile arrange/track window Slick graphic interface Available for Mac and PC Extremely flexible environment and workflow organization	Could be more stable Audio editing could be improved
Digital Performer 4.12 by Mark of the Unicorn	Very clear graphic interface Great MIDI features Very musical and creative approach Fairly easy learning curve	Score editor needs improvement Track list window too limited Could be more stable Requires a pretty powerful computer for a high number of tracks Available for Mac only

it easier to collaborate with them. Probably you will end up with at least two sequencers at the top of your list, which means that if your budget is not too high, you will have to choose one for now and buy the other one (or ones) later. In fact, be prepared eventually to learn, use, and master more than one program, since you might want to take advantage of the key features of each system to bring your sequencing skills to the next level.

In order to have a clearer idea of the pros and cons of the four audio sequencers presented in this book I have summed up their strengths and weaknesses in Table 1.7.

1.7.5 What about the computer?

OK, everything is almost set. But there is one more piece of equipment in the studio that we have to analyze before getting into sequencing in the next chapter: the computer. Even though not strictly a musical instrument or an audio device, the computer plays a crucial role in your productions. We know by now that it serves as both MIDI and audio recorder/playback device, and therefore we can think of it as the "brain" of your studio. Keep in mind that we could devote hundreds of page to this topic, but this would go beyond the scope of this book. Here I want to give you some practical information that will come in handy when you have to decide which computer to get for your studio.

First of all let's take a look at the two main platforms available: Apple computers (which run the Mac OS) and PCs (which run Microsoft Windows). The debate on which one is better has been going on a long time, basically since the mid-1980s, when both operating systems (or OS) become available to the general public. As you have probably learned by now, there is no perfect OS. Each one has advantages and disadvantages that are really up to the end user to consider. Apple machines have the reputation of being more expensive, while PCs are believed to be cheaper. Even though to a certain extent this is true, you have to consider that Apple computers are slightly more expensive for a reason, mainly because they usually come equipped with features you might have to add as options to a standard PC, such as FW 800 ports, dual video outputs, and dual internal HD. Another advantage that can be attributed to Apple computers is their stability. No matter how much your next-door neighbor brags about his superfast PC that can show you 200 frames per second (fps) playing Unreal Tournament 2004, for your studio you need a machine that is reliable, stable, easy to configure, and troubleshoot. Macs here have a clear advantage, especially with the introduction of OS X.

Remember that you are going to use your computer for music *only*; do not try to run games, Internet surfing, e-mail, bank account management, shopping list, and more on your main machine. You must have a clean, simple, and well-organized computer to run the audio sequencer in your studio, so leave the other tasks to a second machine.

The latest version of Windows XP is a greatly improved OS, and it has become a stable alternative to Mac OS X. The timing of its data transfer paths has come up to standard for MIDI and audio applications, while its stability is solid. Nevertheless one of the main advantages of Apple computers is that the same company makes both the machines and the OS. This guarantees a tighter and more stable interplay between the two. In the case of PCs we have thousands of different designs made by thousands of companies over which Microsoft has no direct control. This comes down to several incompatibility issues that can ruin your creative day more often than you would like. Remember that all the equipment discussed so far (computer included) are only tools that allow your creativity to develop and grow. As any other tool you don't want them to get in the way of your compositional process. If you were writing for orchestra you wouldn't be happy if your first violinist came up to you every time he or she broke a string, right? Well, since the MIDI studio is your modern orchestra, in the same way you don't want to be interrupted by technical problems. Therefore my advice is always to choose the more stable, safe, and sound options, even though they can sometimes be a bit more expensive.

When considering a computer for your studio think in terms of the best machine you can afford. While for MIDI-only sequencing you don't need a very fast CPU, for audio data handling you need the fastest CPU available, especially if you plan to have a high track count with many plug-in effects. It is hard to give some reference about CPU speed, since the rate at which computers evolve in terms of features and options is extremely high. Besides considering the clock speed of your main processor (or processors in the case of a multiprocessor machine), you have to make sure you have the fastest bus speed supported by your motherboard. The bus system is the connection path over which the data are transferred from and to the different components of your motherboard, such as the I/O interfaces, CPU, and memory. Having a fast bus speed allows your computer to perform better in all the tasks involved with digital audio recording and playback. Always try to install as much memory as your computer allows. Random access memory (RAM) is the temporary storage space your CPU uses to store big chunks of data in transit to and from the I/O ports; if the CPU runs out of RAM, then it will use the hard disk (permanent storage space) instead, as virtual memory. Remember that the hard disk handles information much more slowly than RAM. This option therefore would work fine for applications where the timing of data fetching is not an issue. But with MIDI and audio you need very tight time management, and therefore more available RAM allows your computer to handle more data with tighter timing.

Permanent storage space devices, such as internal and external hard disks (HDs), are also a very important aspect of your computer system. When you record audio sessions in your audio sequencer, the data converted from the AD converters are stored on the HD of your computer as a sequence of 1s and 0s. The higher the sampling frequency and bit resolution of the digital conversion, the more data the computer needs to store on the HD. You will soon realize that those 80 GB of hard disk space you thought would last forever won't be enough. While I will discuss space and file management of your sessions later in the book, for now keep in mind that, as with the memory issue, for HD the bigger the better. Not only can a bigger HD hold more data, but it is also usually faster, since the computer has more space and freedom to organize the data in a rational way. In terms of speed, when buying an HD for audio applications, get a model that uses mechanisms with at least 7200 rpm, 10 ms or better access time, and at least 8–10 MB/s transfer rate.

Always have at least three HDs in your studio. The first one is the internal one that holds the OS and the applications (or programs) you run, such as sequencer, notation program, and utilities. The second HD, which can be either internal or external, is the main session disk, where you store your projects during the recording sessions. The third HD is the "overnight" backup (more on backing up and archiving in later chapters), where you make a copy of your project as soon as you have a minute, even during the session itself. For no reason should you allow even one day to go by without making a copy of your current projects to the backup device! Catastrophic events always seem to happen right before the deadline of your important project, so you *must* have a plan B (as in backup!).

Another feature you should consider equipping your computer with is a multiple video monitor setup. The regular one-video configuration is good enough for word processing and Web surfing applications, but it is way too limited for MIDI and audio composing/sequencing. Imagine having to write a score for studio orchestra on a 2-inch by 2-inch piece of paper. Well, if you have only one monitor, that's pretty much the feeling you would get. On the other hand, if you have a two (or more) monitors, you increase your desktop "real estate" without affecting considerably your budget. What you need is an additional video card for your computer—nothing fancy, a card that supports at least a video resolution of 1024 × 768 will work fine—and of course a second monitor. The advantage of having a second monitor is that you will be able to organize the edit windows of your sequencer across both desktops and drag elements and windows of your current project across the two monitors, making recording and mixing on your computer a much more enjoyable experience. What I usually do is to set a template so that when I open my audio sequencer I immediately have the track/arrange window displayed on the main monitor and the mix window on the second. This setup allows me to do quick edits and to apply changes to my mix without opening and closing different windows. Because of the low cost of monitors (especially CRT monitors) and video cards, this is something you should seriously consider.

1.8 Final considerations and budget issues

Well, now you should have most of your pieces together to move on to the next chapter and finally start sequencing and using your equipment. We added different sections of your project studios together little by little, starting from the MIDI gears, moving to the MIDI interface and the computer, to the audio gear, such as the mixing board and the speakers, to the audio interface. Your studio now should look like the one in Figure 1.57.

While the studio shown in Figure 1.57 is a theoretical project studio, you can use it as a starting point to design your own. When planning a new MIDI/audio setup you should always start with a sketch where you carefully plan how your devices will be connected, which type of connections are required, and if possible also the configuration of the MIDI and audio

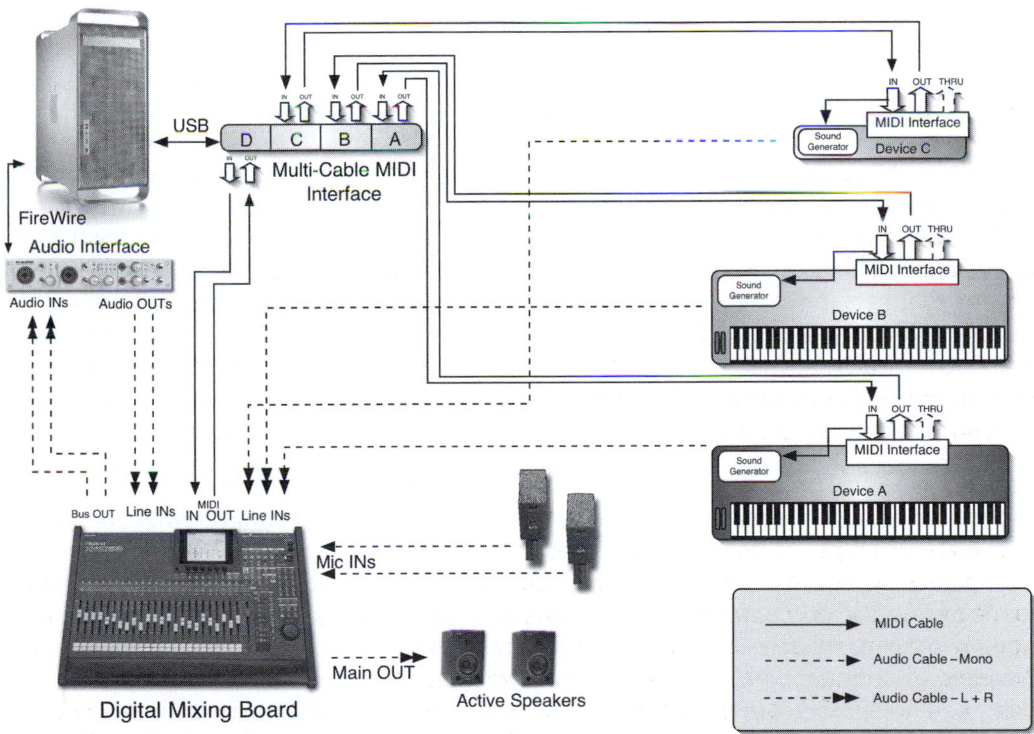

Figure 1.57 The project studio (Courtesy of Apple Computer, Roland Corporation U.S. M-Audio, and AKG).

devices. The creation of a blueprint allows you to plan your budget precisely. I highly recommend when building a project studio to start from the budget available to you and buy the equipment that fits it. These are the steps to build a balanced and efficient project studio. Start with a budget, then make a sketch of the equipment you need in your studio, and finally do some research with different music dealers and find out the prices for each device. Remember that you have to stick to the plan you designed on paper. If you did things right, you will need every device you see on the blueprint, and this will allow you to keep your focus on the pieces of equipment you really need. One of the most common mistakes is to spend most of your budget on one device (let's say your computer) and then not have enough money to buy your MIDI interface, which makes your studio unusable. Therefore be careful when planning and balancing your budget and the equipment. Remember that your final goal is to have a fully functional, versatile, and, as much as possible, trouble-free working environment so that you can focus on the creative process of writing music.

1.9 Summary

In this first chapter you learned how to prepare and set up your working environment for your music productions. The devices in your studio fall mainly into three categories of equipment: MIDI/electronic instruments, sound/audio equipment, and the computer. Two main path signals connect your devices: the MIDI network and the audio network. The former uses 5-pin DIN MIDI cables to connect all your MIDI-equipped devices through the use of built-in MIDI interfaces. Each device has an IN, an OUT, and, on most devices, a THRU port. The computer needs an external or a PCI internal MIDI interface to be able to connect to the MIDI network. This interface can be a single cable (meaning with one set of IN and OUT only) or multiple cables (meaning with multiple sets of INs and OUTs). If you have the latter, then yours is a much more advanced studio that allows you to connect your MIDI devices in parallel (Star network), reducing the transmission delay to a minimum and taking advantage of all 16 channels available on every device. If you use the former, then you have a basic MIDI setup, called Daisy Chain, where multiple devices are connected in a serial chain using the MIDI THRU ports. Remember that this configuration is limited to only 16 channels in total and that they need to be split among all your devices. The MIDI data transmitted over the MIDI network are sent to the computer that, through a software sequencer, is able to record, play back, store, and edit them. The data transmitted on the MIDI network do contain no audio information. The sound generator of each MIDI device, by reacting to incoming MIDI messages, such as Note On, is responsible for producing the audio signal that is eventually sent to the mixing board and to the speakers.

The audio network has the mixing board as the central hub of its audio connections. It collects the audio outputs from the sound generator of your MIDI devices. The audio from the mixing board is sent to the speakers through the main outputs and to the audio interface connected to the computer through the bus outputs. These outputs, usually two, four, or eight in small to mid-size boards, allow you to send to the audio interface inputs several channels of audio separately. The audio interface is connected to the computer usually using FireWire, USB, or PCI interfaces. Through the use of AD/DA converters, the audio interfaces allow the computer to record analog audio in digital binary format. The computer functions therefore as both MIDI and audio central hub of your studio, and, along with the software sequencer, it constitutes the heart of your music production environment.

1.10 Exercises

Here are some exercises to help familiarize you with the material presented in the chapter. Use them as a starting point not only to review the concepts of MIDI and audio network but also to design and create your project studio.

Exercise 1.1

Identify the MIDI devices and their components shown in Figure 1.58.

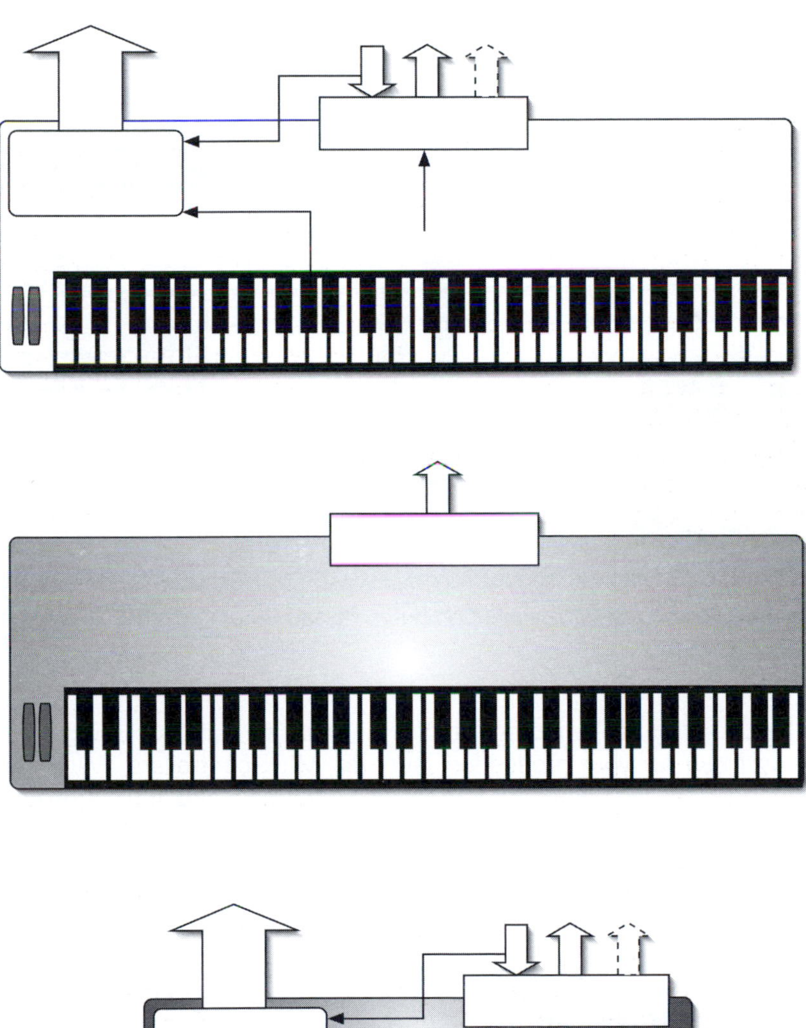

Figure 1.58

Exercise 1.2

Connect the MIDI and audio devices shown in Figure 1.59, indicating clearly the MIDI and audio cables.

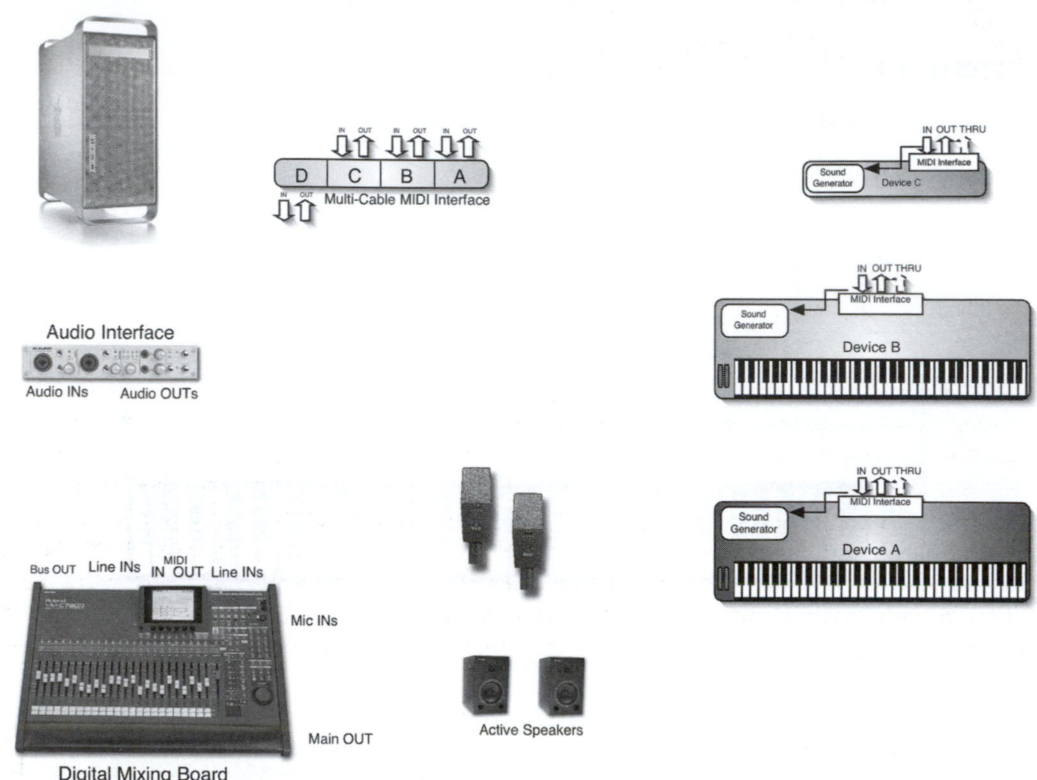

Figure 1.59 (Courtesy of Apple Computer, Roland Corporation U.S. and AKG).

Exercise 1.3

a. Create the blueprint of the project studio you want to build, including all the devices that you will use for your production, such as MIDI controllers, MIDI synthesizers, sound modules, mixing board, computer, MIDI interface, and audio interface.
b. Connect the MIDI and audio devices you designed in your blueprint, indicating clearly the MIDI and audio cables.

Exercise 1.4

Using the resources on the Web or catalogs from different music stores, replace the blueprint from Exercise 1.3 with real devices available on the market that appeal to you. Come up with three different studio scenarios that can fit the following three budgets:

1. Studio A: budget of $25,000
2. Studio B: budget of $12,000
3. Studio C: budget of $8000

2 Basic Sequencing Techniques

2.1 Introduction

Now that you have designed and set up your studio, it is time to start sequencing. If you read carefully the preceding discussion on studio setup, you should be familiar already with the equipment you are going to use in the following pages. In this chapter you start learning the concepts of sequencing. Let me be very clear from the beginning: Sequencing is a lot of fun! Making music while having almost no limitations in terms of instruments, sound palette, editing features, and effects is an amazing experience; at the same time, it can be a bit overwhelming if not approached systematically and with organization. You are going to learn how to set up a sequencing session from different points of view in order to minimize the problems and maximize the fun of writing and producing your own music.

In the first part of this chapter you learn how a sequencer works. How you organize your session is crucial in order to have a smooth sequencing experience. We go through all the necessary steps to avoid mistakes that could cause problems down the road.

In the second part of the chapter we start sequencing, using MIDI tracks first and then including audio tracks with some basic editing techniques. At the end of the chapter is a quick review, some tips, and some exercises you can use as a starting point to experiment on your own.

2.2 The sequencer: concepts and review

As explained in Chapter 1, a MIDI sequencer (software or hardware) allows you to record, play back, edit, and store MIDI data. From now on we are going to learn sequencing techniques on the four top-of-the-line software sequencers: ProTools, Digital Performer, Cubase SX and Logic Pro. While some of the techniques can also be applied to hardware sequencers, in order to take full advantage of the MIDI and audio features available nowadays you really should have a computer-based studio running a software sequencer. It is also important to recognize that as a result of the evolution in technology we have experienced since the mid-1990s, modern computer-based sequencers have the ability to record not only MIDI data but also audio information. This means you will be able to create complex arrangements that include synthesized MIDI tracks and acoustic tracks at the same time, raising the entire sequencing experience to a much higher level. The ability to merge MIDI and audio tracks in a mix not only makes it possible for you to improve the sonority and realism of sampled and synthesized sounds, but also enables you to record or import any audio material (such as vocal tracks, acoustic instruments, lead parts) and include it in your sequence. Of course, the path signals that MIDI and audio data follow are very different, and it is very important to understand the difference between the two. As we learned in Chapter 1, while the MIDI data reach the computer through the MIDI interface (which allows the computer to communicate and exchange MIDI data with other MIDI devices), audio data are recorded and played back through the use of an audio interface, which mainly converts the analog audio signal into digital and vice versa. These two main types of data constitute the basic elements of your sequences, and that is where you are going to start your first project. It is time to begin sequencing!

2.3 How a sequencer works and how it is organized

The first time you start a new project in your sequencer you are presented with the main track list window. Keep in mind that this option can be changed depending on the sequencer and its preferences. For example, you can create templates that have different windows and editor layouts, or you can have the sequencer customized to start up with an "Open" dialog window so you can load a preexisting project. There's no perfect setup, but I usually like to have the sequencer set to a "do nothing" option and choose each time I launch the software which action to take. The track list window is where all the tracks available in your project are organized and presented in a horizontal way. The layout of the tracks may change slightly among the four sequencers analyzed in this book, but the concept is the same for all of them. There are two main types of tracks: MIDI and audio. The MIDI tracks receive the data from the MIDI interface, usually through the USB port, while the audio tracks receive the audio from the audio interface connected to the computer through an FW, USB, or PCI connection. In Figure 2.1 you can see a comparison between the track list windows of the four sequencers.

Tracks, of course, can be added, in order to accommodate the most complex projects. It is usually a good idea to create one or more templates with the instrumental ensembles you use the most. In order to set up a song as a template in DP and CSX, you can open a new project by selecting "New" from the File menu of your application. Create and rename your MIDI tracks according to your ensemble, and then select "Save as Template." Now every time you start a new project you will be able to select the template that better matches its instrumentation. If your sequencer doesn't have an option to save templates directly from the application, there are ways around. You can create a folder on your HD where you save empty projects that act as templates. These are only "empty" sequences that you name according to their ensemble setups. In order to work with a certain template you simply have to open the project-template you previously created and then start sequencing. Make sure that the first time you save the project you select "Save As" in order to avoid writing over and replacing your template with the current project. In Mac OSX a good way to avoid this mistake is to alter the status of your project-templates files by getting the Information Window from the Finder (*Command-I*) and make sure that the

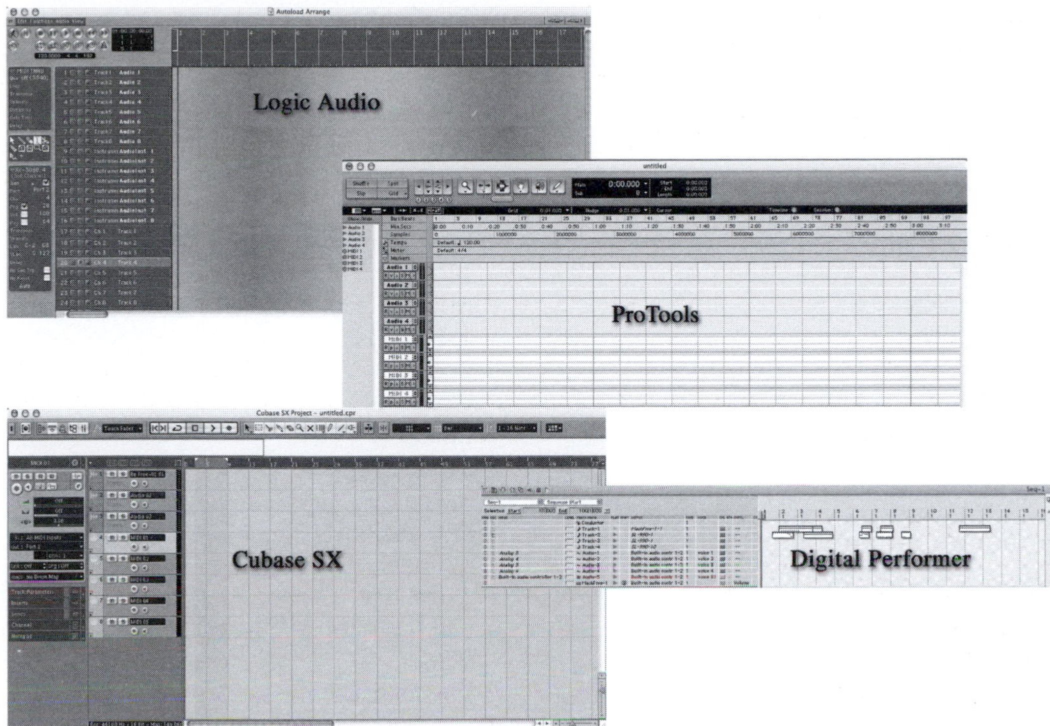

Figure 2.1 The main track list window in PT, DP, CSX, and LP.

"Stationary Pad" option is checked. Now every time you open the template from your sequencer you will work off a copy of the template. If you wish, rename the project and save it in the folder of your choice. You should have a variety of templates available in order to speed up your setup work. Here's a list of suggested templates.

Jazz Ensemble:
- Trumpet
- Sax
- Drums
- Percussion
- Piano
- Bass

Rock Ensemble:
- Distorted Guitar 1
- Guitar 2
- Lead Guitar
- Drums
- Percussion
- Keyboards
- Bass

Latin Ensemble:
- Trumpet
- Trombone
- Sax
- Drums
- Percussion 1
- Percussion 2
- Keyboards
- Piano
- Bass

Symphony Orchestra:
- Piccolo
- Flutes
- Oboes
- Cor Anglais

- Clarinets
- Bass Clarinets
- Bassoons
- Double Bassoons
- Trumpets
- Trombones
- Bass Trombone
- Tuba
- Horns
- 1st Violins
- 2nd Violins
- Violas
- Cellos
- Double Basses
- Harps
- Percussion
- Timpani

Studio Orchestra:
- Piccolo
- Flutes
- Oboes
- Cor Anglais
- Clarinets
- Bass Clarinets
- Bassoons
- Saxophones
- Trumpets
- Trombones
- Bass Trombone
- Tuba
- Horns
- 1st Violins
- 2nd Violins
- Violas
- Cellos
- Double Basses
- Harps
- Percussion
- Timpani
- Drum Set
- Electric/Acoustic Bass
- Piano
- Keyboards/Synthesizers

These are just a few ideas that you can expand and adapt according to your style of writing. It is advisable, though, to have a comprehensive palette of templates in order to be ready for any situation. On the included CD you can find the foregoing templates for DP, CSX, PT, and LP.

2.4 MIDI tracks

Each MIDI track needs to be assigned to a MIDI output. Depending on your setup you will have either a single-cable or multicable MIDI interface (refer to Chapter 1 for a review of these two configurations). If you have a single-cable interface, then you have to set up the output of each track to a MIDI channel between 1 and 16; this will tell the sequencer to which channel to send out the MIDI data. The receiving channel will play back the MIDI messages sent in real time from the controller or recorded on the sequencer track. Remember that the sequencer and the computer act as the central hub of the MIDI network, so it is crucial to assign the track to the right output. If you have a multicable interface, then you have to specify for each track the cable (meaning the MIDI device) and the channel to which the MIDI data will be output. Depending on the software and the interface, each cable will be identified with a letter, a number, or the name of the device connected to it (Figure 2.2).

In Figure 2.2 the different cables of the MIDI interface are named with the model number of the device connected to it. The computer and your sequencer know this information because it was entered in the MIDI setup application that is part of the operating system of your computer. In OS X this application is called *Audio-MIDI setup*. It is reminiscent of

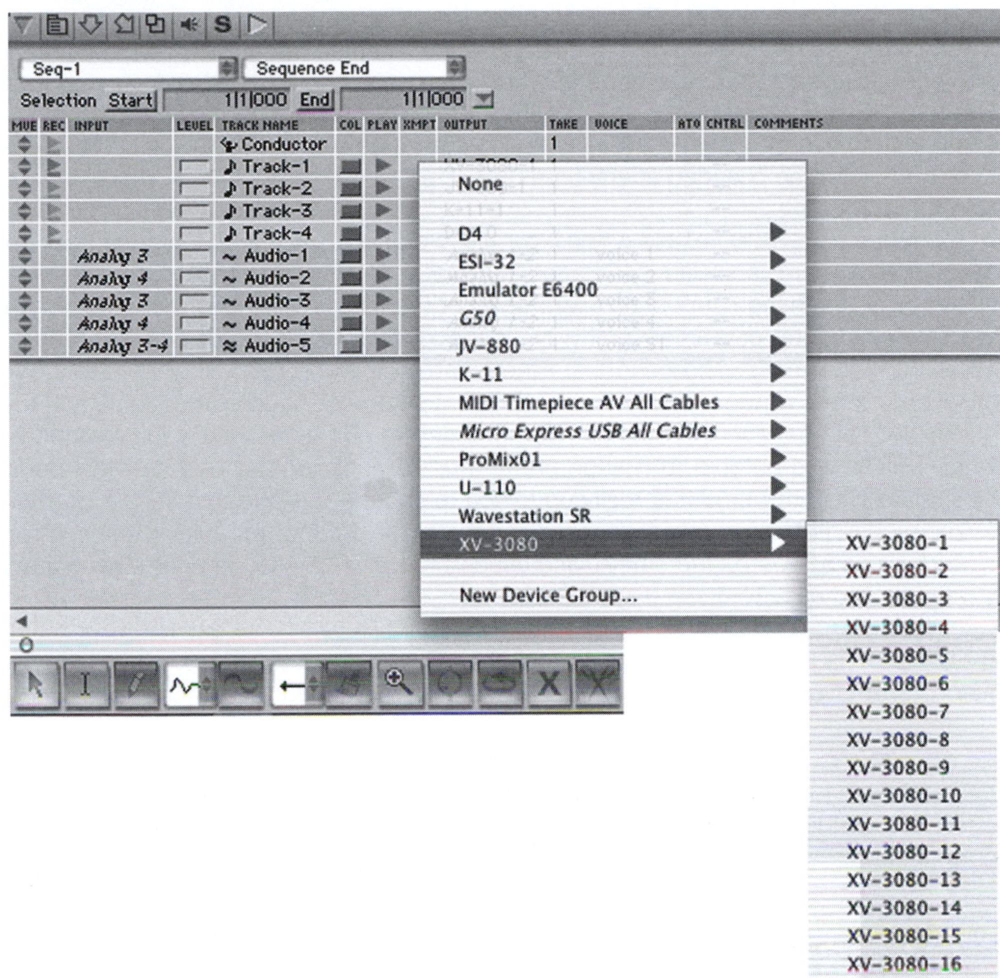

Figure 2.2 MIDI output channel/cable assignment in DP.

Free MIDI and OMS in OS 9, and it makes it very easy to set up your MIDI cable assignments. To configure your MIDI setup, launch the Audio-MIDI setup, which is located inside the Utility folder inside the Applications folder on your HD. After launching it, select the "MIDI devices" tab at the top of the window. Make sure that your MIDI interface is connected to your computer. Hopefully your MIDI interface will appear in the center of the window with all the I/Os "open" (as a quick troubleshooting step in case the MIDI interface is not recognized by the application, try to disconnect and reconnect its USB cable from the computer). At this point you have to add every MIDI device that is present in your studio by clicking on the "Add Device" icon located at the top of the window. You can specify the brand and model of the device by *double-clicking* its icon. The last step of the MIDI configuration consists in connecting the I/Os of each device to the I/Os of the MIDI interface in the same way they are connected in your studio. To do this simply *click and drag* the arrows at the edge of each device to the corresponding arrows located on the MIDI interface icon. Once you have made the right assignments for all MIDI tracks, you are ready to start sequencing. Remember that the I/O assignments are also saved with your templates, so use this feature to customize the templates as much as you want. For example, if you know that your oboe part will always be played by the device on cable 2/channel 1, then you can assign the MIDI out of the oboe track to cable 2/channel 1 so that you won't have to assign it every time you start a new orchestral project. Of course this parameter can be changed at any time.

MIDI tracks are extremely flexible when it comes to editing. Since in this type of track only performance data are recorded (as learned in Chapter 1), it is very easy, with the help of the sequencer, to change almost any parameter of the MIDI messages recorded. First of all we can look at the data in several ways, depending on the type of edits we need to make. Among the most used editors are the graphic (or piano roll) window, the list window, and the score window. These three represent the most used edit options, and they are available in almost every sequencer. Remember that no matter which edit method you use, you will always be able to change pretty much any parameter of your MIDI performance, so most of the time it is really up to you to choose the one that suits better your style and working habits. I usually recommend the graphic editor for quick fixes, common edits such as copy, paste, and delete, and especially for CC message edits. This editor is very intuitive and easy to grasp (Figure 2.3).

The list editor is another well-known tool that finds its roots in the early hardware sequencers. It is usually the favorite tool of more seasoned MIDI composers/producers

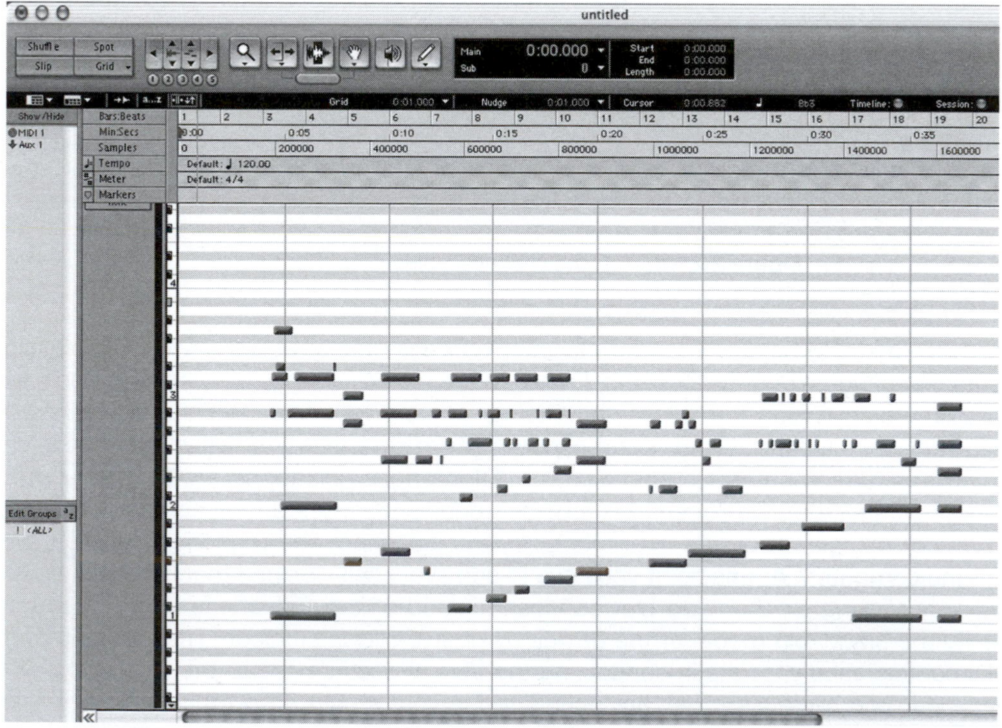

Figure 2.3 Graphic edit window in PT.

because it was really the only editing option in the early days of sequencing. It has the advantage of being very specific and selective: All the data are categorized by event, which makes changing a single parameter quick and effective (Figure 2.4). The list editor in PT is similar to the one found in DP.

The score editor is the most popular among trained musicians, since it allows you to see the flow of MIDI data in a more traditional notation view (Figure 2.5). From this editor not only can you edit the MIDI data but you can also organize the score in a page layout mode to print out professional scores and parts for your projects. CSX and LP especially are excellent in terms of flexibility and quality of the notation printouts.

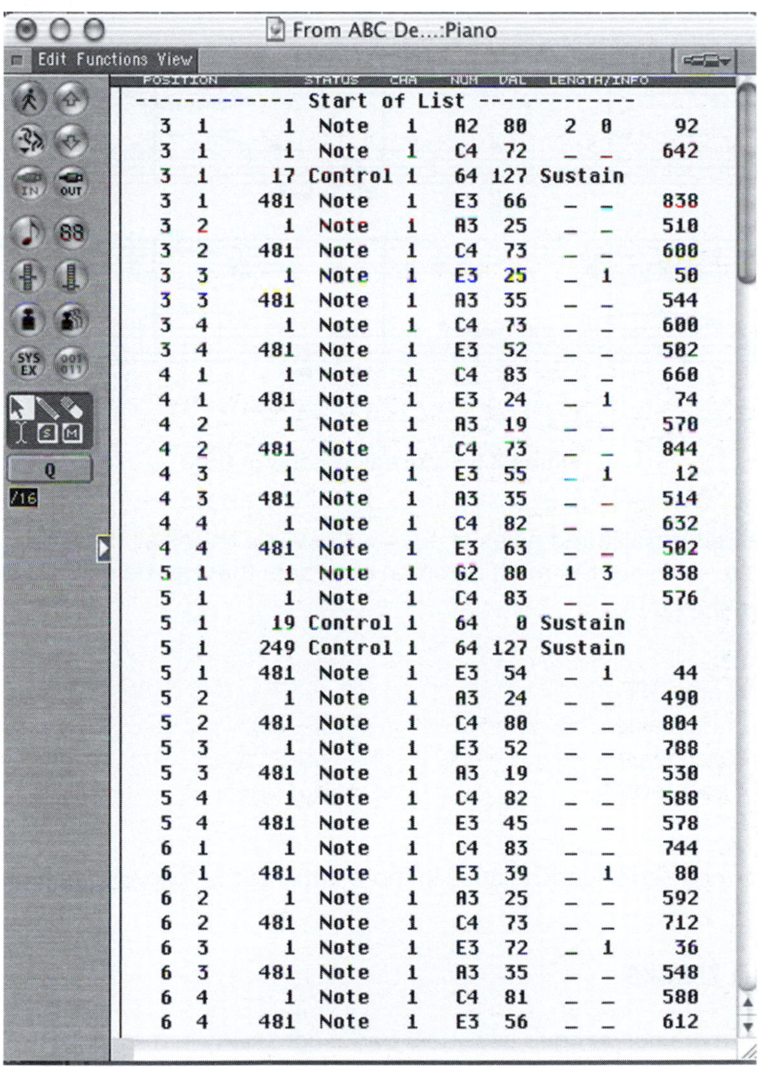

Figure 2.4 List edit window in LP.

Figure 2.5 Score edit window in CSX.

As mentioned before, all three types of editors allow you to edit all the parameters of your MIDI data. Here's a list of the most common elements that can be edited from the aforementioned windows:

- Note number
- Velocity ON and OFF
- CC number and value
- Position of each event
- Duration of each event
- Pitch Bend

I will describe these editing techniques in more detail in the following paragraphs.

2.5 Audio tracks

Even though we are not going to use audio tracks right away in this chapter, it is important to understand how these tracks differ from the MIDI tracks you just learned to set up. In an audio track the I/O labels are related not to the MIDI interface but to the audio interface.

The input and output setup depends on the type of audio interface you have and on the number of physical inputs and outputs available. Another difference between MIDI and audio tracks is that the latter can be set up as mono or stereo, depending on the material you are going to record and play back. In your templates it is also a good idea to have at least four mono tracks for solo instrument recording and two stereo tracks for mix-downs and loop import. Figure 2.6 shows the I/O audio track option in LP.

Audio tracks can be edited in two ways: destructively and nondestructively. While the former alters directly the original audio file and its data that were stored on the HD after the analog-to-digital conversion, the latter doesn't alter the original file but stores the edits as a list of events that points to areas (regions) of the original audio file without in fact altering

Figure 2.6 I/O audio track option in LP.

it. This technique allows the sequencer always to recall the original file without the risk of losing part of it by committing to permanent edits. All four sequencers have options to edit both ways. Most of the time you will want to make audio edits in a nondestructive way, but keep in mind that there are situations in which you will have to use the destructive editors.

In addition to the regular audio tracks just described there are three other main types of tracks that can be related to this category: auxiliary, virtual instrument, and master tracks. These types of tracks are different from the audio tracks because they don't contain any audio data and they don't show any waveform in the edit window. Nevertheless these are very important tools for your final productions. Let's examine them one by one.

Auxiliary tracks: These have the same function as the aux channels on a mixing board. They collect signal from an input (which usually is a bus) and send it to an output (which can be another bus, another track, or the master output). Auxiliary tracks are very versatile. They can be used for assigning multiple audio tracks to a single fader or to share a common effect (such as reverb) among several audio tracks. The application of auxiliary tracks can also be extended to multiapplication signal routing. In this case, if the programs used allow it, signal from one application can be sent to another running at the same time through the use of virtual auxiliary channels. Usually the interapplication signal exchange is controlled by software protocols that take care of the synchronization and data flow between applications. One of the most widespread protocols is "Rewire," which allows programs such as DP, CSX, LP, and PT to communicate in real time with other applications, such as Reason and Live. The ability to route signal between applications opens up infinite creative mixing and signal processing opportunities, such as adding effects to a signal generated by one application with plug-ins inserted in another. The signal coming in from another application not only can be routed through plug-ins and effects but in some cases also recorded to an audio track, making the exchange of audio material among different software fairly easy.

Virtual instrument tracks: Certain sequencers, such as DP, LP, and CSX, use dedicated tracks to insert software (or virtual) synthesizer modules that can be triggered by a MIDI controller. Virtual instrument tracks do not contain any waveform (like the audio tracks) but in fact are more similar to the aforementioned auxiliary tracks. In order to use a virtual instrument track in PT, a software synthesizer plug-in is usually inserted in one of the insert slots of the track in the mixing board window. In LP and DP a software synthesizer plug-in needs to be inserted on the I/O slot (or "Insert" in the case of DP) of an Audio Instrument track. In CSX you can add a software synthesizer through the VST Instrument options found in the "Devices" menu. The software synthesizer tracks act as additional mixing board channels that receive the input signal not from the sound generator of an external MIDI device but from the virtual output of the inserted software synthesizer.

Master track: A master track has the same function as a master fader on a hardware mixing board. It receives the input from all the active tracks (audio, auxiliary, and virtual instrument) in your sequence. It is useful if you want to insert mastering effects on the entire mix, for example, or if you want to generate an overall fade without having to work on the individual channels. All the main sequencers allow you to create one or more master faders.

2.6 Organizing your projects

Now that you know how to organize your tracks and have your templates set up, we should take a look at how to organize your projects and your session files on your HD. Even though each sequencer has a slightly different approach to storing session data, the overall techniques are very similar. For a problem-free composing experience it is absolutely crucial to understand how your sequencer stores the data and the hierarchical structure of the files associated with your project. As we learned previously, MIDI and audio tracks store two very different types of data: performance data in the form of MIDI messages in the former, and digitized audio renderings of an analog waveform in the latter. These two types of information, plus other data, are stored on your HD in different files. Let's take a look at the file/session organization of a project.

When you first launch your sequencer and choose to start a new project, the program asks you to choose in which folder of the HD to store it. Notice that this operation needs to be done before you start recording any audio data, since the audio information coming in from the audio interface are stored in real time during the recording process. The only exception is LP, which asks you to choose the folder where to record the audio files as soon as you record-enable an audio track. Your new projects should always be saved in separate new folders. Keep your folders very well organized; this is crucial for a problem-free working environment.

After naming your new project you will be brought back to the main window of your sequencer. What it is important to understand is the structure of the folder of your projects and where the different files are stored in your project folder. The two most important items of every project are the *session file* and the *audio files folder*. The session file stores all the information about your MIDI tracks, MIDI data, channel assignments, links to the audio files, automation data, preferences, etc. Inside the audio files folder only the actual audio files data are stored. Remember that those two items, session file and audio files folder, should never be separated. If you misplace the audio files folder, then all your audio information will be lost, resulting in empty audio tracks when you try to open the session file. Depending on the software you use in your project folder, you can find several other items that also should never be separated from the session file. In Table 2.1 you can see a description of the file structure for DP, PT, CSX, and LP.

The main purpose of having a well-organized project folder has to do with backup and archiving procedures. If all your files are kept constantly in the same folder, then your backup sessions will run smoothly and quickly. If your project files are scattered around your HD, then you increase the chances of deleting or forgetting some of the files of your project, resulting in a waste of time, energy, and creativity.

In order to be able to obtain the best performance from your system it is highly recommended to use a dedicated HD for your recording sessions and projects. Since the main internal HD hosts the operating system and the programs, it is usually busy taking care of the tasks related to the normal computer activities. If we would record the audio and session data to the same internal HD, the overall performance of the system would be affected, resulting in lower track and plug-in counts. By using one (or more) dedicated HDs you ensure that your recording system runs smoothly and with the highest performance

Table 2.1 Project file structure in DP, PT, CSX, and LP

Program	File structure: folder and subfolder
Digital Performer	*Project folder* Session file: contains the MIDI data, automation, etc. Audio Files folder: contains the audio files only Undo folder: contains the list of edits and undos Fades folder: contains the list and record of fades applied to audio files Analysis Files folder: contains data regarding the overview of waveforms
ProTools	*Project folder* Session file: contains the MIDI data, automation, etc. Audio Files folder: contains the audio files only Fades folder: contains list and record of fades applied to audio files Session File Backups Folder (.bak): contains backups of previously saved sessions using the Auto Save function from the Preferences menu
Cubase SX	*Project folder* Session file (.cpr): contains the MIDI data, automation, etc. Session file (.csh): contains image information for edited audio clips Audio folder: contains the audio files only Edits folder: contains list and record edits Images folder: contains data regarding the overview of waveforms
Logic Pro	*Project folder* Session file: contains the MIDI data, automation, etc. Session Files Backup folder (.bak): contains backups of previously saved sessions (Notice that the audio files in Logic Pro can be saved anywhere on your HD, but it is recommended to store them in an Audio Files folder created inside the main project folder)

available. Partitioning your HD doesn't achieve the same result as having a physical second HD; buy a new HD instead of partitioning your existing one.

2.7 Flight-check before takeoff!

If you followed carefully all the preparation steps up to this point, you should be in good shape to start sequencing and finally having fun with your studio. In this section we are going to make sure that everything is in working condition, meaning that all the signals are routed following the right paths and that you won't have any technical problem to stop your creative flow.

Since in this chapter we are going to use MIDI tracks, first let's focus on the MIDI signal path. To check whether your connections are working, try to record-enable a midi-track, meaning make sure that the record button of the MIDI track you are going to test is checked or highlighted. By doing this you tell your sequencer to reroute the MIDI messages coming in from your controller through the computer MIDI interface to the cable/channel assigned to the output of the MIDI track record-enabled. Try to send some MIDI data by pressing the keys of your controller. You should hear sound coming from your speakers. If not, then it's time to learn some troubleshooting techniques that can help you fix technical problems related to wrong signal path assignments.

There are two different areas to check when you don't get any signal from your MIDI tracks. First check that the MIDI data flow is set up right. Check if by pressing a key on the controller you see MIDI activity on the sequencer. Depending on the software you use, you can monitor the messages received by the sequencer in several ways. In Cubase and Logic there's a MIDI activity meter in the transport window. In PT you can record-enable a MIDI track and check whether in the mix window you see activity in its view meter. In DP there's a useful dedicated window called MIDI Monitor (accessible from the Studio menu) that shows messages received for every cable and channel independently. If your sequencer doesn't have this function, you can check if MIDI messages are received by simply hitting record, press on some keys, and see if the data were recorded on the MIDI track selected.

If no data are received by the sequencer, then check whether you record-enabled a track, check the MIDI cable connection, check that the MIDI interface is connected in the right way to your computer, and also double-check your MIDI setup in the Audio and MIDI setup control panel of the OS.

If your sequencer is receiving MIDI messages but your device is not, then check that the track you record-enabled has a cable/channel assignment. You can check whether the device is receiving data by looking at the "MIDI activity" LED located on the front panel of most MIDI devices. If the LED is not reacting when you press a key on your controller, then check for the cables connecting the receiving device to the MIDI interface of the computer. If the LED shows activity by blinking, then the problem probably lies in the audio path. Make sure that the audio cables connecting the MIDI devices to your mixing board are well plugged in and that the faders on your board are up. If you see signal activity on the viewmeters of the board, then the problem is probably part of the bus assignment or the speaker system. In general the best way to fix signal problems (both audio and MIDI) is to narrow down the areas that can cause the problem. By isolating different regions of the signal path, you can fix your studio problems quickly and easily so that you can move on to the creative aspect of music production. If everything fails, try shutting down your system, computer included, wait a minute, and then restart from fresh; sometimes this is the only solution to the mysteries of electronic equipment! If everything is in working condition, let's move on to the next section.

2.8 The first session: click track and tempo setup

Depending on your style of writing, the way you set up the tempo and meter of your compositions can vary significantly. The type of project (e.g., a groove-based song, a jazz ballad, an orchestral piece, a movie score) has an impact on how you set the tempo, or tempos, of your sequences. One thing to remember, though, is that no matter what style your project is based on, it is always recommended to sequence to a *click track*, even if you plan to have passages or entire sections as rubato. It is crucial to play to a click track because it is the only way the sequencer can assist you later during the edit and quantize operations. In this chapter you will learn how to set up the click track, the metronome, and the meter of your sequence using a steady tempo mark; in the following chapters I will discuss how to create tempo/meter changes and some advanced tempo-related sequencing techniques that allow you to reach more fluid and realistic results.

2.8.1 Who plays the metronome?

The first thing to set up when starting a new project is the metronome and in particular its sound. Keep in mind that the settings you select for the metronome sounds are shared by all projects, so it is necessary to go through this process only once, unless you decide to change them later. The metronome can be played by two different sources: the internal sound generator of the computer and any MIDI device connected to the interface. It is usually wise to choose only one of the two, even though technically you could choose both options at the same time. These settings are selected from the "Click and Count Off Options" menu (sometimes referred to as "Metronome settings" in LP and CSX or as "Click Options" in PT). In Figure 2.7 you can see the metronome settings window in CSX.

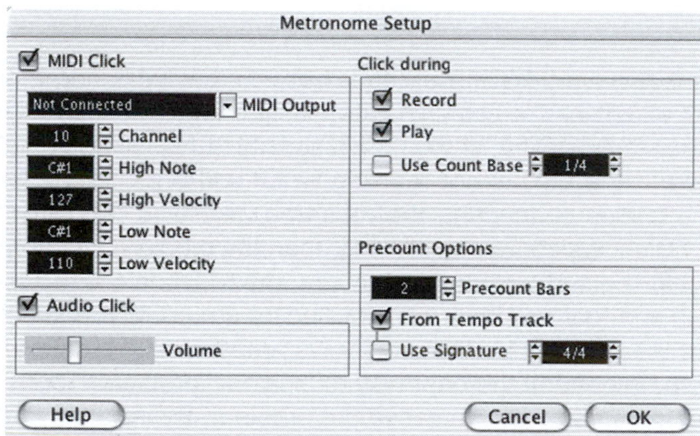

Figure 2.7 Metronome settings window in CSX.

I recommend using a MIDI device as the audio source of your click track. The main reason is that the timing of the internal sound generator of your computer usually drifts a few milliseconds in relation to the MIDI data flow, creating a discrepancy between the MIDI click and the audio click coming from the audio output of your computer. Even though it is not a huge difference, that your audio click doesn't represent exactly the MIDI click can be misleading when sequencing. Therefore I suggest assigning a MIDI device to play the click track. Once you make this decision, then it is up to you to choose to which device (cable number) and MIDI channel to assign the metronome. After assigning the cable/channel, you have to choose the patch and note that will be assigned to your metronome. While this is entirely up to you, I suggest you use a drum patch and in particular to pick up a sound that stands out clearly in the mix and that you won't use in your sequence as an instrument. For example, you might want to stay away from hi-hat or rim shot sounds, which you will probably use in your sequence, and choose instead high-pitched woodblocks or cowbells, which are less likely to be included in your production. Please notice also that you can choose two different notes for the "high note" (beat 1, or "strong beat") and for the "low note" (all the other beats depending on your time signature, or "weak beats"). You can have different velocities for high and low notes. Usually you want to have the strong beat slightly louder than the weak beats. I have found that velocity 127 and 110, respectively, work fine in most situations. If you prefer to have the all the clicks with the same intensity, then just leave velocity 127 for high and low notes.

You also have control over when the click track will be played and over the count-off options. The click can be set up to play always (when the click is on), only when you record, or only during count-off. I recommend having the click always on and turning it off manually when you don't need it. The main reason is that while the click is a valuable tool during recording, it is also essential during playback in order to check the rhythmic accuracy of your performances and recorded takes. Regarding the count-off, you can choose the number of precount bars and other options, depending on the sequencer you use. For example, in DP you can also choose to have the count-off playing only when recording (which is a very useful feature), or in CSX you can have the count-off set at a different time signature than the one set for your project.

No matter how you set up your metronome options, one crucial aspect to consider when sequencing is to select the right tempo at which to perform and compose your music. The tempo of your sequence can be changed from the transport bar of your sequencer. In this chapter we are going to deal with a steady tempo throughout the sequence. The first time you play to a click track is going to be a challenge, whether you are inexperienced or an accomplished musician. During many sessions I have run into professional musicians who had a hard time getting used to playing musically against the metronome. You will get used to it—don't worry; it is just matter of practice. The first thing I recommend before playing anything on your controller is always to keep in mind the tempo at which you want your sequence. If the tempo is well established in your mind, then it is going to be much easier to follow it in your performance. If you have a groove, a melody, or a rhythmic phrase in your head, try to find out its tempo first. Do not try to match your musical idea to a specific tempo simply because it is already set to your sequencer. A composer doesn't write a piece of music at a certain tempo because the orchestra can play only at one tempo. Therefore keep in mind that the technology you have in front of you is only a tool. You, not your computer, are in charge of the creative process. Sing the musical idea in your mind or out loud first and snap your fingers or tap your foot in order to have a preliminary idea about the beats per minute (BPM) at which you will set your metronome. Then press Play on your sequencer and start changing the tempo slowly until it matches the tempo you have in mind. You will see that now it will be very easy to play your musical idea over the metronome. Of course when you are more experienced you will be able to skip this process, but if this is your first real sequencing experience, I highly recommend following the aforementioned procedure.

2.9 Recording MIDI tracks

The moment you were waiting for has finally arrived. You learned all the components that form your studio, you learned how to connect them and how to have them interact nicely and (hopefully) smoothly with each other. You just decided on and set up the tempo of your sequence. Now it is time to record and to put to use your knowledge and creativity.

As I explained earlier in this chapter, you must have a track record-enabled in order to have MIDI data moving from the IN to the OUT of your computer's MIDI interface. Make sure the local control of your controller is set to OFF (in case you have a MIDI synthesizer as controller), as explained in Chapter 1. You can change the sound assigned to the channel to which the track is directed either by selecting a bank and patch from the front panel of the receiving MIDI device or by sending a Patch Change message from the sequencer. If you choose the latter, your options will vary depending on the setup and sequencer you use. In the worst-case scenario when clicking on the patch menu of a track in the sequencer you will see only a list of numbers, indicating the corresponding patch numbers on your device (Figure 2.8). If your sequencer was programmed with the names of the patches

MIDI 1

Patch: - Controller 0: - Controller 32: -

none	9	19	29	39	49	59	69	79	89	99	109	119
0	10	20	30	40	50	60	70	80	90	100	110	120
1	11	21	31	41	51	61	71	81	91	101	111	121
2	12	22	32	42	52	62	72	82	92	102	112	122
3	13	23	33	43	53	63	73	83	93	103	113	123
4	14	24	34	44	54	64	74	84	94	104	114	124
5	15	25	35	45	55	65	75	85	95	105	115	125
6	16	26	36	46	56	66	76	86	96	106	116	126
7	17	27	37	47	57	67	77	87	97	107	117	127
8	18	28	38	48	58	68	78	88	98	108	118	

Patch name file: None

Increment Patch Every 3 Sec

Change... Clear Cancel Done

Figure 2.8 Generic patch list table in PT.

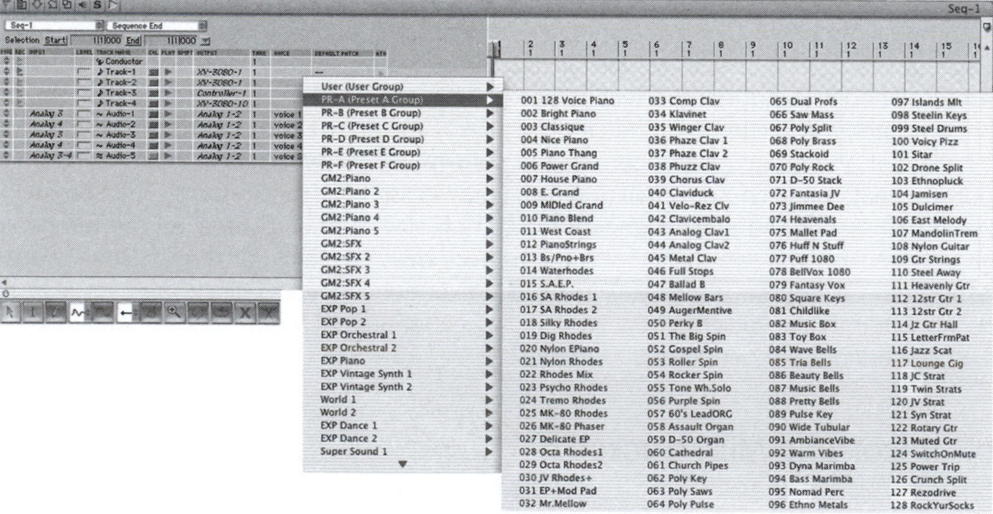

Figure 2.9 Detailed patch list table in DP.

stored in your device, then instead of a meaningless list of numbers you will see a list with the patch names (Figure 2.9). To choose your patch in DP, simply click on the column named "Default Patch" in the track list window; in CSX select the patch in the "Inspector Window" on the left side of the Project Window; while in LP you can click on the Patches field to the left of the Arrange Window. The same result can be achieved in PT by clicking on the small icon on the channel strip of the track in the mix window (Figure 2.10).

Click the Record button (Record and Play in PT) in the transport window of your sequencer and start playing your part. While you are sequencing you will see the MIDI data being recorded by the computer displayed on the record-enabled track. If you don't like what you

Figure 2.10 Patch list icon in PT.

played, you can just go back to the beginning of your part, delete the data by selecting them with the mouse and pressing delete, or simply undo your last take by choosing *Undo* from the Edit menu (*Ctrl-z* in Windows or *Command-z* on a Mac).

If you like the part you just recorded but think you can do a better job, it is a good idea to keep the old take and try a new one. This technique is called multiple takes. In a MIDI sequencer it is possible to achieve the multiple-takes feature in several ways. The simplest one, and sometimes also the fastest, is to record-enable a different track, mute the one you just recorded, and start recording again. You can repeat this procedure until you are satisfied with your last take or you know you can compile the perfect take by copying and pasting sections from several previous takes. This technique has some advantages: It can be used on any sequencer, and it allows you to see all the different takes at the same time, which makes editing easier. Another option is to use the "multiple-take" feature present in some sequencers, such as DP and PT. Virtual tracks, called takes in DP or playlists in PT, are manually selected for each track (Figure 2.11). After you record, if you want to try a different take all you have to do is to switch the selected track to another take and rerecord the part. This process can be repeated as many times as the memory of your computer allows. In DP if you want to compare two takes, you can simply switch back to any previous take by clicking on the Take column. You can also copy and paste between takes. This technique has the advantage of keeping the track count down, since each take is virtually "superimposed" on the others (keep in mind, though, that you can listen to only one take at a time for each track), but it has the disadvantage of not showing the data of more than one take for each track at the same time.

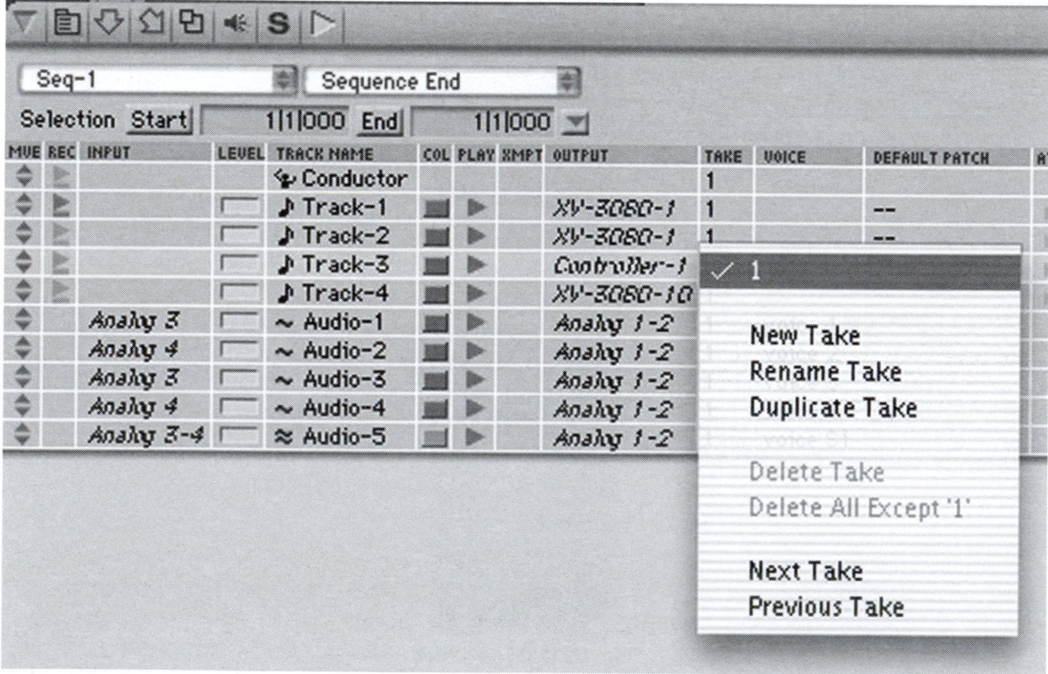

Figure 2.11 Multiple-take selection in DP.

A similar approach is used by PT, where different takes can be stored in different playlists. To create a new playlist in PT simply click on the "double arrow" button next to the track name of a track. From the list that appears select "New" for a new playlist or "Duplicate" to create a new playlist with the same content as the current one. By choosing "Delete Unused" you can purge all playlists that are not currently utilized in the project for that particular track. Use the same technique to switch among playlists. Other sequencers, such as CSX and LP, use a slightly different technique, called *cycle recording mode*, that can be as effective as the one based on multiple takes. In CSX this feature is activated from the transport window and it allows you to create a loop between two locators (named "Left" and "Right" locators). When you hit the Record button, the sequencer will keep cycling between the two locators until you press Stop. You can have the sequencer set to erase the old data at every new pass ("overwrite" mode), to keep the last complete pass ("keep last" mode), or to create a new part for each pass ("stack" mode). These options are available directly from the CSX transport window. Each new part will be stacked over the previous one without erasing it. In LP, after selecting the Cycle button from the transport window, you can choose to have the sequencer create a new track automatically after each pass, and you also have the option of muting the previous tracks (takes). You can change and activate/deactivate these options from the "Recording Options" found in the "Song Settings" submenu, which is located in the File menu.

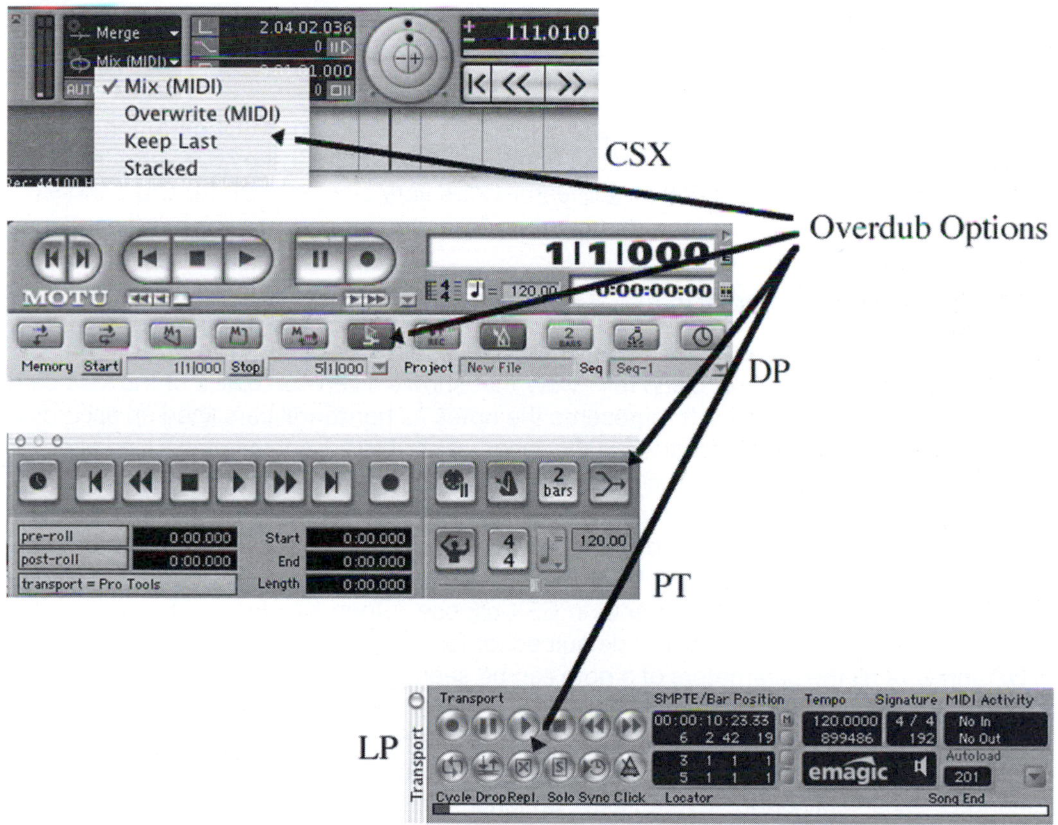

Figure 2.12 Overdub options in CSX, DP, PT, and LP.

Another aspect of recording MIDI tracks that is important to recognize is that you can keep overdubbing a MIDI track without erasing the data that were already stored on that track. This feature is called the *overdub* or *merge function*. In most cases you can choose from the transport window of your sequencer whether you want to merge the new data to the preexisting ones on the record-enabled track. If you prefer, you can opt for a more traditional approach (more similar to a regular tape recorder), where, if you record over a track, you automatically erase the preexisting data. I recommend having the overdub option always on in order to avoid deleting precious takes. If you want to delete them, it is easier simply to select the MIDI data in any editor window and manually erase them. In DP and PT the overdub function is accessible directly from the transport window. In LP the merge function is controlled from the transport window using the Replace button. If the button is highlighted, then the merge function is off and the preexisting data will be replaced by the new ones recorded. In CSX you can choose from the transport window between "merge" or "normal" mode (which overdub the MIDI data) and "overwrite" mode (which erases the preexisting data) (Figure 2.12).

Now that you have recorded your first track, keep recording a few more in order to have enough material to learn the basic editing techniques of the next section. You can open some of the examples found on the included CD to practice your editing skills.

2.10 Basic MIDI editing techniques

In this section I discuss the three most popular edit windows for MIDI data: the graphic (or piano roll) editor, the list editor, and the notation editor. These three constitute the most used type of editors for MIDI data. Their features are very similar in DP, LP, CSX, and PT.

2.10.1 The graphic editor

The graphic editor provides a very intuitive and quick way of accessing the MIDI data you want to edit. It is based on a "piano roll" view that shows a vertical replica of the keyboard on the left side of the window. It represents the notes as horizontal bars lined up according to their pitch (*y axis*) and their positions in the song (*x axis*). The advantage of this kind of editor is that it appeals to both the musician and the nonmusician MIDI programmer. In order to edit notes and data in this editor you don't need classical music training, since the notes and their parameters are easily read and changed in a very intuitive way. To open the editor in DP, select the option "Graphic Editor" from the Project menu, in LP select the option "Open Matrix Edit" from the Windows menu, in CSX choose "Open Key Editor" from the MIDI menu. PT uses the Edit window as a default editor for the MIDI tracks. In the graphic editor in DP and CSX all the parameters of a note can be seen, and if necessary changed, by simply clicking on a note and looking at the top of the main window. Here all the data of the selected note (such as velocity, start position, end position, length, and MIDI channel) are displayed and can be changed by simply clicking on each field. The same technique can be used in LP by selecting the "Event Float" option in the "View" menu of the Matrix editor. This will open a floating window that displays all the parameters of a selected event. The graphic editor layouts and features are very similar in DP, CSX, and LP. In PT the piano roll editor is built in the main edit window and it is the only way to edit MIDI data in PT. Another advantage of the graphic editor is that it allows you to edit not only note data but also any other MIDI data, such as CC, Pitch Bend, and velocities of the notes in a graphic way. These additional data are shown at the bottom of the graphic editor window in an expandable frame. Figure 2.13 shows the graphic editor and the editable parameters in CSX.

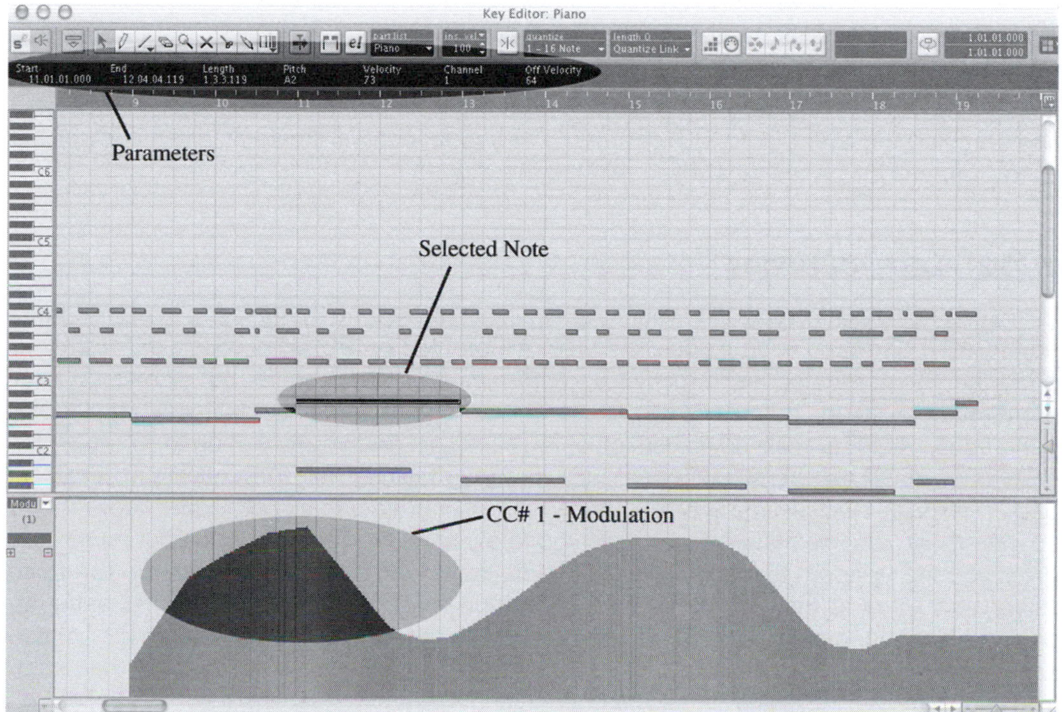

Figure 2.13 The graphic editor in CSX.

In addition of being able to change notes' parameters by using the information area at the top of the window, you can change the notes' position, start/end points, and length by using the mouse. To move a note, click in the middle of it and drag it horizontally and vertically as needed. To change its length, move the cursor toward the beginning or end of the note until you see the shape of the cursor change (into a small hand in DP and LP, a double arrow in CSX). In PT you can change the start, end, and length parameters manually by punching in the data. A series of additional graphic tools also allows you to create, change their shape, and delete several events. The most common tools are the *pencil*, the *eraser*, and the *reshape* tool. The pencil allows you to insert new data, such as notes, CC and Pitch Bend. In LP it also allows you to extend the beginning of a note. With the eraser you can delete several data at the same time, including CC and Pitch Bend. The reshape tool in DP (*Command-click* in CSX) is designed to change the parameters of existing data without inserting new ones. It is particularly useful in situations where you want to quickly alter a volume fade, modulation, or Pitch Bend data without playing the part again or inserting new messages. In the graphic editor all the usual edit functions (accessible from the Edit menu) and shortcuts apply, as shown in Table 2.2.

Table 2.2 Edit functions available in the graphic editor

Function	Mac OS	Windows
Copy	*Command-C*	*Ctrl-C*
Cut	*Command-X*	*Ctrl-X*
Paste	*Command-V*	*Ctrl-V*
Select All	*Command-A*	*Ctrl-A*
Undo	*Command-Z*	*Ctrl-Z*

A quick way of copying data in the graphic editor is to press the *option* key while clicking and dragging the event (note, CC, or other). The sequencer will create an exact copy of the event that will be placed where you release the mouse.

2.10.2 Level of Undos

Every editing operation you do in your sequence can be undone by selecting the Undo function in the Edit menu (*Command-Z* on Mac or *Control-Z* on PC). It is possible not only to undo the last command, as in many low- and mid-range sequencers, but also to go back numerous operations before the last one and restore your sequence as it was prior to the edits. While PT LE (a scale-down version of PT) allows an undo history of up to 32 steps, PT, CSX, LP, and DP feature an unlimited undo history, meaning that theoretically you can go back in time as much as you wish and bring back your project to any previous stage. This is an extremely valuable feature that allows you to experiment creatively with your music and gives you the freedom to make mistakes and fix them with the touch of a button. The *undo history* is a list of all the actions and commands you executed on your sequencer since the beginning of your project or since the last time you "purged" the list. Clicking on the list you can jump to any point in time in your project history and revert all the action up to that point. The undo list in LP, CSX, and DP can be opened and altered by selecting the "Undo History" option from the Edit menu (called simply "History" in CSX). In PT the undos are controlled using the *Command-Z* keys (Undo) and the *Shift-Command-Z* key (Redo).

2.10.3 The list editor

The list editor is probably one of the oldest editing tools for MIDI data. It finds its roots in the first hardware sequencers and it has been a valuable tool ever since. The philosophy behind the list editor is very simple: the MIDI data and messages are listed in chronological order and can be seen, edited and deleted by clicking on each event. While this method might look a bit basic and unsophisticated compared to the graphic editor, nevertheless it can be very useful and speed things up especially when used in association with a view filter (more on this in a little bit). It can be used to quickly delete certain data in your track or take a quick look at which data are quantized and which are not. The parameters displayed in the list window are the same as the ones we learned in the graphic editor (note, velocity, position, length, etc.) and they can be edited by clicking on the value field of each event (Figure 2.14).

A very helpful tool that can be used in conjunction with any editor but is particularly so with the list editor is the *view filter* option. It allows you to decide which data are displayed in any window of your sequencer. This is a feature that can be handy when you are looking

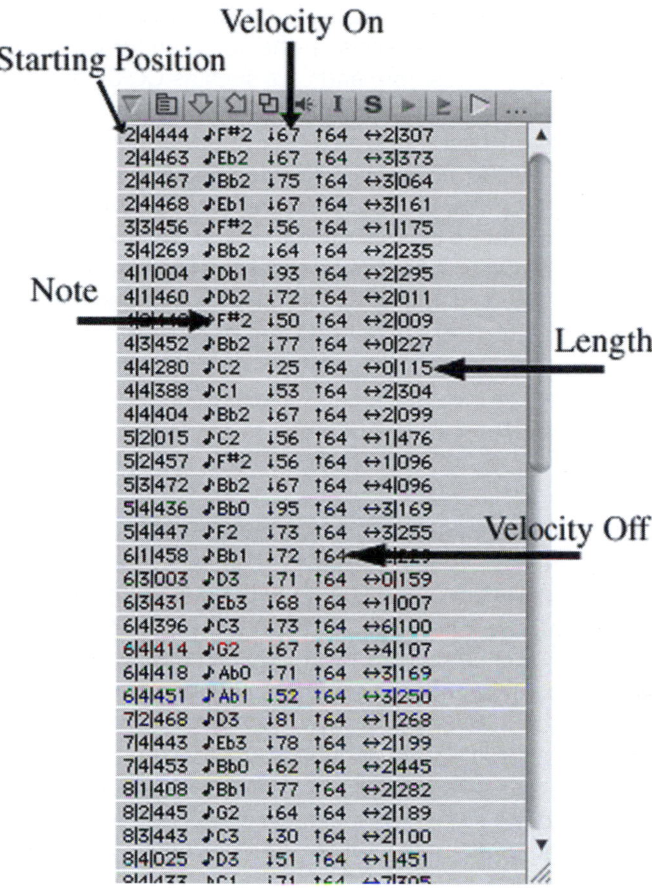

Figure 2.14 The list editor in DP.

for a particular MIDI message or CC. Let's say that you want to delete all of CC 7 from the bass track. One option would be to open the graphic editor, set the lower part of the window to display CC 7, and then graphically delete the data. But another way (probably faster) would be to set the view filter of your sequencer to show only CC 7 and then to select all (*Command-A* or *Ctrl-A*) and choose *cut* (*Command-X* or *Ctrl-X*). In DP the view filter can be found in the "Set View Filter" option under the "Setup" menu. In LP you can filter data by selecting the messages you want to see on the right side of the track list window (you can choose between notes, program change data, Pitch Bend, CCs, and System Exclusive messages). In CSX the filter is activated by clicking the "Show Filter View" in the top left corner of the List Editor window and enabling the events you want to hide. In PT you can filter the messages when they are being recorded (Input Filter).

2.10.4 The score editor

The MIDI messages recorded in your sequencer can also be displayed and edited as regular notation in the score editor. This is a very useful tool, not only to check and edit in a more traditional way the MIDI parts you recorded but also to prepare and print out professional scores and parts of your compositions. In the score editor you can edit MIDI events such as notes, velocity, duration, position, and CCs such as CC 64 (sustain pedal). While it is a good idea to use it for checking and inputting your parts, I discourage using it as your primary editing tool. The main reason is that all score editors use automatic quantization correction in order to display the notes in a readable way. If you played your part fairly freely (and did not quantize it), the sequencer in the notation window would try to make some educated guesses in order to display the part in the clearest possible way. This could mislead you while you are trying to edit the data, since what you see is not precisely what you recorded. Therefore I suggest using the notation editor mainly for scores and part preparation. LP and CSX feature very advanced and complete notation editors. They can be used for both quick parts checkup or in professional scoring situations. DP has a decent notation editor that can be used for quick parts checkup and some basic parts and score preparation, but it lacks the advanced layout features of LP and CSX. PT does not provide a score editor at the moment.

The notation editor has two main views: the edit view and the page layout view. The former is used mainly to edit the score and the parts and gives you a scroll view of the tracks. In edit view the song position is constantly updated on screen, making it very easy to keep track of errors and changes you wish to make to your parts. Page view allows you to see the score and parts as they will be printed. Here you can freely add layout symbols, markers, dynamic markers, annotations, bar numbers, etc. Every item or graphic object can be moved or placed freely around the score. Usually the edits and settings applied to a part or track in the score editor do not affect the playback of MIDI messages in the track list window. In CSX, for example, they affect only their appearance. Nevertheless you can, if necessary, apply those edits to the actual MIDI data by selecting the "Scores Notes to MIDI" function in the "Global Functions" menu under the Score main menu.

2.11 Basic principles of MIDI note quantization

One of the biggest advantages that MIDI and sequencing offer to the modern composer is the ability to quickly "fix" performance mistakes and the option to tie up the parts in a precise and quick way. The idea of quantization was introduced since the first hardware sequencers, but it has been definitely brought to perfection and to a higher level of sophistication with the introduction of the latest generation of software sequencers. To *quantize* literally means "to limit the possible values of a quantity to a discrete set of values." This concept, applied to a MIDI sequencer, describes a function that allows you to rhythmically correct the position of MIDI data in general, and notes in particular, according to a grid determined by the "quantization value." In practice, when you quantize a part, a track, or a region of your sequence, you are telling the computer to move each piece of MIDI data to the closest grid point specified in the quantization value. To better understand this concept, compare Figure 2.15a with Figure 2.15b.

In (a) you see a sequenced part that has not been quantized. The notes were not played accurately in rhythm with the metronome. While this sometimes can create a nice natural and smooth effect, most of the time it results in a sloppy sequence and can lead to a poorly produced project. In (b) the same part has been quantized with a 16th note grid.

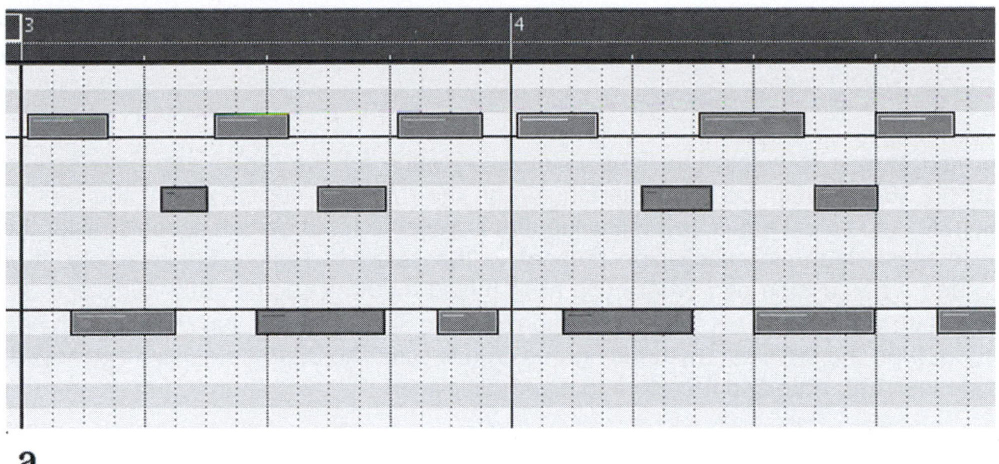

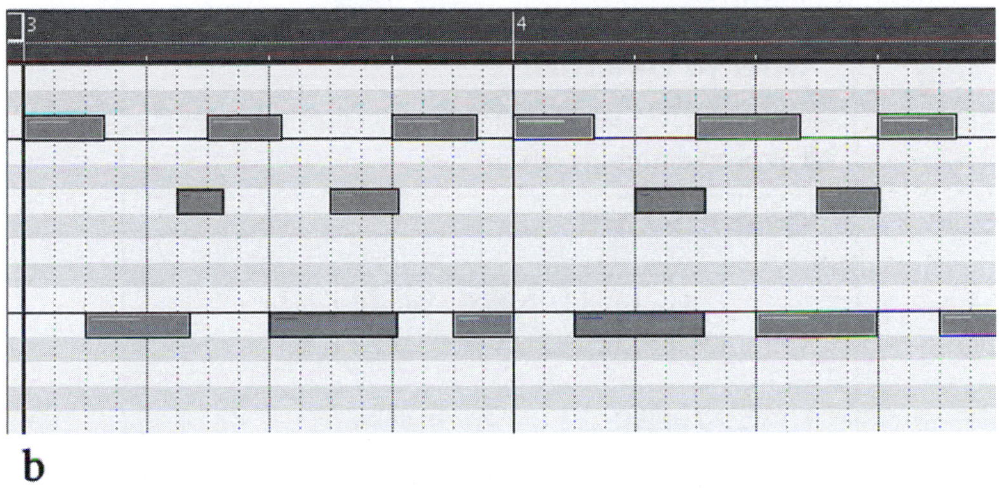

Figure 2.15 (a) Unquantized part and (b) quantized part, in 16th notes.

As you can see, every event has been moved to the closest 16th note, rhythmically "cleaning" the part, which now plays in perfect sync with the metronome. Listen to audio Examples 2.1 and 2.2 on the audio CD to hear the difference between quantized and non-quantized parts.

For basic quantization you must first select the part, region, or notes you want to quantize. You can do so from any editor window of your sequencer. If you want to quantize a large section, I suggest selecting the region from the track list (or arrange) window. If you want to quantize only a section or few notes, then it is better to open the track in the graphic or list editor and specifically select the events you want to quantize (remember that you can select noncontiguous events by holding the *Shift* key and clicking on the events). At this point you have to select the quantization option. In DP go to the "Region" menu, in LP choose the quantization value in the "Object Parameters Box," in CSX use the "Over Quantize" option found in the "MIDI" menu, while in PT you have to select "Quantize" from the "MIDI" menu.

The next step is to select the right quantization value for the region you have selected (Figure 2.16). This can be tricky, depending on the rhythmic complexity of the part. The main rule is to select a quantization value that is equal to (or in certain cases smaller than) the smallest rhythmic subdivision present in the selected region. For example, if you recorded a percussion part where you have mixed rhythms such as 8th, 16th, and 32nd notes, then you have to set the quantization value to 32nd notes. If you choose 16th notes instead, then the smallest rhythmic figures (in this case the 32nd notes) will be moved to the closest 16th note and therefore lost and misplaced.

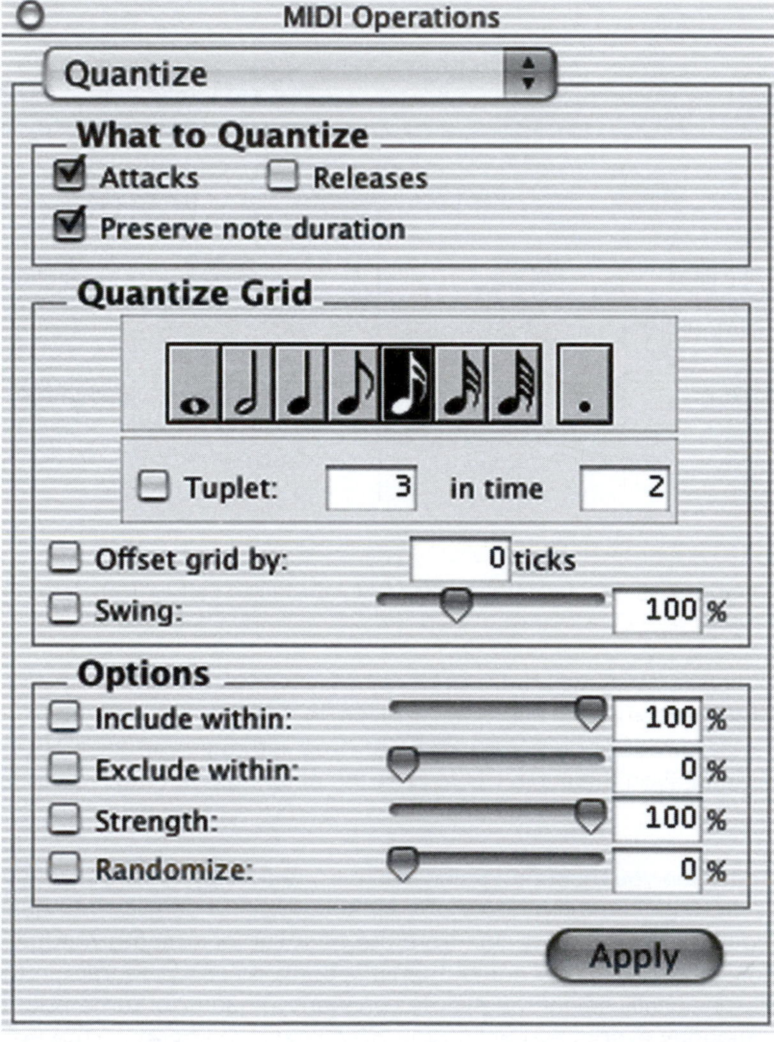

Figure 2.16 Quantization options window in PT.

Listen to Examples 2.3, 2.4, and 2.5 on the audio CD and compare, respectively, the non-quantized version, the quantized part with the right value, and the quantized part with the wrong value. While this procedure looks fairly simple, things get a bit more complicated when your part features mixed rhythmic figures, such as straight 8th notes and 8th note triplets (or more generally, any kind of straight notes versus tuplets). In this case you have two options. The first is to selectively choose and quantize regions that have similar rhythmic features—for example, open the graphic editor and select and quantize all the events that share a common rhythmic subdivision (i.e., 8th notes). Then do the same with 8th note triplets, and so on. Even though this technique can be a bit time consuming, it will guarantee you the best and most accurate results. Another technique is to use the so-called "smart quantize" feature present in some sequencers, such as DP (accessible from the "Region" menu). This function creates a floating grid that automatically adjusts itself depending on the notes recorded. This way you don't have to choose a quantization value. It is ideal for quantizing a long section of a track where mixed rhythms are most likely to be found.

Quantization is almost a "magic tool" that can make your parts sound exactly in time with the metronome. While this feature can definitely improve the general feel of the production, at the same time it can take away the life of your music by flattening the groove of your performance and making it sound very stiff. But don't worry, the techniques that I just described are only a small part of what a sequencer can do in terms of quantization. In succeeding chapters you will learn more advanced techniques that will help to compensate for the stiffness introduced by straight quantization.

The quantization action, like any other editing action, can be undone using the undo history in the Edit menu. You can also apply quantization by inserting a MIDI filter on the track or tracks you want to quantize. This feature gives the option of turning quantization on and off at any time without making a real commitment to a certain quantization value. In DP, for example, you can insert a quantization plug-in on the mixer channel of a MIDI track by simply selecting it from the list of plug-ins that shows up when clicking on an insert slot in the mix window (Figure 2.17).

A similar result can be obtained in CSX by inserting the MIDI effect called "Quantizer" in the mixing board channel of the MIDI track you want to quantize. In LP you can quantize a part directly from the *sequence/audio region parameters* window located at the top left of the *arrange* window. An entire track or each part individually can have separate quantization values that can be changed or even turned off at any time. In PT the quantization features are less sophisticated than in the other three sequencers, but the techniques and the procedures for quantizing a region are the same. The quantization option in PT can be accessed from the MIDI menu. In succeeding chapters you will learn more advanced quantization techniques, such as "groove quantization," and more detailed parameters, such as swing feel, sensitivity, and strength, that you can use to avoid the mechanical groove caused by straight quantization.

2.12 Audio track basics

So far in this chapter you have learned how to record, edit, and quantize MIDI tracks. While the MIDI standard is an extremely valuable music production tool, the integration of MIDI

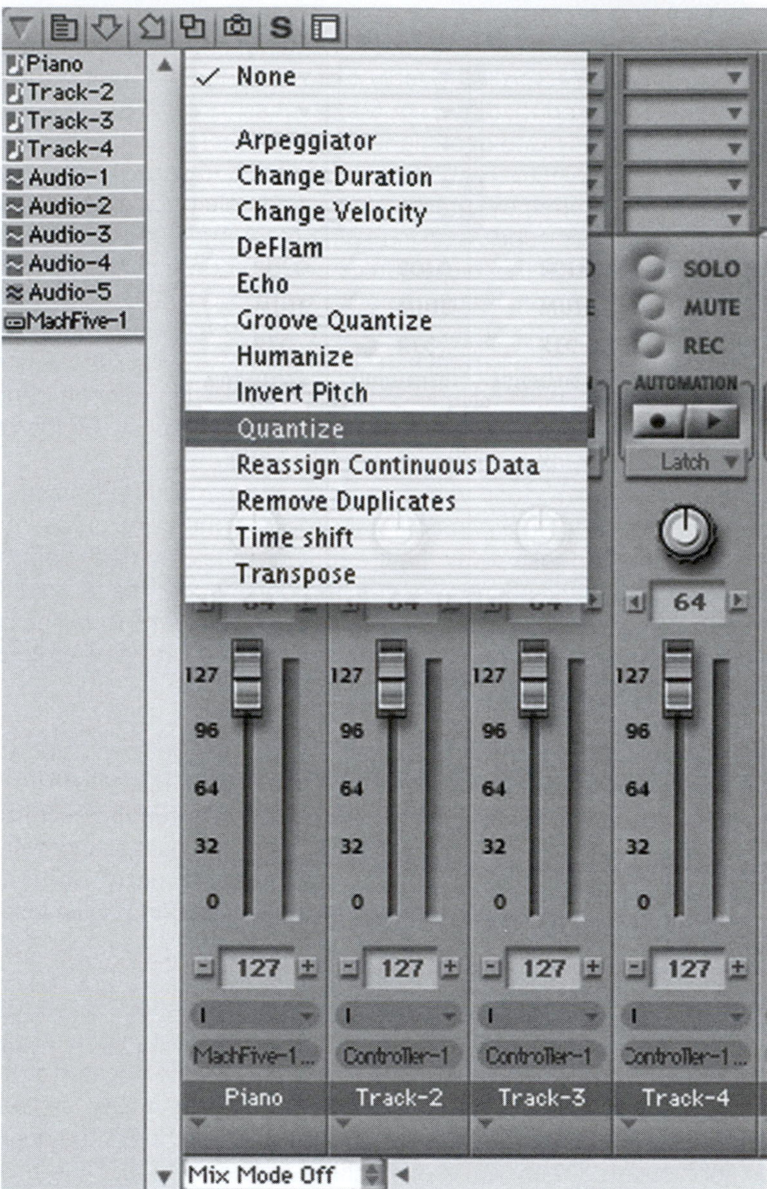

Figure 2.17 The quantization MIDI plug-in in DP.

tracks with audio tracks has brought the sequencing experience to a much higher level. Therefore it is time now to start exploring the interaction of MIDI and audio tracks.

Audio tracks can be of two types: mono (meaning tracks that contain and play back only one channel of audio) and stereo (two channels of audio, usually codified as left and right). Before recording and/or importing an audio file, you should set the track to the right type, depending on the type of audio material you will be working on (mono or stereo). Let's create a new stereo track.

In DP, select "Add Track" and then "Stereo Audio Track" from the Project menu. A new track with a *double waveform* symbol will be added to the track list window. In PT, simply select "New Track" from the File menu and change the default "Mono Track" option to "Stereo." In LP, first create a new track by double-clicking on an empty field in the track list column, and then select the type of track you want to create by *clicking and holding* on the track name (you will see a hierarchical menu showing different options, among which is "audio track"). After creating the audio track, you can change the type (mono or stereo) by clicking on the respective symbol at the bottom of the channel strip for the track you just created. In CSX you can create a new track by choosing "Add Track" from the Project menu, selecting "Audio" from the menu, and then clicking on "Stereo."

Once you have at least one stereo audio track ready, import some audio loops to play along with the MIDI tracks in the project. You can import almost every type of file you want in the sequencers presented in this book. They are all professional tools able to deal with pretty much any format. On the included CD you can find a series of professionally recorded loops and grooves inside the folder named "Audio Loops." Feel free to use them to get familiar with importing and editing audio files. Most of the audio applications use the following file formats to record and store the audio information on the HD: AIFF (audio interchange file format), wave (WAV), SDII (Sound Designer 2), and BWF (broadcast wave format). While these are the most popular in the industry at the moment, other formats are supported, and they may vary depending on the sequencer you are using. Table 2.3 shows a summary of the formats and the applications that support them.

Table 2.3 Audio file formats supported by the four main sequencers

	Sequencer			
Audio file format	**DP**	**CSX**	**PT**	**LP**
AIFF	Yes	Yes	Yes	Yes
WAV	Yes	Yes	Yes	Yes
SDII	Yes	Yes	Yes	Yes
BWF	Yes	Yes	Yes	No
Audio CD	Yes	Yes	Yes	Yes
MP3	Yes	Yes	Yes	Yes
MPEG2	No	Yes	No	No
Wave 64	No	Yes	No	No
REX 2	Yes	Yes	No	Yes
Ogg Vorbis	No	Yes	No	No
QuickTime	Yes	Yes	Yes	Yes
AVI	Yes	Yes	No	No
AIFC	No	Yes	No	No

Logic version 7 also supports AAC and Apple Loops formats.

The fastest way to import an audio loop into your sequence is to drag the audio file directly into the arrange/track list window of your current project. This procedure works in LP, DP, and CSX, while in PT you can import audio files by selecting the option "Import Audio to Track" from the File menu. The file will be copied into the working directory of the project. The only exception is LP, where the imported files are left in their current directory as default. In order to copy or move them to the project's directory, use the options in the Audio

File menu accessible from the Audio Window located in the Audio menu. After the overview of the waveform is calculated in the background, it is going to be placed in the arrange/track list window (Figure 2.18). Try to import one of the files you find in the CD, and make sure that it plays back.

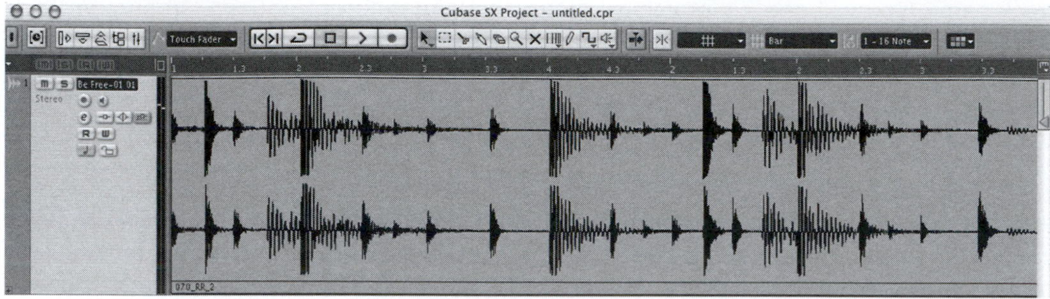

Figure 2.18 Waveform overview in the arrange window of CSX.

The file you just imported can be edited and manipulated in several ways, but while some of the techniques are similar to the ones you learned for the MIDI track, others differ drastically. Even though this book focuses mainly on the MIDI aspect of music production, it is important to know the editing techniques available for audio tracks. In this chapter I describe the basic graphic editing procedures available for audio tracks and sound bites, while in subsequent chapters you will learn more advanced options and techniques.

2.12.1 Destructive and nondestructive audio editing

In an audio sequencer, the audio files can be edited in two different ways: destructively and nondestructively. To understand the difference between the two techniques we have to become familiar with the way audio data are stored by the sequencer on the hard disk. When you record or import a file, the data are stored in audio files generated or copied into the working directory of the current project. These files usually assume the names of the track on which they were recorded or the name of the files imported. Any nondestructive edit that is done doesn't actually change the contents of the original file that was recorded or imported but instead changes virtual pointers that point to different parameters of that file.

For example, when you change the beginning or end of a sound bite (also called *audio region*), you don't trim the original file on the hard disk but trim a virtual copy instead. This technique has many advantages. First you can always go back to the original file in case you decide that the edits you made were not suitable for the project, and second it saves a lot of space on the HD, since when you copy a region you don't actually copy the audio file but just a new set of pointers that still point to the original file on the HD. Destructive editing, on the other hand, is less forgiving, in the sense that the edits you do are permanently incorporated in the original file (or in an entirely new one). This technique is more dangerous because it gives you less space for error. Both editing options are nevertheless available and used in the four main sequencers in one way or other. In this chapter we discuss the nondestructive ones, which usually are the most common.

In most cases, all nondestructive edits can be done from the arrange/track list window, making it really easy to adjust audio sound bites, depending on your needs. This is specially true for CSX and LP. It is possible, directly from their arrange windows, to trim, separate, cut, copy, paste, repeat, and fade audio regions without having to open a separate editor. To trim the beginning or end of a sound bite (called "part" in CSX and "sequence" in LP), simply move your cursor to the beginning or end of it and, with the *arrow tool*, click and drag. You will see your region's boundary move with it. When you release the mouse, the sound bite is trimmed. Remember, though, that, as I mentioned before, this a nondestructive edit, so you can also revert the region to its original status by clicking, dragging, and trimming back its boundaries. To trim a sound bite in DP you first have to double-click on it from the track list window in order to open the graphic editor for audio tracks and then follow the same procedure as explained for CSX and LP. In PT all the audio edits are done from the main edit window. To trim a sound bite, simply select the *trimmer tool* (Figure 2.19) from the main edit tools in the top part of the edit window, and click and drag from the beginning or end of the audio region.

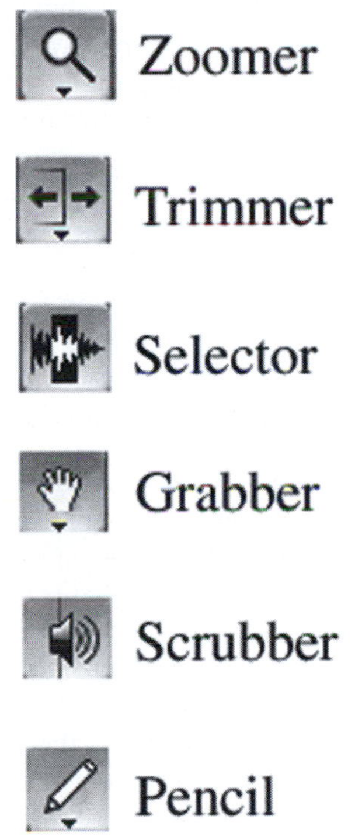

Figure 2.19 PT tool palette description.

The basic operations of cut, copy, and paste also apply to audio regions and follow the regular procedures of editing. You can split a region by selecting the *scissor tool* from the tool palette in DP, CSX, and LP and click wherever you want to insert the split. This action will cut the region in two, creating two completely independent bites (but not a new audio file, since it is based on nondestructive editing). In PT you would have to choose the *selector tool* (Figure 2.19), click on the point where you want to cut the region, and select "Separate Region" from the Edit menu (or use the shortcut *Command-E*). There are several techniques to join back together two or more regions that were previously split. In PT simply select the regions with the *grabber tool* by *Shift-clicking* on all of them, and then select "Heal Separation" from the Edit menu (*Command-H*). In CSX and LP simply select the *glue tool* and click (*Shift-click* in LP) on the parts you want to join together. In DP you have to select all the sound bites you want to join and choose "Merge Sound bites" from the Audio menu (keep in mind that this action will create a new audio file and therefore will increase the size of the project).

2.12.2 Playing it nice with the other tracks

Now that you have learned the basics of handling audio tracks, let's deal with some practical issues that very often puzzle beginners as well as advanced MIDI producers. When you import a loop from an audio file, most likely its tempo won't match perfectly the tempo of your sequence, which can be a real disaster. Fortunately this problem is easily fixable in several ways. The first thing that probably comes to your mind is to adapt the tempo of the sequence to the loop. This works fine if you are flexible with the tempo of your composition. But in most cases composers don't like to make adjustments (read "compromises") when dealing with their compositions. Also keep in mind that if you import more than one loop and they all have different tempos, then you are in trouble. Therefore the goal is first to find the tempo of the imported loops and then to time-stretch them (meaning change their tempos without affecting their pitch) in order to have them match the tempo of your sequence.

The first thing you have to do is find out the exact tempo of the loop (or loops) you imported in your sequence. While in some cases the tempo is indicated on the booklet of the loop library you bought, there are other situations on the file name of each loop where we don't know for sure the tempo of an audio loop. There are several ways to determine it.

The most generic technique involves a simple equation that can be applied to any digital audio file. By knowing the sample rate, the number of beats, and the number of samples of a loop, you can find out the tempo, in beats per minute, by applying the following formula:

$$(NB \times SF \times 60)/NS$$

where NB is the number of beats of the loop, SF is the sampling frequency in hertz, and NS is the number of samples of the loop. For example, if you have a two-bar loop in 4/4 meter (a total of 8 beats), recorded at 44,100 Hz, that contains 200,540 samples, then the tempo will be

$$(8 \times 44{,}100 \times 60)/200{,}540 = 105.5\,\text{BPM}$$

To find out the number of beats in your loop, simply listen to it. Information on the sampling frequency and number of samples can be collected in different windows, depending on the sequencer you use. Table 2.4 sums up where you can find them on each sequencer.

The technique explained in the preceding paragraph can be used from any audio sequencer. Each application, on the other hand, has specific procedures and tools for finding out the

Table 2.4 Location of sampling frequency and number of samples on the four main sequencers

	Sampling frequency	**Number of samples**
DP	Open the Soundbites window from the Project menu. Select the audio file you are working on, and select the Info option at the top of the Soundbites window. On the right side of the window you will see a list of all the information about that particular audio file.	Follow the same procedure as for finding the sampling frequency.
LP	Open the Audio window. At the top of each sound bite you will see information about the sample rate and whether the file is stereo or mono.	Open the destructive audio editor (called Sample Editor) by double-clicking on the audio region of your loop from the main arrange window. Use the View menu of the window, and select "Samples" as a view option (this will guarantee that the length of the audio files is shown in samples and not in other formats). Select the entire loop (*Command-A*), and read the number of samples in the upper left corner of the destructive audio editor window.
PT	The sample rate of any audio files in a session matches the one of the session, since the sample rate of every audio file that was imported is converted into the one at which the session was created.	Use the selector tool to highlight the region for which you want to find the tempo. Make sure you have selected the main counter to show the number of samples by choosing the "Samples" option in the Display menu. Read the number of samples of the area you selected in the Event Edit Area next to the Location Indicators.
CSX	Open the Audio Pool window. You can see all the information regarding any particular audio file, including the sample rate.	Open the destructive audio editor (called Sample Editor) by double-clicking on the audio region of your loop from the main arrange window. Select the entire loop (*Command-A*), and read the number of samples in the field named "Selection Range," located in the upper right corner of the window.

tempo of a loop that are worth considering because in most cases they can help speed up the process. You will find a brief explanation of these techniques in the following paragraphs.

In DP a quick way to find out the tempo of a loop is to trim the beginning and end of the sound bite to the length of the loop so that you are able to select the exact number of measures that form the loop. Select the trimmed sound bite and choose "Set Soundbite Tempo" from the Audio menu. Punch in the time length of the loop in beat. The tempo of the sound bite will be calculated automatically and listed next to the "Tempo" field in the lower right corner of the active window. By clicking the OK button, you will "stamp" the selected sound bite with the tempo that was calculated by the computer. Now, to simply change the tempo of the loop to match the tempo of the song, select the function "Adjust Soundbites to Sequence Tempo" from the Audio menu. If you want to change the tempo of the sequence in order to match the tempo of the selected loop, choose "Adjust Sequence to Soundbites Tempo" from the same menu instead.

A similar technique is available in PT. Use the *selector tool* to highlight an exact number of beats of your loop. Zoom in if necessary to be as precise as possible. Select the function "Identify Beat" from the Edit menu. Make the necessary changes to the *start* and *end time signature* fields, depending on the time signature of the loop. Insert the end location value, which will be equal to the start location plus the number of beats of the loop you selected. After you click the OK button, you will see that a new tempo change marker has been inserted in the tempo ruler at the top of the edit window. That tempo is the tempo of your loop. Now that you know its exact tempo, you can adapt it to the tempo of the sequence by selecting the loop in the Edit window and choosing Time Compression Expansion from the AudioSuite menu. Look at the Tempo fields: In the "Source" field insert the original tempo of the loop, and in the "Destination" field insert the tempo to which you want to convert the loop (probably the tempo of your sequence). Press the Return key and then click on "Process." Your loop now will match the tempo of the sequence.

In LP, if you want the tempo of the loop to match the one of the sequence, trim the loop to an exact number of bars (2, 4, 8, or similar) and then set the locators to match the same number of bars of the trimmed loop. From the Functions menu of the arrange window select "Object" and then "Adjust Object Length to Locators." The loop will be time-stretched according to the tempo of the sequence. To adapt the tempo of the sequence to the tempo of the loop follow the same procedure, but press the "T" key instead.

In CSX, open the sample editor by double-clicking on the audio loop from the main arrange window. In the sample editor turn on the "Hitpoint Mode" by clicking on the Hitpoint icon in the top center of the sample editor window. The "Hitpoint Detection Window" will appear on the screen. Adjust the bars parameter according to the length of your loop, and set the minimum and maximum of the BPM range if you have a vague idea of the tempo of the loop (if you don't, you can just leave it very wide and let the computer do the calculation). The "beats" field is set according to the overall rhythmic feel of the loop (this will help the computer to detect and place hitpoints according to the type of loop you are analyzing). Click on the "Process" button. Select an exact number of bars (usually 2, 4, or 8) using the two markers in the ruler above the waveform. Insert the information about the loop in the fields located at the right of the Hitpoint Sensitivity slider (Bars, Beats, and Time Signature). The value you insert must equal the number of bars you selected in your loop. The tempo of the loop will be shown in the "Original tempo" field. If you want to change the tempo of the sequence to match the tempo of the loop, simply select the function "Set Tempo from Event" in the "Advanced" section of the Audio menu. If, on the contrary, you want the loop to adjust to the tempo of the sequence, select "Create Audio Slices" from the same menu. From now on the loop will follow the tempo changes of the sequence.

Listen to Examples 2.6 through 2.9 to compare the same loop time-stretched to accommodate three different, new tempos.

2.13 Basic automation

Another big advantage of using an audio sequencer to record, mix, and produce your projects is that almost every parameter of a track can be automated either graphically or in real time. In this section we focus on the automation of basic parameters, such as volume, pan, and mute. The automation of volume and pan can contribute greatly to the enrichment and improvement of your production. Through automation you can create crescendos, decrescendos, and fades independently for each audio track and MIDI channel, or you can create stereo effects by having the signal moving from left to right, and vice versa. Listen to Examples 2.20 through 2.22 on the audio CD. The automation can be applied in similar ways to both MIDI and audio tracks. In the case of MIDI track automation it is achieved through the use of CC messages (CC 7 controls the volume of a MIDI channel, while CC 10 controls its pan). The automation of audio tracks is controlled by sequencer internal protocols. Depending on your background, you might find yourself more comfortable inserting automation data in real time through the use of a hardware control surface or a digital mixing board, by recording a fader's movement controlled by the movement of your mouse, or by inserting automation data graphically through the graphic editor of the sequencer. No matter which option you choose, the type of data inserted will always be the same, and the automation information can be edited in all the aforementioned ways at any time. I usually recommend doing a rough automation mix with the mouse or with an external control surface and then going back to polish your final mix through graphic editing of the data previously inserted.

There are two main types of automation: static and dynamic. Static automation is achieved through the use of *snapshots* taken at different locations in your project. The parameters of your virtual mixing board will change according to the data inserted on each snapshot. This is a pretty basic way of automating your mixes. Its advantages are that it can be a quick way to insert basic ideas that can be developed later and that it uses very little processing power and MIDI bandwidth. A dynamic mix involves continuous changes over time of different mix parameters at the same time. In a dynamic mix scenario you can have the volume of track 1 slowly fading in while the pan of track 2 moves continuously from left to right. This type of mix gives much more flexibility, versatility, and control over your mixes.

2.13.1 Static automation

Static automation can be recorded in two different ways: In DP you can use the *snapshot* function available from the mix window, while in PT, CSX, and LP you can manually insert data either in the graphic editor or in the list editor. To insert a snapshot in DP simply open the mix window, set the locator at the point in time where you want to insert the snapshot, make the necessary changes to your mix (set the volume, pan, and mute assignments for your tracks), and click on the small camera icon in the top left corner of the mix window. A menu will appear with a few parameters regarding the snapshot you are going to take. In Table 2.5 you can see a brief explanation of these parameters.

By clicking on the OK button you take a snapshot, and all the automation data for all the tracks specified in the previous settings window will be inserted. In order for the sequencer

Table 2.5 Snapshot mix parameters in DP

Parameter	Description
Time range	Allows you to specify an exact time range in which the snapshot settings will be inserted
Tracks	Allows you to specify which tracks will be included in the snapshot
Data types	Allows you to specify which automation data (i.e., volume, pan, etc.) will be included in the snapshot

to follow and play back automation data, you need to make sure that each track on which you want to have the automation executed has the automation playback option enabled. In DP you do so by clicking on the automation playback button of each track from either the track list window or the mix window (Figure 2.20a). In CSX click on the "R" (Read) button next to the faders on the mix window or in the inspector window (Figure 2.20b). In LP and PT make sure you select "Read" from the menu located above the fader of each channel in the mix window (Figure 2.20c and 2.20d).

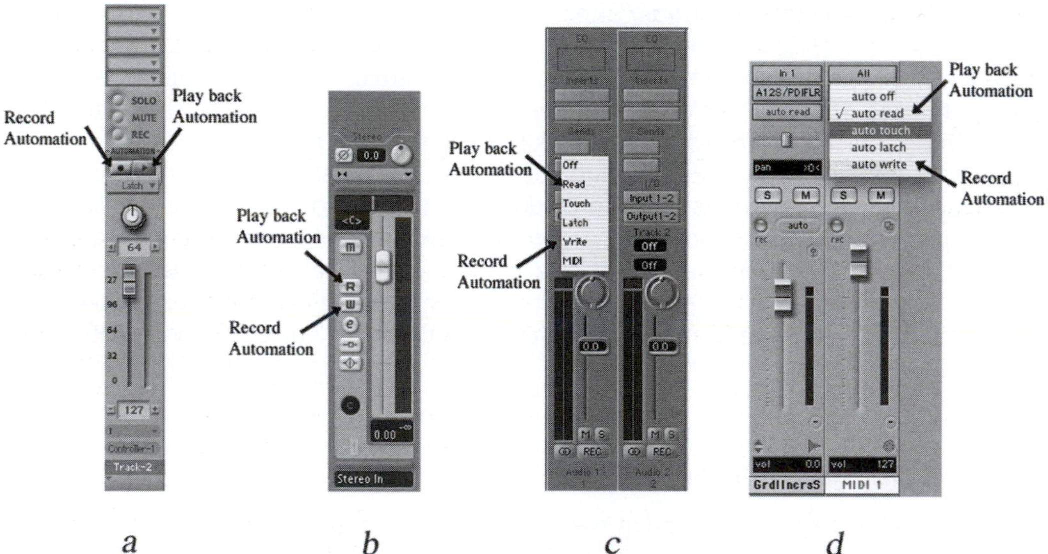

Figure 2.20 Real-time playback automation enabler in DP, CSX, LP, and PT.

While LP, CSX, and PT don't have a snapshot function built in, you can still take advantage of static automation by manually inserting automation data for each track and parameter you want to automate. You can do so in the graphic editor, the list editor, or, in the case of LP, in the "Hyper Edit" window. I am going to talk specifically about these techniques in the next few paragraphs, where I describe how to edit automation data.

2.13.2 Dynamic mix: real-time automation

The first step to record automation data for your project in real time is to record-enable the automation for the tracks on which you want to record the data. This can be set from the mix windows. In DP, click on the record button in the "Automation" section of the track's channel strip in the mix window. Be careful not to click on the actual record-enable button of the track (Figure 2.20a). In CSX, click on the "W" (Write) automation button, placed to the left of each channel in the mix window (Figure 2.20b). In LP and PT select one of the automation recording modes by clicking on the automation field right above the view meters of the track in the mix window (Figure 2.20c and 2.20d).

The next step is to choose the record-automation mode you are going to use. There are three main modes with which automation can be recorded: overwrite (called *write* in PT and LP), touch, and latch. The main difference between these modes is the way the automation data are written on the track. Take a look at Table 2.6 to get a better understanding of the differences between the modes.

In addition to the settings mentioned in Table 2.6, each sequencer has its own variations of automation modes that simply expand the basic options explained earlier. For example, in CSX the *X-latch* mode has the same features as *latch*, with the only difference being that the sequencer stops recording automation data when it encounters preexisting data. In PT, DP and CSX the *trim* mode allows you to reset the offset of preexisting automation data and therefore to modify their values without changing the original shape of the automation.

Table 2.6 Record automation modes

Mode	Description
Overwrite	If you select *overwrite*, the automation data will be recorded as soon as you press the playback button on your sequencer, even if you don't move any faders or knobs. The existing automation data present on the track will be replaced by the new settings. The recording of automation data will stop only by stopping playback. This mode can be a bit dangerous if you are not careful, because you might end up deleting some prerecorded automation data if you don't stop the sequencer in time. It is useful, though, when you insert automation for the first time on a track, since it allows you to insert data without moving any object in the mix window.
Touch	The *touch* mode allows you to record automation only when you click on an object of the mixing board and change a parameter. When you press the play button, no automation will be recorded until you click and move a fader or a knob. If you release the controller, no automation data will be recorded and the controller will automatically go back to its previous position. This is the safest option because you will be recording data only when you move a controller on the mix window. It is the best option to replace only specific passages of automation.
Latch	The *latch* mode is a combination of the two previous modes. Like *touch* mode, after you press the play button, the sequencer starts recording automation data, but only when you click and move a fader or change a parameter. Like *overwrite*, though, the recording of automation data will stop only by stopping playback, meaning that even after releasing the mouse the current settings will be recorded until you press the stop button.

2.13.3 Editing automation data

Automation data inserted manually, through snapshots or through real-time recording techniques, can be edited in two ways: graphically and in the list editor. Even though the latter option could work for a small amount of data, it would quickly become overwhelming if used to edit a large number of data for several tracks. A much easier way of editing and inserting automation data is based on the use of the graphic editors available in every sequencer. Automation data on MIDI tracks can be edited using the techniques you learned in the first part of this chapter. Volume data are represented by CC 7 and pan data by CC 10. The automation data for audio tracks and the editing techniques associated with them are represented using proprietary methods of each sequencer, and, even though they are fairly similar, the operational differences can be substantial. Therefore I am going to list the different ways of editing automation on a "per sequencer" basis.

In DP you have plenty of ways to edit the parameters and data related to automation. For MIDI you can edit automation data in the graphic editor and filter the data you want to work on by simply selecting the right CC number from the "Continuous Data Filter" available in the lower left corner of the graphic window. If you choose CC 7, then on the lower part of the window you will see only volume MIDI data. At this point, by using the mouse, you can select, delete, copy, cut, and paste the automation events previously recorded.

Another way to edit the automation data is through the Sequencer Editor window (*Shift-S*) shown in Figure 2.21. This is probably the best option since from this window you can edit MIDI and audio tracks at the same time, including their automation data. In order to edit the automation, simply open the Sequencer Editor and select the track you want to edit by clicking on the track names to the left of the window (if the list is hidden, click on the "Expand"

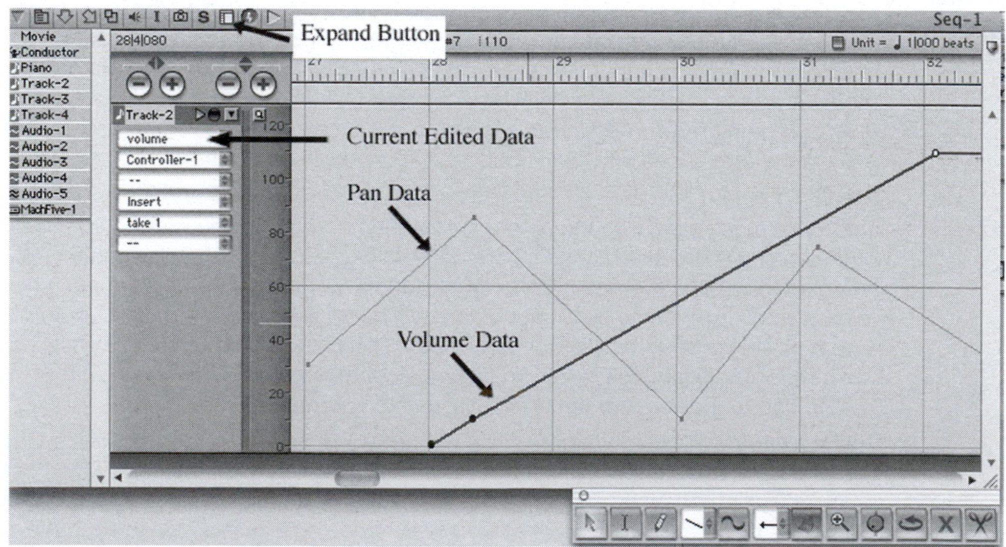

Figure 2.21 Graphic editing of automation data in DP.

button in the menu bar of the window). Select the data type you want to edit (i.e., volume, pan, mute, etc.) by choosing the parameter from the first drop-down menu below the track name. Use the *pencil tool* to insert new data, the *reshape tool* to alter existing data, and the *marquee tool* to select, insert, and move individual data. You can also use the "reshape flavor" button to choose between different shapes when you use the pencil or reshape tool.

In CSX the automation data can be inserted and edited directly from the main arrange window, making it a very easy, intuitive, and quick process. For each track you can look at the automation data in the form of additional track layers. You can make changes at any time and in real time using the tools in the tool palette. To open the different automation layers for a track, click on the small "+" sign located in its lower left angle (Figure 2.22). A new track layer will show a grayed-out version of the MIDI data or waveform plus a line with the automation data for a particular parameter (such as volume, pan, mute). Using the *draw tool* (pencil) you can insert new automation data. With the *line tool* (line) you can change or insert data with different preprogrammed shapes, such as line, parabola, sine, square, and triangle. Use the *Object selection* (arrow) tool to change or insert single points on the data curve.

A similar approach is used in LP, where automation can be edited and inserted from the main arrange window. To reveal the automation parameter layers, choose the "Track

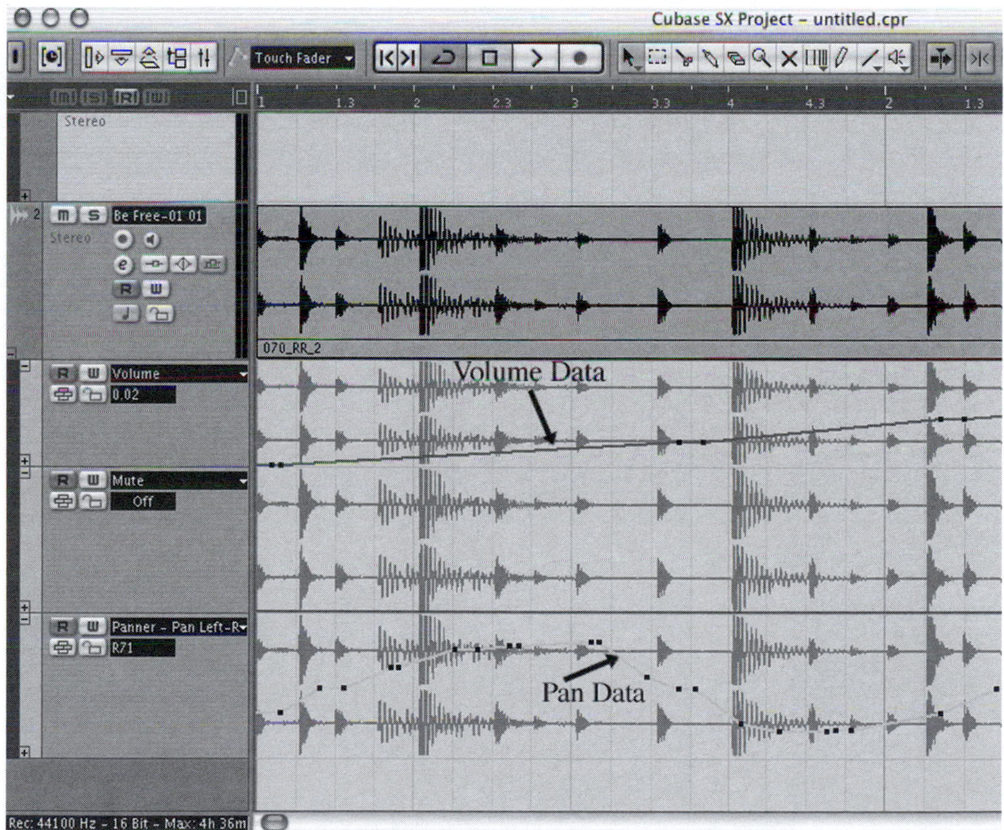

Figure 2.22 Graphic editing of automation data in CSX.

Automation" option in the View menu. The default parameter set is going to be Volume for both MIDI and audio tracks. You can see several parameters at the same time on several automation layer-tracks (each displaying a different controller) by clicking on the small arrow pointing left in the last automation track available (Figure 2.23). Use the *pencil* to

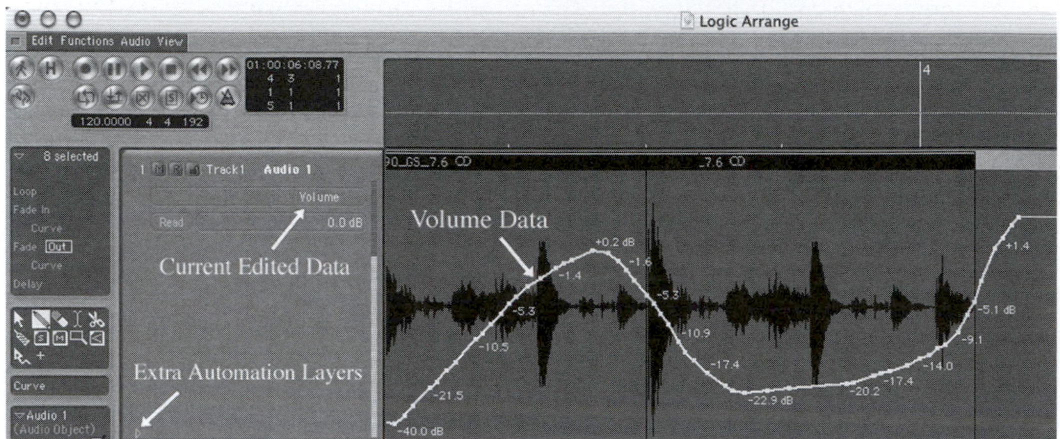

Figure 2.23 Graphic editing of automation data in LP.

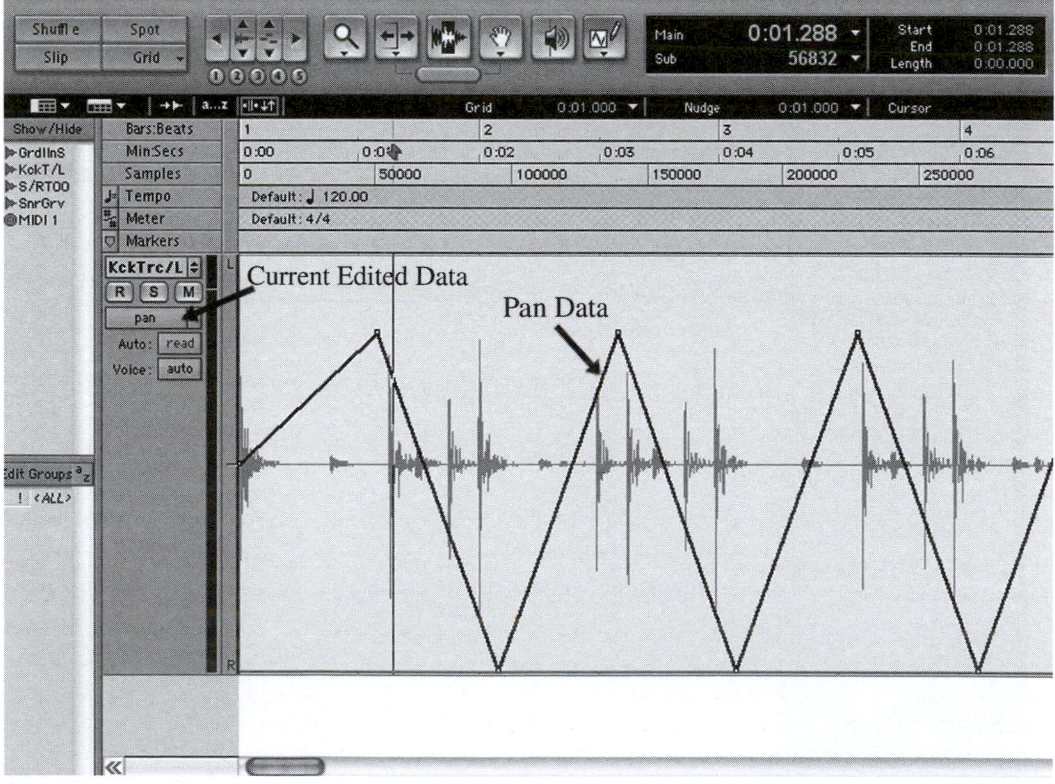

Figure 2.24 Graphic editing of automation data in PT.

insert or modify data, the *eraser* to delete data, the *arrow* to insert or change single data, and the *automation tool* to select or change the curvature of existing automation (four types of curve are available: concave, convex, and two types of S-curve).

In PT all the automation edits are done in the edit window (Figure 2.24). Each track has different layers, which can be activated by clicking on the drop-down menus below the track names. As a default you can access volume, pan, and mute parameters. Eventually, as with the other sequencers, you can automate any parameters of any plug-in and effect. To edit the automation parameters of a track, you can use the main editing tools. Use the *pencil tool* to insert or change data. As with the other sequencers, you have several preset curves, such as line, square, triangle, and random shape, to facilitate repetitive edits. With the *grabber tool* you can insert single data or change existing ones, while with the *selector tool* you can select and delete multiple data at the same time.

2.14 Practical applications of automation

The actual uses and practical applications of automation can have a big impact on the final result of your projects. While you will discover new frontiers in automation in every project and sequence you will work on, there are some very useful applications I would like to share with you. Use them as a starting point for your sequencing activities.

2.14.1 Volume automation

The automation of the volume parameter has some common applications, such as fade in and fade out of single track, multiple tracks, or entire sequences if applied to a master fader track. These can be applied to both MIDI and audio tracks. Keep in mind, though, that for MIDI tracks you are allowed only one volume per MIDI channel, *not* one per track. If, for example, you have a drum part spread among several MIDI tracks (let's say bass drum, snare drum, hi-hat, and toms), all sent to the same device and MIDI channel, then you will have only one common volume available for all of them, making it fairly impractical to fade one drum track at a time.

There are several workarounds, though. The first solution would be to assign different tracks to different channels or devices if your setup allows it. Otherwise you can record the MIDI tracks as individual audio tracks by routing the output of the sound generators of the MIDI device to the inputs of your audio interface. Once the tracks are recorded as audio, you will have independent control over any parameters, including volume, pan, mute, and effects. Do not delete the original MIDI tracks, in case you want to change some of the parts at a later time. While this is a good workaround, it will increase considerably the size of the project in terms of hard disk space. A third solution would be to use the velocity parameters of each part to achieve different volumes, fade-ins, and fade-outs. Use the graphic editor to open the track on which you want to create the volume change, select the notes for which you want to edit the velocity, and, by using either the *pencil* or the *reshape tool,* change the velocity according to your needs. If you use this procedure, make sure to create backup copies of the original MIDI tracks in case you want to restore the original velocities of your performances.

Volume automation can be a solid tool to improve MIDI performances of acoustic samples and synthesized sounds. The main problem of sequencing acoustic instruments such as string instruments, woodwinds, and brass consists in reproducing a realistic attack and release for different passages and dynamics. In slow passages and long sustained sections I recommend using volume automation to achieve a much more realistic reproduction of the string instruments' dynamics. Listen to Example 2.10 on the audio CD. It features a string section played without the use of automation. As you can hear, the parts sound unrealistic and choppy. If you use a bit of volume automation at the beginning and end of each sustained note to fade in and out each bow movement slightly, you will achieve a much more realistic effect. This is because string instruments, when played with the bow, have the natural tendency to slowly fade in at the beginning of the bow movement before reaching full volume when the string is completely set in motion and reaches full amplitude. Listen to Example 2.11 on the audio CD for a more realistic string effect. You can add the automation fade-in and fade-out for each sustained note either while playing or graphically after you record the part. If available, I recommend using a fader from your keyboard controller (digital mixing board or control surface), assigned to CC 7, to record the volume automation while performing the part. It will give you a better idea of when and how to use the volume to increase the realism of the sampled string section.

The same idea can be applied to wind instruments. This is particularly true for sections in which you alternate fast passages with slow and sustained notes. The use of the regular attack and release parameters on your synthesizer wouldn't be able to accommodate both playing styles (fast attack and release for staccato parts and slow attack and release for slow passages). The best option is to keep a short attack and release and to use automation for smoothing the beginning and end of sustained notes. Listen and compare Examples 2.12 and 2.13 on the audio CD. Notice how the oboe part sounds much more realistic in the second example, where the beginning and end of the sustained notes are edited with volume automation.

The automation of volume can also be used to simulate dynamic compression on both MIDI and audio tracks. If you feel that the performance you just recorded has abrupt changes in dynamics that seldom occur in the track, you could adjust the volume to correct and smooth the highest peaks using a compressor. Listen to Examples 2.14 and 2.15. The first one is a guitar track without compression that has big jumps in volume. The problem was fixed with volume automation in order to have a smoother overall performance.

2.14.2 Pan automation

The automation of the pan parameter can be used to create interesting stereo effects to help widen the stereo image of synthesized pads, leads, or acoustic instruments for both MIDI and audio tracks. For example, by choosing a predefined shape, such as triangle, sine, or square, and using the *pencil tool* to draw pan automation, you can create the effect of a constantly moving instrument on the stereo image. This will help to add stereo image depth, clarity in the final mix, and interest for the listener. Use a small grid (16th note or 32nd note) for fast and rhythmic parts such as synth arpeggios (Examples 2.16 and 2.17 on the audio CD), or use slower resolution (whole note or half note) to open the stereo image of sustained pads and sustained parts in general (Examples 2.18 and 2.19 on the audio CD).

Another creative use of the pan is to apply it to cymbal swells, mark tree passages, or harp glissandos and guitar arpeggios, where the pan of the track is automated from one side to the other following the glissando. This technique adds a dramatic effect to the part, shifting the attention of the listener from the orchestral foreground (melody and harmony) to the background (percussion and sound effects). Listen to Examples 2.20, 2.21, and 2.22 on the audio CD to compare pan automation techniques applied to percussion and harp parts.

If used on acoustic harmonic instruments such as guitars, harps, and keyboards, pan automation can help increase the stereo image and the depth of their sounds. In this case you want to use a very mild and subtle automation in conjunction with a slow rhythmic setting in order to simulate the movement of the performer in front of a stereo microphone. Listen to Examples 2.23 and 2.24 on the audio CD and compare a guitar track without and with pan automation.

2.14.3 Mute automation

Even though most of the effects that can be achieved with the automation of the mute parameter can be accomplished with regular editing techniques, I usually like to experiment with this type of automation, especially on synthesized pads and sustained parts, to create rhythmic effects that would be too complicated and tedious to achieve through regular editing. You can, for example, model interesting rhythmic patterns on a sustained pad by simply selecting a random pattern curve and by using the *pencil tool* to set mutes on the track. It will bring the sound in and out at randomly selected intervals (Example 2.25 on the audio CD). Or you could use a more regular pattern, such as the square curve, and a medium-to-fast grid and insert mutes every other subdivision in order to create an arpeggio effect (Example 2.26 on the audio CD). By applying all three types of automation together (volume, pan and mute) you can achieve some very interesting and original effects, such as the one in Example 2.27 on the audio CD.

2.15 Summary and conclusion

The sequencer is the central hub of your MIDI and audio network. It allows you to record, play back, edit, and store MIDI and audio data. In order to achieve the best results in a "problem-free" and smooth working environment, you have to organize your sessions and projects in the right way. The use of templates can improve considerably the organization and the speed of your sessions. I recommend having a series of premade templates for different ensembles and project situations that you can recall when necessary. After organizing your MIDI and audio tracks, make sure all the connections in your studio work properly. Always do a "preflight" check before every session, especially if you expect to work with a client or a colleague. Check that the MIDI tracks are working and that every device in your studio is receiving and sending MIDI data. Do the same for your audio tracks. Try to record and play back a few seconds of audio to test the audio connections. Make sure before starting your project that you have a clear idea where all the session files are located on your HD, that you have enough space to record, and that you have a quick backup plan to use during session downtimes. Before starting recording, also make sure that the metronome is set up right, that it is played by the right instruments (a MIDI track), and that it is set at the tempo you feel most comfortable. Remember that the click track is your best friend in a sequencing environment, not your worst enemy, so play with it and do not fight it.

When recording MIDI or audio tracks, make use of multiple takes so you have as much material as possible to edit later. Record different passages, choose the best takes, and compile the final track from several passages.

When editing a part, use the edit window that best suits what you are trying to achieve. Each editor has pros and cons. Use the one you feel most comfortable with and that you think is going to help you complete the editing task quickly and easily. Quantization can highly improve the final result of your projects if it is used in the right and wise way. Remember that quantization is not a magic tool; it can only fix what was already good enough to be interpreted in the right way by the sequencer. Choosing the right quantization value is crucial: It should be equal to or smaller than the smallest rhythmic subdivision present in the region you are quantizing.

The addition of audio loops and audio parts in general can greatly improve the overall feel of your project. When dealing with rhythmic loops, make sure they match the tempo of your sequence. After importing the loop, if you don't know its tempo you can use the following formula to calculate the tempo, in beats per minute: $(NB \times SF \times 60)/NS$, where NB is the number of beats of the loop, SF is the sampling frequency in hertz, and NS is the number of samples of the loop. You can also use specific shortcuts available on each sequencer to make the tempo of the loop match the tempo of the sequence, or vice versa.

The automation of parameters such as volume, pan, and mute is a fantastic tool for improving and expanding the boundaries of your projects. An automated mix can be *static* (with the insertion of snapshots of single data) or *dynamic* (with the insertion and recording of continuous movements and data). The latter can be achieved in three different ways: by recording automation data through a control surface, though the movement of the mouse, or through the insertion of data in one of the editors. Once the data are inserted, they can be edited either graphically or though the list editor. The same principles for the editing of MIDI data are applicable to automation data. Practical applications of automation include fades (in and out), crescendos and decrescendos, simulation of attack and release of acoustic instruments such as strings, woodwinds and brass, and more creative applications, such as creative panning and mute.

In this chapter you learned the basic methods that constitute the building blocks for more advanced sequencing techniques. If you think what you learned was exciting, wait until you read the next few chapters, where we will experience more detailed and advanced sequencing tools.

2.16 Exercises

Exercise 2.1

Set up a sequence with the following features:

- Fourteen MIDI tracks
- Five audio tracks

- Instrumentation for MIDI tracks: violins, violas, cellos, basses, drums (with one track each for bass drum, snare drum, hi-hat, toms, and cymbals), electric bass, woodwinds, synthesizer pad, rhythmic guitar, and tenor saxophone.
- Tempo: 110 BPM

Exercise 2.2

a. Using the template created in Exercise 2.1, sequence 16 bars using at least 10 MIDI tracks (at least three must be drums). Make sure to use the multiple takes feature of your sequencer.
b. Quantize each track using the right quantization value.
c. Copy and paste material from the first 16 bars in a creative way in order to have a total of 64 bars.

Exercise 2.3

Find the tempo, in beats per minute, of the loops listed in Table 2.7 according to their sampling frequency (SF), number of samples (NS), and number of beats (NB).

Table 2.7 List of loops for Exercise 2.3

	SF	NS	NB	Tempo
Loop 1	44.1 kHz	100,340	8	
Loop 2	96 kHz	400,567	16	
Loop 3	48 kHz	220,100	8	
Loop 4	44.1 kHz	150,320	8	

Note: 1 kHz = 1000 Hz.

Exercise 2.4

Using the sequence created in Exercise 2.2, import at least two audio loops found on the CD or from your own library. Find the tempo of each loop, and adapt their tempos to the tempo of your sequence using the time-stretch feature of your sequencer. Insert the loops in your project as you wish.

Exercise 2.5

Insert automation in order to have the following data:

a. Fade the strings in with a slow fade 4 bars long.
b. Automate the pan for the synth pad track moving constantly from left to right, and vice versa, at a pace of 16th notes throughout the entire 64 bars.
c. Mute the bass from bars 17 to 32.
d. Program a crescendo of one track of the drum kit using one of the techniques explained in this chapter.

3 Intermediate Sequencing Techniques

3.1 Introduction

What you learned in the first two chapters has given you a solid background and a basic knowledge of sequencing. By now you should have a good idea about the potential offered by the tools you are using and how they can inspire your creativity and improve your productions. If what you learned in Chapter 2 provided you enough information and skills to start sequencing, in this chapter you will gain a deeper knowledge of more advanced sequencing techniques and tools, such as "groove" quantization options, MIDI channel layering, audio and MIDI track layering, editing techniques, and how to use and program tempo changes. In the second part of this chapter you will learn about the practical aspects of sequencer-based production and project management, such as the synchronization of nonlinear machines and backup and archive techniques.

Before continuing in this chapter, make sure you are familiar with all the concepts presented in Chapter 2. Master the basic techniques introduced in that chapter, and use the exercises listed at the end of it to familiarize yourself with how a sequencer works and how to start a project with both MIDI and audio tracks. Since each chapter of this book builds on the previous ones, it is very important to be familiar with the material presented earlier. If you "did your homework," then you are ready to continue discovering more advanced and exciting sequencing techniques. Let's go!

3.2 Groove quantization and the "humanize" function

In Chapter 2, when I discussed the basic quantization options available in your sequencer, we learned that by quantizing your MIDI parts, the computer can check and correct the timing (rhythmic position) of the MIDI events you recorded by moving and aligning them according to a grid. The quantization value you select determines the grid setting. As you probably experienced by quantizing the parts in the sequences you recorded so far, this basic quantization technique seems to flatten out the original groove of your performance. Since every event is moved to a position determined by the quantization value, the original rhythmic "freedom" that was part of your performance is lost. This can lead to very stiff, cold, and impersonal parts and sequences. Up to very recently, the terms *sequencer* and *quantization* were synonymous with "mechanical," "stiff," and "drum-machine-like" music. In fact this preconception has for years kept many composers away from the MIDI environment. Unfortunately it is really a misconception that sequencers can be used only for mechanical and loop-based composition. With the latest innovations introduced by the most advanced sequencing software, the modern composer is able to achieve levels of smoothness, precision, and overall rhythmic feel that were impossible before. Gone are the days when sequencing meant boring and rigid rhythms. These improvements are possible mainly through the increased sophistication of the quantization algorithms implemented in the sequencers. The options available to composers are countless and include quantization parameters such as swing percentage, sensitivity, strength, randomization, iterative quantization, groove quantization, and MIDI-to-audio quantization. Let's analyze all the different quantization options provided by the four sequencers we are learning about, and let's examine their practical applications.

3.2.1 Quantization filters

In Chapter 2, when applying quantization, we assumed that all the notes and events in the selected region needed the same quantization settings in terms of particular notes/events and how they were going to be affected. This approach implies a very basic type of quantization. It assumes that all events need the same amount of quantization and that all notes need to be quantized. The idea of "quantization filters" can be used to gain more control over which parameters, which notes, and how much the events will be quantized. The main idea is that instead of having only the option to fully quantize an event or not quantize at all, we can create several variations between the two extreme settings (100% to 0%) that allow us to be more precise and to decide how much an event will be quantized. Here, 0% corresponds to no quantization and 100% corresponds to full, or "straight," quantization. The same idea can be applied to which events will be quantized or to the amount of, say, swing that will be applied to the quantization (more on this later). Let's learn these techniques.

One of the most basic quantization filters you can set up allows you to determine which parameters and events will be affected by the quantization action. You can choose to quantize MIDI or audio events, MIDI CCs, or any other events. By choosing one or more parameters you instruct the sequencer to disregard all the others that were not included in the list. The quantization therefore will be applied only to those events selected. You can also choose to quantize the *attack point* of the MIDI note (which is usually the case) or the *release point* (meaning the end of the note). In the latter case the computer, instead of aligning the beginning of the note to the grid, will align the end of it (this option is rarely used in a practical sequencing situation). The quantization filter options vary from sequencer to sequencer. In DP, for example, you access these settings by opening the quantization menu located in the Region menu (Figure 3.1) and clicking on the events you want to be affected by the quantization. To see more options click on the "What to quantize" drop-down menu. A similar menu can be accessed in PT by selecting "Quantize" from the MIDI menu.

Although a generic quantization filter is not directly accessible from LP, you have two other options: You can select the events you want to quantize manually from any editing window,

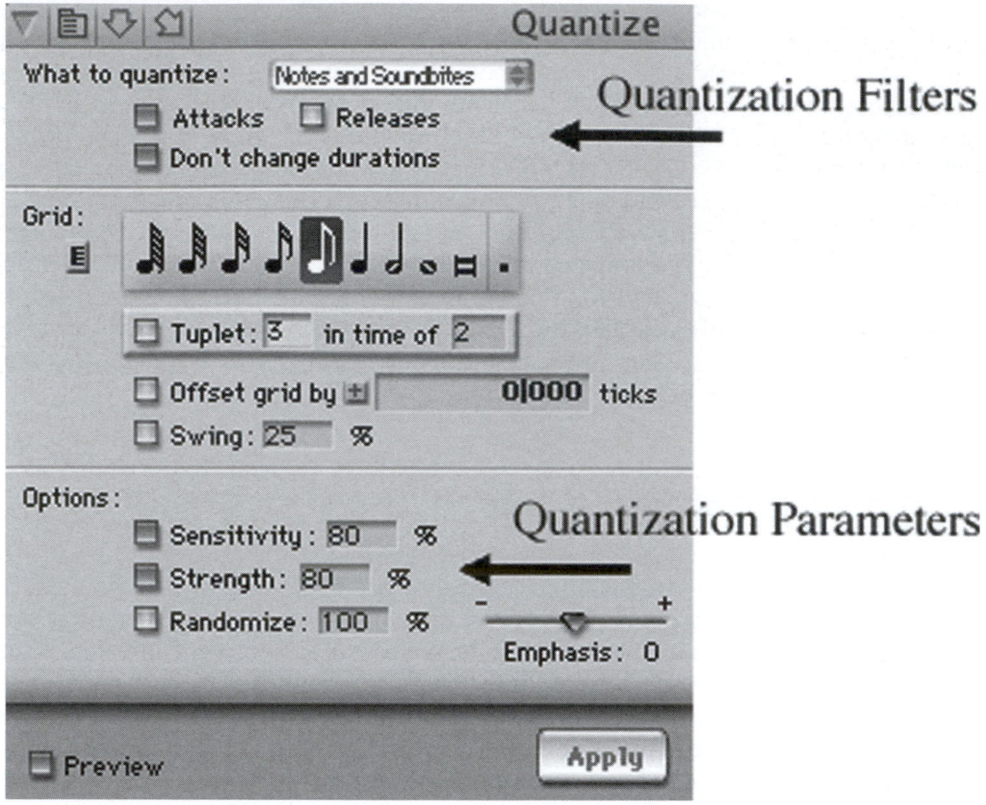

Figure 3.1 Quantization parameters and filters in DP.

or you can use the very sophisticated "Transform editor," which allows you to choose any filter events based on several criteria assignable to different parameters. Keep in mind that these filters can be used for many other situations. This editor can be opened by selecting "Open Transform" from the Window menu. The window is split in two sections. The top part sets the "conditions," meaning which events will be affected by the "operations," which are set in the lower part of the window. For example, you could set the conditions and operations as "apply 16th note quantization only to notes between bar 1 and bar 5 on MIDI channel 1" (Figure 3.2). This allows you to be very specific in terms of which events will be quantized and how they will be affected by the quantization.

In CSX you can set the quantization filters in several ways. If you select the option "Advanced Quantize" from the MIDI menu, then you can choose to quantize either the end of the notes ("Quantize Ends") or their lengths ("Quantize Lengths"), which means that lengths of all the notes selected will be readjusted according to the quantization value selected. Remember that the quantization value in CSX is set at the top right of the Arrange window using the "Quantize-Type" drop-down menu. The "Undo Quantize" function allows you to restore the original timing of the performance you recorded no matter how many quantizations you applied to the part. Use "Freeze Quantize" to make the quantization permanent for the selected part. The "Part to Groove" option allows you to create custom quantization templates (more on this in Chapter 4). Other, more advanced quantization

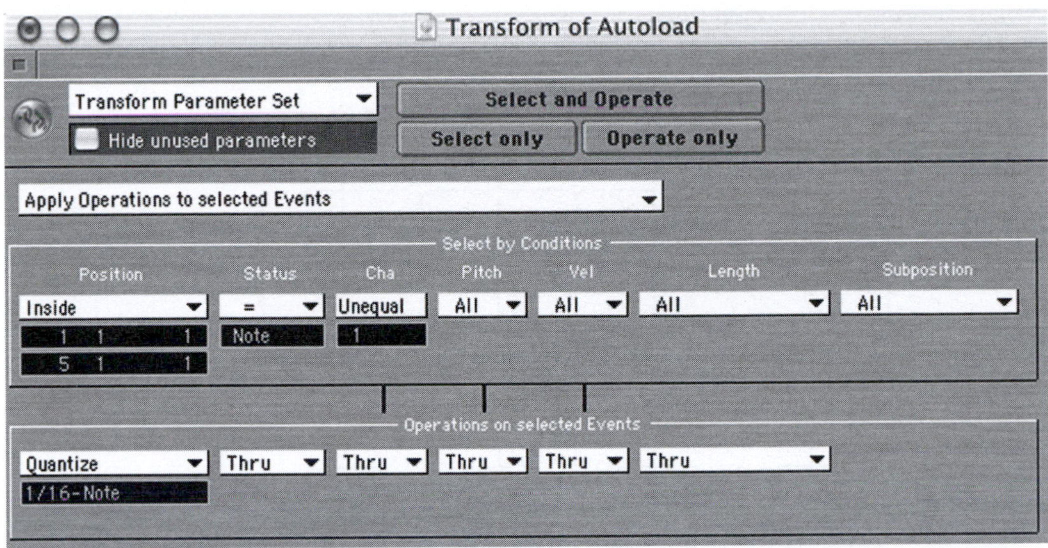

Figure 3.2 The Transform window in LP.

parameters and filters are available from the "Quantize Setup" submenu accessible from the MIDI menu, which is discussed later in the chapter.

One of the most important features of modern quantization algorithms is the ability to choose the events you will quantize based not only on their type (e.g., notes, CCs, audio sound bites), but also on their position relative to the grid. This allows you to preserve the original groove of your performance without compromising its tightness to the click track. In other words, by controlling the *sensitivity* of the quantization algorithm, you can choose which events will be quantized and which will be left unquantized, based on their position and not on their type. In Figure 3.1, the *sensitivity* parameter in DP ranges from 0% to 100%. With a setting of 0% no events will be quantized, while with a value of 100% all the events will be affected. Any value in between will allow you to extend or reduce the area around each grid point influenced by the quantization action. With a setting of 100% or with the regular quantization options, each grid point has an area of influence (a sort of "magnetized" area) that extends 50% before and 50% (a total of 100%) after the point itself. Events that fall in these two areas will be quantized and moved to the closest grid point. By reducing the sensitivity, you reduce the influence area controlled by the grid points. Therefore if you choose a sensitivity of 50%, each point will "attract" only notes that are 25% ahead of or 25% behind (a total of 50%) the grid points. This setting is used mainly to "clean up" the events around the grid point and leave the natural rhythmic feel of the other notes. If you choose a negative value for the sensitivity parameter, you will achieve the opposite effect: Only the events that were played farther from the grid points will be quantized, leaving the ones that were fairly close to the click in their original position (Figure 3.3). This setting is perfect to fix the most obvious mistakes but leave the overall natural feel of your performance intact. On a practical level I recommend using a sensitivity value between −50% and −80% to fix major rhythmic mistakes but keep the overall feel and groove of your performance. Listen to Examples 3.1, 3.2, and 3.3 on the audio CD to compare different sensitivity settings.

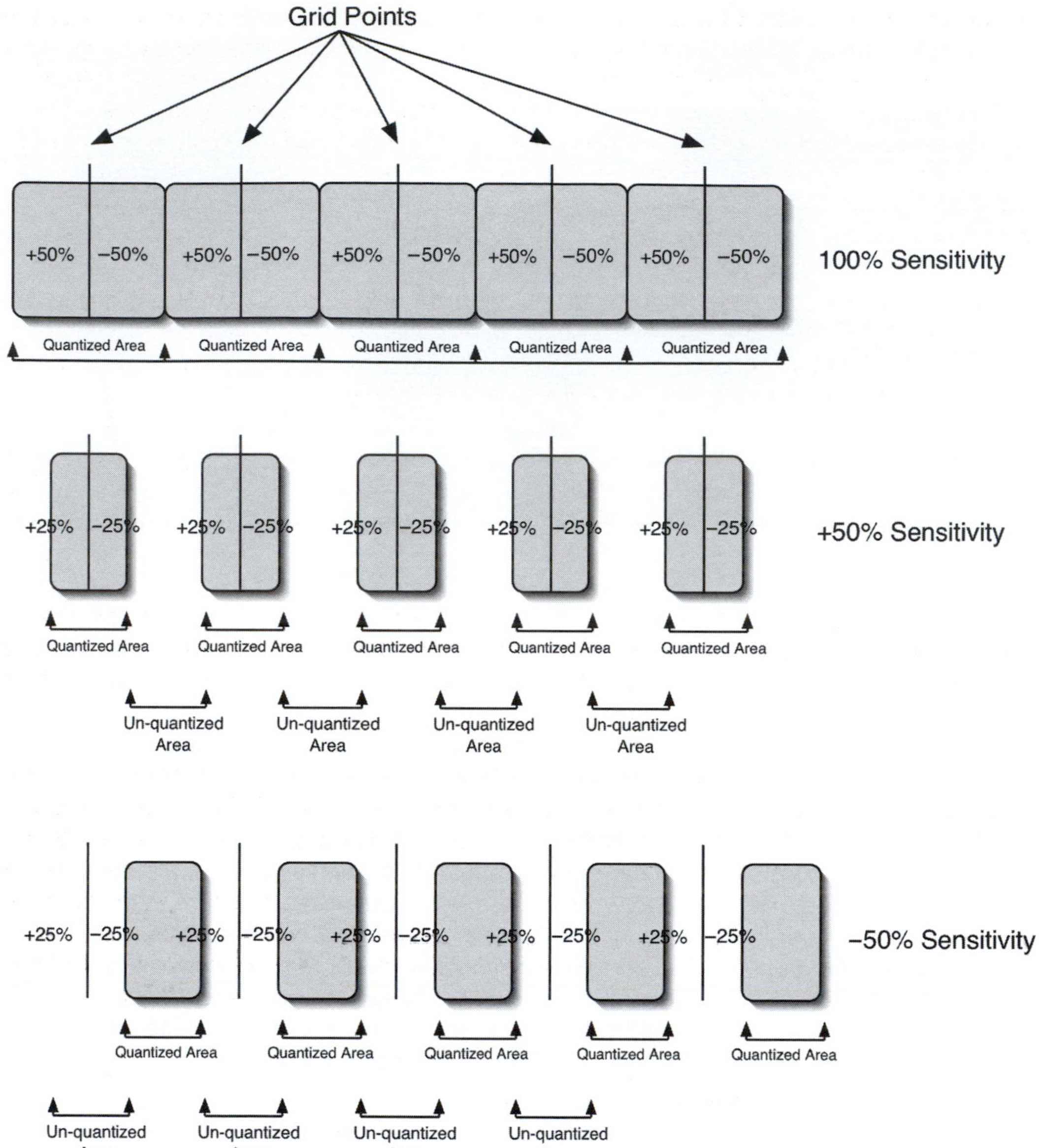

Figure 3.3 Quantization parameters: sensitivity.

Another parameter you can use to improve the quantization of your parts is the *strength* option. While the sensitivity parameter has an impact on which events will be affected by the quantization, the *strength* allows you to control "how much" the events will be quantized. By choosing a value of 100%, the events will be moved all the way to the closest grid point. On the other hand, if you choose a value of 0%, their original position won't be changed. If you choose a 50% value, the events will be moved halfway between their original position and the closest grid point. This option gives you great control over the "stiffness" of your quantization. While with a 100% strength (usually the default), your parts will sound very rigid and mechanical, choosing a value between 50% and 80% will help maintain the original smoothness of the parts and at the same time correct the major mistakes. Listen to Examples 3.4 through 3.7 on the audio CD to hear the difference between different amounts of strength quantization. A third parameter, called *randomize*, allows you to place the events randomly inside the grid area of each point. If you keep its percentage under 50%, it can generate enough random data to simulate a more loose performance. Sometime this function is also referred to as *humanize*. I find this option pretty useless on a practical level since the random nature of the algorithm doesn't match any real performance situation. You can experiment with it, though, and see if something useful "randomly" comes up.

In DP and PT these settings are laid out exactly as explained in the preceding paragraph (Figure 3.1). In PT the *sensitivity* parameter, instead of having a positive and a negative value, is set using the "include within" option (same as positive sensitivity values) and the "exclude within" option (same as negative sensitivity values).

In CSX you can control the sensitivity of the quantization by selecting the "Quantize setup" in the MIDI menu. The parameter called "Magnetic Area" controls the positive values of sensitivity, while the parameter labeled "Non Quantize" controls the negative sensitivity. The strength of the quantization is set through the parameter called "Iterative Strength." The iterative quantization (accessible from the MIDI menu) can be repeated several times in a row in order to move the events closer and closer to the grid points each time by the amount specified in the "Iterative Strength" field. This technique allows you to quantize your part slightly more each time you apply it, giving you very precise control over the rhythmic feel of your performance.

In LP the additional quantization features are accessible from the "Extended Sequence Parameters" floating window found in the "Options" menu. The *Q-Range* has the same function as *sensitivity*. The value, instead of being expressed in positive or negative percentages, is inserted in *ticks* (subdivisions of a beat). If the *Q-Range* is set to positive values, the quantization will affect only events for which the distance in ticks is smaller than the value inserted (same as having a positive sensitivity percentage). If the *Q-Range* is set to negative values, the quantization will affect only events for which distance in ticks is greater than the value inserted (same as having a negative sensitivity percentage).

3.2.2 Swing quantization

One of the most controversial uses of a sequencer in the past has been the production of genres that don't require a "straight"-note feel, such as jazz, hip-hop, and R&B. The ability to reproduce a convincing and realistic swing feel has been (and in a certain way still is) one of the most challenging tasks when sequencing. This is due mainly to the intrinsic feature of the swing feel, which is based on a constant and subtle variation of the rhythmic relationship between the notes in every beat. In fact while straight quantization can be applied to different styles and genres without affecting the overall groove too much, when it comes to a swing feel there are almost infinite possibilities regarding how the notes can occupy the space inside a single beat.

Let's take 8th notes as an example. Straight 8th notes are represented in notation as in Figure 3.4a. When you quantize without swing, this is the rhythmic feel you get. The 8th

Figure 3.4 (a) Straight 8th notes. (b) Swung 8th notes.

note swing feel has a much "rounder" and smoother groove; this is represented in the notation of Figure 3.4b.

Listen to Example 3.8 on the audio CD and compare the difference between the two types of rhythmic feel. While these two rhythmic subdivisions can easily be recreated by using, respectively, straight 8th note quantization (Figure 3.4a) and 8th note triplet quantization (Figure 3.4b), through the use of the *swing* parameter available in the quantization window of your sequencer you can create any variation between these two extremes. In CSX, DP, and PT with the *swing* parameter set to 0% you will get the same result as using a straight quantization; with the *swing* parameter set to 100% you will get the same rhythmic feel as an 8th note triplet (in DP values between 100% and 300% move the notes closer to the next beat). What's interesting, though, is that you can now choose any value in between to control "how much your computer can swing"! With values below 50% you will get a slightly looser feel than with straight quantization, providing a more relaxed and slightly "rounder" rhythmic feel to the quantized part. On the other hand, if you use values between 70% and 120% you can achieve a more natural swing feel, avoiding the unnatural and sometimes mechanical 8th note triplet feel (audio CD Examples 3.9 through 3.11). In LP you can apply different swing settings by using the swing presets available in the Parameters Box located on the left of the Arrange window. Here, in addition to the straight quantization options, you can find values with the addition of swing. Their amount varies from "light" (i.e., "Swing A") to "extreme" (i.e., "Swing F"). An additional swing percentage parameter can be selected from the Extended Sequence Parameters box accessible from the Options menu. Using the Q-Swing parameter you can fine-tune the amount of swing applied to the selected sequence: With a 50% setting the part is quantized with no swing, with values above 50% every second note will be pushed closer to the next grid point; while with values below 50% every second note will be pushed closer to the previous grid point. In Figure 3.5 you can see a graphic representation of the impact that the swing settings have on an 8th note pattern.

To improve the swing groove of your parts even faster, you can use different percentages in different sections of your projects. This helps to recreate a more human and variable rhythmic feel. Try to vary the swing parameter values by ±10% among different sections of your project. This creates a variable swing feel that will improve the overall groove of your parts. The swing parameter can also be used to improve the rhythmic interaction between tracks. By quantizing different tracks with slightly different swing values, you can create a more coherent and realistic "virtual" band. The final result will be just as if every "musician" in your studio had his or her own rhythmic personality and groove. The interaction between all the different grooves and swing settings will generate a more convincing interplay effect among the tracks (audio CD Examples 3.12 and 3.13).

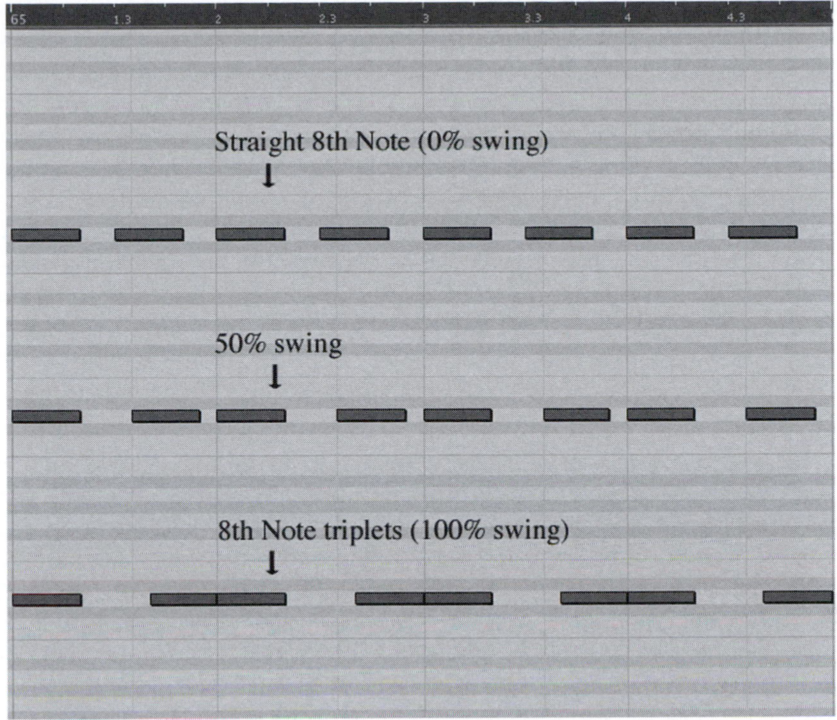

Figure 3.5 Different swing-feel quantization settings in the CSX graphic editor.

3.2.3 Groove quantization

So far we learned how to improve the basic quantization of your parts through the use of filters such as sensitivity, strength, and swing. But what if you want to recreate a much more realistic and cohesive groove based on what a real musician would be capable of? The answer is a function called *groove quantize*, which can be found in some of the top sequencers. The principle behind groove quantize is simple: Starting from templates provided with your sequencer or that you can buy as expansion packs, you can apply several styles, such as "laid back," "push forward," or "shuffles" to your parts. The grooves control primarily the timing, velocity, and lengths of the notes in the MIDI tracks or parts you are quantizing. According to the template you choose, the MIDI events will be "shaped" to give you a more realistic rhythmic feel. Keep in mind that this type of quantization shouldn't be used to correct rhythmic mistakes but to create more realistic and natural-sounding performances. Therefore I recommend quantizing your part first with the regular techniques explained earlier and then applying groove quantization in a second pass in order to loosen up the stiffness of straight quantization.

All four main sequencers allow you to apply predefined groove templates to MIDI parts. Some of them go even further by giving you the ability to create your grooves from audio or MIDI parts. DP is bundled with some DNA Groove Templates (created by Canadian company WC Music Research). The templates can be accessed from the "Region" menu and the "Groove Quantize" submenu. Here you can choose the general groove categories (e.g., "Pushing," "Laid Back," "Shuffle") and a specific groove (e.g., Medium Shuffle, Soft Shuffle, Hard Shuffle). You can alter the parameters of each template by clicking on the "More Choices" button. This will bring up a window with sliders controlling three parameters: timing, velocity, and duration (Figure 3.6). If you increase the percentage of each parameter, you will amplify the impact that this particular parameter has on the groove. For example, choosing 100% timing means that the timing of your part will be highly affected by the predefined groove you have chosen. If you select a small percentage, then your part will be only marginally affected by the groove quantization. Usually I recommend using only a mild setting (50% and lower). High values can alter the timing and the velocity to an extreme, making the quantization process useless. Use the "Beat Division" drop-down window to choose the basic rhythmic groove of the part. The same rules listed for the "quantization value" can be applied in this case. Listen to audio Examples 3.14 through 3.17 to hear the difference between straight quantization and groove quantization.

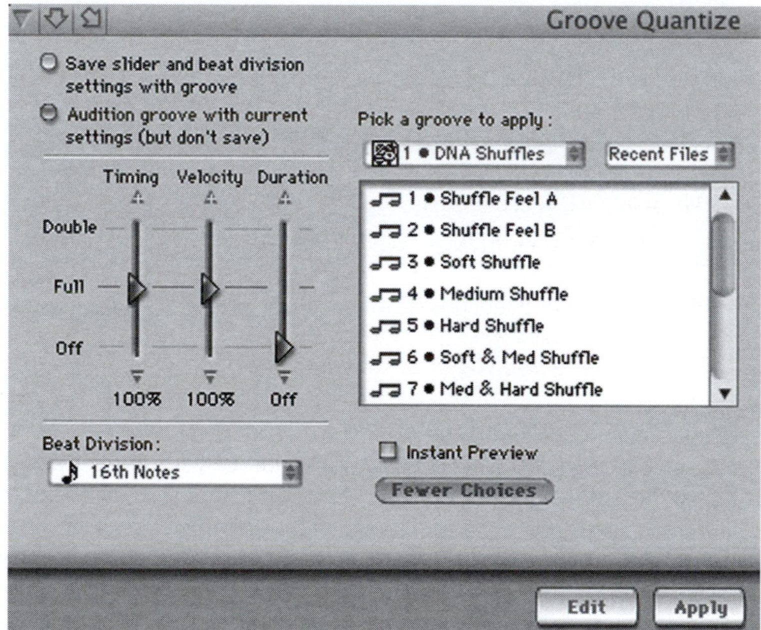

Figure 3.6 Groove Quantize window in DP.

In PT the parameters are the same as in DP. Select the "Groove Quantize" option from the MIDI menu, select the groove template, and adjust the parameters following the directions listed in the preceding paragraph.

LP can import and use predefined DNA groove templates created by companies such as WC Music Research. To apply a groove template, you will first have to import the grooves into LP. Create a folder called "Grooves" in the same folder as your LP application and copy the DNA grooves into this folder. Launch LP and select "Import DNA groove templates" from the submenu "Groove Templates" in the "Options" menu. Your new groove templates will be listed in the regular quantization list and can be used for any regions or sequence. You can also create your own templates based on preexisting MIDI parts (more on this in Chapter 4).

In CSX you can apply grooves generated either from MIDI or audio tracks. To generate a groove from a MIDI track, select the part with the groove you want to extract, and choose "Part to Groove" from the Advanced Quantize submenu in the MIDI menu. If you want to apply a groove to a MIDI track, select Quantize Setup from the MIDI menu, choose the groove you want to use from the drop-down menu called "Presets," and click on the "Apply" button.

In addition to the predefined templates that are either bundled with your sequencer or can be bought separately, you can create your own templates from audio files. This can yield an incredible number of options. Imagine, for example, importing a CD track from your favorite album featuring "Philly" Joe Jones or Dave Wackel, capturing their groove, and applying it to your own parts. In LP, CSX, and PT TDM this is possible, and we will analyze it in the next chapter, where you will discover more advanced sequencing techniques.

3.3 Layering of MIDI tracks

Now that you have mastered the quantization techniques and feel comfortable organizing and moving around your session, we are going to focus on some new procedures and techniques to improve the quality of your productions in terms of sound palette. In Chapter 2 we learned how to assign a MIDI track to a certain device and MIDI channel. While sometimes you are going to use a single patch for each track, very often you will find yourself looking for a new sonority or for a richer texture that is not directly available to you from a single device or channel. In this case you can use a technique called *track layering* (sometimes referred to as *patch layering*). Through this technique you can output a single MIDI track to several devices and/or MIDI channel at the same time. Its advantage is that you can create complex patches and sounds without altering and reprogramming the actual MIDI synthesizer or sound module. Instead you use several MIDI channels that receive the MIDI data from a single track of your sequencer. For example, if you want to create a "supersynth" sound that is a combination of a total of four patches, two from your Roland XV-3080 module and two from your trustworthy Korg Wavestation, you can assign the output of one of your sequencer tracks to two available MIDI channels on each device. It is true that you could achieve the same result by copying the MIDI data from one track to other tracks and assign each of them to a different device and sound module, but that would result in two major problems: (1) If your sequencer doesn't allow you to "multirecord" (meaning to simultaneously record two or more tracks at the same time) when you record your part, you will be able to hear only one sound of your layer at a time. (2) When you edit the tracks, you will have to repeat the edits on each track. With track layering, on the other hand, you can simply assign the output of one track to multiple devices and MIDI channel and therefore eliminate these problems.

Almost every patch of your synth palette will sound richer and fresher if layered with other patches from other devices. Here are some suggestions and recommendations for a creative use of the patch layering technique. When dealing with acoustic sounds such as stringed instruments (strings, guitars, acoustic and electric basses), try to layer two or more patches from different devices made by different manufacturers. Try to mix and match between different textures. For example, if you have a nice and edgy electric bass that is missing some depth in the lower range, try to layer it with a deep acoustic bass. The result will be a completely new patch that has the edginess of the electric bass patch but the low end of the acoustic bass patch. Another practical application of patch layering is the combination of sampled and synthesized sounds. A good way to make your string patches more robust and natural sounding is to layer a sampled string patch with a synth string patch. The former will give you the edginess and realistic feel, while the latter will provide adequate lower end support. Of course, you can apply the same technique to constructing and building your own synthesized patches by combining as many layers as you want. This works particularly well in the case of synth pads, where the nature and texture of several individual patches really enriches the final quality of the sound.

To construct layers in DP is very simple. In the Output column in the track list window, instead of selecting a single device and MIDI channel, select "New Device Group." From the new window that appears (Figure 3.7) you can select any device and MIDI channel available in your studio. From now on any MIDI data played or recorded on this track will be output to all the devices and the MIDI channel assigned to its device group. To toggle the checkbox view shown in Figure 3.7, use the *expand/compress* icon located below the small keyboard sign next to the group name. To rename a group, simply *option-click* on the group name from the Device Groups window. If you need to edit a group from the track list window, *option-click* on the group name of the track.

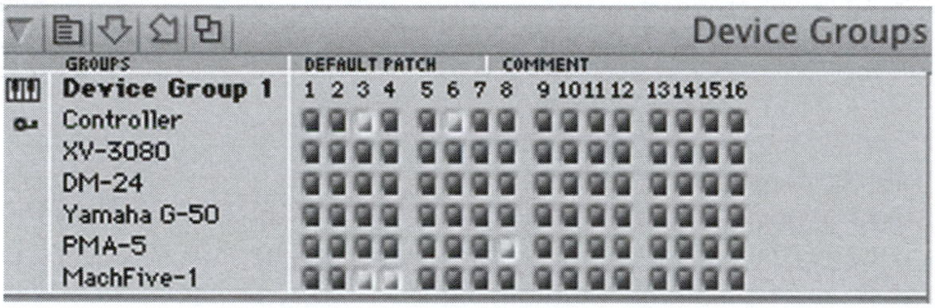

Device Groups
GROUPS DEFAULT PATCH COMMENT
Device Group 1 1 2 3 4 5 6 7 8 9 10 11 12 13 14 15 16
Controller
XV-3080
DM-24
Yamaha G-50
PMA-5
MachFive-1

Figure 3.7 The Device Groups window in DP.

In LP you can create MIDI track layers by using the "Environment Editor." Open the "Environment window" (located inside the Windows menu) and create a new *instrument*, either single or multi, by selecting the "Instrument" option from the New menu. Assign the output of the instrument to different devices and a different channel by *option-clicking* on the connection cable of the instrument (the small arrow that points toward the outside from the newly created instrument), as shown in Figure 3.8. From the hierarchical multi-menu, select the device and channel using the list of available devices in your studio. You can repeat the same procedure for all the instruments you want to layer. The newly created "instrument" will be added to the list of available MIDI outputs. From the Arrange window, select a MIDI track (or create a new one) and assign its output to the new instrument. All the data recorded on this track will be output to the cables and MIDI channel you specified in the Environment window. These outputs can be changed at any time in order to experiment with different layers.

In PT you can assign the MIDI output of a track to multiple devices and MIDI channels by simply *control-clicking* on the device and channel assignment drop-down menu of your MIDI track.

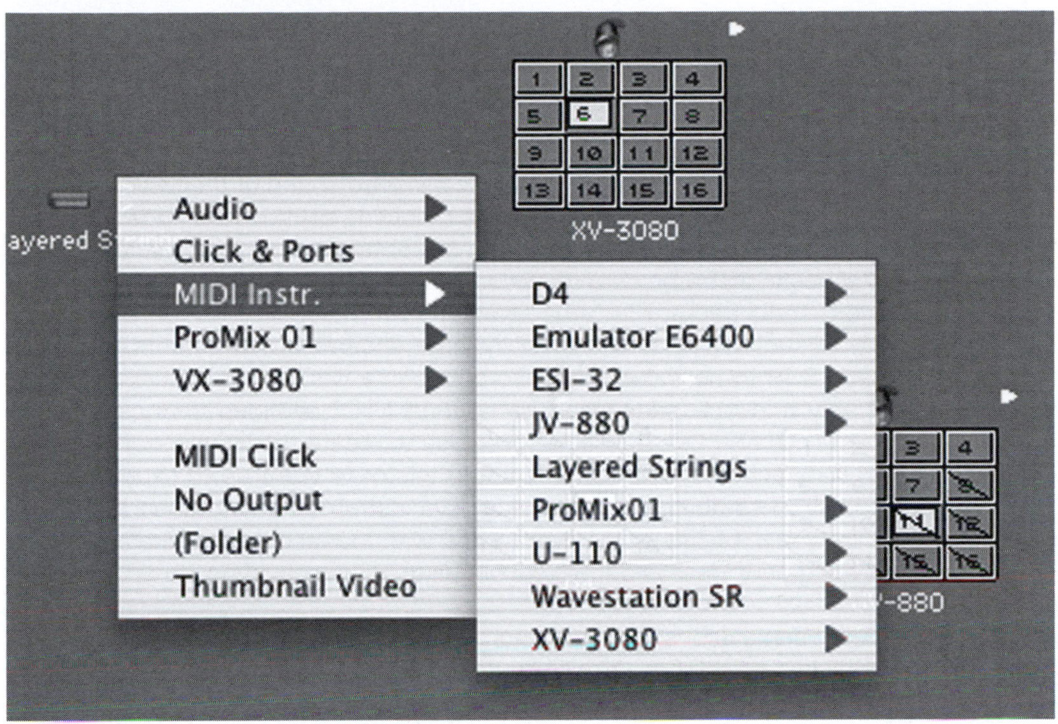

Figure 3.8 MIDI track layering in LP.

In order to achieve a higher control over the blend of your layers it is essential to be able to control the volume of every single MIDI channel that is part of the combined sound. Let's say you just created a layer where you combined a synth pad from a Yamaha Motif set on MIDI channel 1 and a synth bass from a Korg Triton MIDI set on channel 3. If you control the volume from the main track that is output to both devices and channels, you will send the same CC 7 to both tracks and therefore won't be able to control the individual volume of each patch. Instead you will have to create two more MIDI tracks, each assigned to the individual device and channel (in our example, one will be assigned to channel 1 of the Motif and the other to channel 3 of the Triton). These two new tracks won't be used to record actual MIDI notes, but they will serve as individual volume for each patch. In this way you will be able to control the blend of the two patches to create the ultimate layered sound. Make sure to label the tracks with the right names. For example, you could call the layered track "Fat Synth Pad" and the two volume tracks "Volume Fat Pad Motif" and "Volume Fat Pad Triton." Of course if your layer involves more than two tracks, you will need volume tracks for each additional layer.

3.4 Layering of MIDI and audio tracks

Not only can the techniques involving the use of layers to produce richer, newer, and fresher sonorities be applied to MIDI tracks, but they can also be used with MIDI and audio tracks together. The layering techniques involving the two types of tracks can be used in a variety of ways and situations to achieve very different results. One of the major applications of these techniques is the combination of live acoustic instruments (such as strings, woodwinds, or brass) and synthesized or sampled sounds. One of my favorite layers consists of using a MIDI track to sequence a large string section and to overdub, using one or more audio tracks, real string solo instruments (such as a violin or a viola) doubling the same parts sequenced on the MIDI track. While the sampled sound of the MIDI track provides the main body of the orchestra, the acoustic recording of the violin adds a realistic touch and a genuine edge to the performance. You can obtain even better results by overdubbing several acoustic instruments (e.g., two violins, two violas, two cellos, and two basses) on different audio tracks. The same technique can be applied not only to strings but also to any acoustic instruments, especially large ensemble instruments. Listen to Examples 3.18 and 3.19 to hear the difference between a MIDI track sequenced using sampled sound only and one sequenced by layering MIDI sampled sounds and live acoustic strings. You can also use the same technique to create totally new sonorities by combining synth pads and acoustic instruments. In this case feel free to experiment as much as you can. Try to combine sharp and edgy live instruments with deep and round synth pads, or vice versa, or use bright synth leads with low and powerful basses. Any combination will bring new ideas and fresh sonorities to your productions.

In general the use of some acoustic instruments in your projects will increase their overall quality. Sometimes, just adding one or two acoustic tracks will bring your MIDI production to life. No matter how simple or complicated your sequence is or how well programmed your MIDI tracks are, without an acoustic touch they will always have something missing. Most of the time, just adding an audio track with an acoustic guitar, an electric bass, or some live percussion will improve your project dramatically. Try to avoid loops for anything but percussion or drum parts (unless the style you are writing in calls for it); they tend to be too repetitive and most of the time will kill the spontaneity of the composition. Keep this as a mantra: Always use at least one live instrument in your production!

3.5 Alternative MIDI track editing techniques: the drum editor

While the editing techniques you learned in Chapter 2 (graphic, list, and score) are the most common ones, they are not the only ones available. In fact each sequencer features some peculiar ways of editing and inserting MIDI data that are worth mentioning and learning.

One of my favorite "unconventional" editors is the *drum editor* available in DP, LP, and CSX. The philosophy behind this editor is to recreate the working environment of a vintage drum machine, such as the Roland TR-808, and expand it to a higher functionality and flexibility. The drum editor can be used in several ways, depending on your needs and skills. It can be used to input rhythmic MIDI parts (such as, but not only, drum parts) or to clean up and edit MIDI parts that were sequenced from a controller but need some minor tweaking, or it can also be used to create arpeggios and repetitive melodic parts. The drum editor can be seen as a subsequencing-environment for a selected track or for a series of tracks. This means that when you open the drum editor for a track, you will be presented with a series of subtracks, each corresponding to a single note or key of your keyboard. That is why I call it a subenvironment, because each MIDI track is split into several subtracks, each assigned to a different note. This configuration is particularly appealing to drum tracks because of their intrinsic nature of multiple sound-based sets. A drum patch is made up of several sounds and sonorities corresponding to several drums and percussive instruments. By choosing the drum editor to record or edit drum tracks, you have a higher control over each individual piece of the drums and percussion set. You have not only individual *mute* buttons but also independent grid-quantization settings for each note/sound. In the case of CSX you can access the drum editor by selecting the "Open Drum Editor" option from the MIDI menu. Here each subtrack can be assigned to a different note number, MIDI channel, and device. These settings can be stored in different "Drum Maps" that can be recalled at any time. The "Drum Map" feature is one of my favorites, since it allows you to create customized drum sets that share different sounds and instruments among different devices and MIDI channels. In CSX you can insert MIDI events using the *drum stick tool* and delete them using the *eraser tool* (both accessible from the tool palette located in the top right corner of the Drum Editor window). In DP's drum editor (located in the Project menu), the layout and concepts are very similar to the ones found in CSX (Figure 3.9). Here you can't assign different subtracks to different devices or MIDI channels but only to different notes. In the top left part of the editor you can view several MIDI tracks at once, and each track can be assigned to different channels and devices in your studio. Keep in mind that each

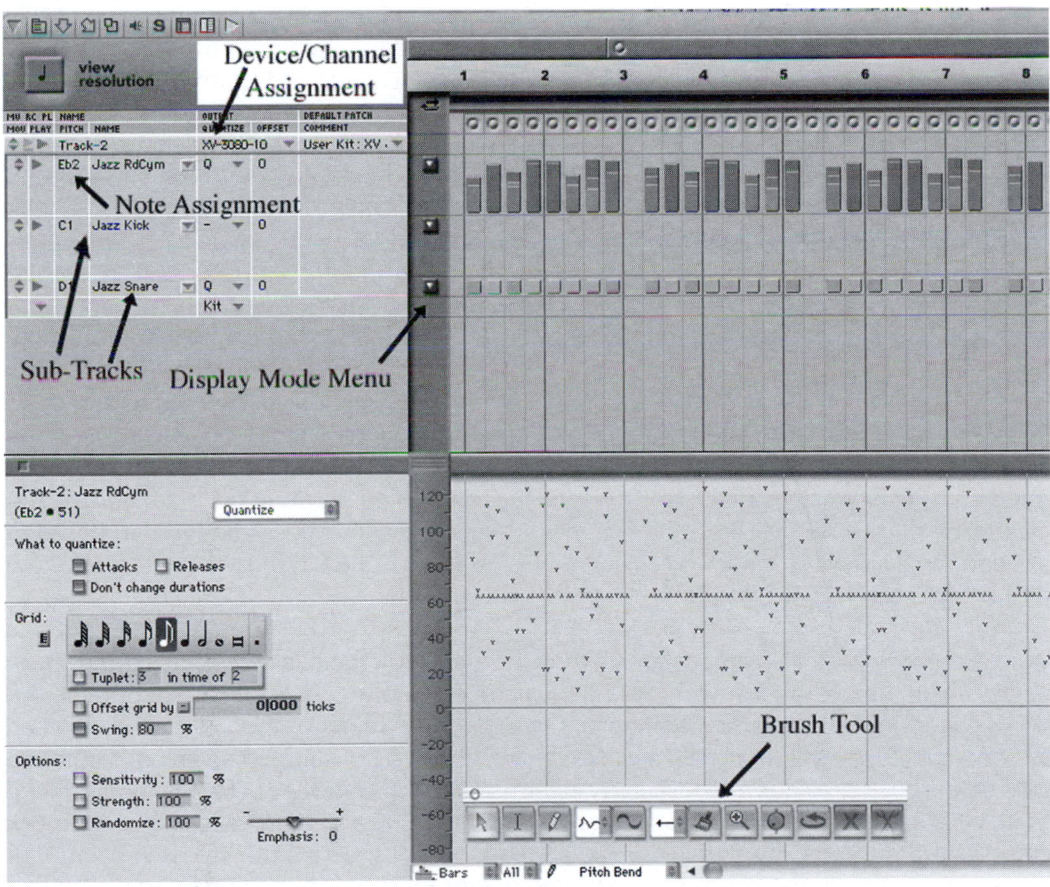

Figure 3.9 Drum Editor in DP.

subtrack needs to be assigned to a note; otherwise you won't be able to add/edit events for that subtrack. To assign a subtrack to a note, click on the "Pitch" field next to the playback button of that track. As in CSX, you can set separate quantization grids for each subtrack by selecting a subtrack with the mouse and changing the quantization settings for that particular subtrack in the lower left area of the editor. This option makes it very easy to create complex rhythmic parts with intricate rhythmic subdivisions. Use the *pencil tool* to insert or edit events in any subtrack. For each subtrack, you can also switch (through the *display mode menu*) between different views that allow you to see notes only, notes and velocity, velocity and duration, and a *free* setting that displays the notes not according to the grid but according to their real location (a sort of a miniature piano roll window). A great feature of the drum editor in DP is the use of the *brush tool*, accessible from the tool palette (*Shift-O*). With this tool selected you can "paint" a predefined pattern on any subtrack. You can choose the pattern from a list of several preinstalled templates, or you can define your own patterns. This feature is very convenient for quick drum part layouts. Even though the predefined patterns are so-so, they provide a good starting point for sequencing quick drum grooves.

In LP there's a similar editor that in fact is much more powerful than a regular drum editor. The "Hyper Editor" (accessible from the Window menu) is an extremely flexible and multifaceted environment that you can customize to edit basically any parameters or MIDI event you wish. The default set allows you to edit common parameters such as volume, pan, modulation, and pressure. Another predefined set, called "GM Drum Kit", transforms the Hyper Editor in a regular drum editor with all the features we saw earlier in the case of DP and CSX. Switch to this mode using the drop-down menu located in the left side of the Hyper Editor, below the tools palette. You will see separate subtracks for each note of the keyboard, and for each note you can change the name, MIDI channel and device assignment, quantization settings, etc. Use the *pencil tool* to insert an event and input its velocity according to the grid value chosen for that particular subtrack (the grid value can be changed using the "Grid" parameters located in the Parameters Box in the left part of the window). You can group instruments together by *Control-clicking* on two or more subtracks. This will set the group tracks not to play events at the same time—very useful when programming hi-hat parts, for example, where you don't want the closed and open hi-hats to be played at the same time. But where the Hyper Editor really shines is when you create your own sets. This feature can be used not only to create custom drum sets but also specifically designed control sets that allow you to program a series of CCs, events, and any other MIDI events you might need. You can choose the type of message assigned to a certain subtrack by selecting the "status" option in the parameters window to the left of the Hyper Editor.

Here are some practical suggestions and applications to consider when using the drum editor. If you are not familiar with playing the parts on a controller (keyboard or others), the drum editor provides a good starting point to sequence drums or monophonic parts such as simple bass lines or melodic lines. Not only is it more advanced than an old-fashioned step editor, but it provides more flexibility in terms of quantization and velocity programming. Start with simple and short patterns, get familiar with it, and then keep exploring more complicated rhythms. Try to avoid using a steady velocity parameter for each event you create. If you do so, then your parts will sound very mechanical, no matter how hard you work on their quantization (audio CD Example 3.20). Try to create variation in the velocity pattern instead, as a real drummer would do when playing a drum set or a percussion set (audio CD Example 3.21). With this editor it is very easy to quickly change the velocity of several events by clicking and dragging across the screen with the *pencil tool* (or with the *drum stick tool* in CSX).

I often like to use the drum editor to create or edit small variations in a drum pattern. After recording, quantizing, and groove quantizing the part, I open the drum editor and insert small variations here and there that would be too fast to play from a keyboard controller. For example, you can create some interesting hi-hat parts by choosing a small resolution for the hi-hat subtrack (such as 32nd notes) and inserting quick passages every so often to create contrast and variation. Listen to Example 3.22 on the audio CD to get an idea about this concept.

As I mentioned before, you can also use the drum editor in creative ways such as arpeggio generator or groove creator. If instead of opening the drum editor for a track assigned to a drum patch you use it to edit a track assigned to a synth bass patch, you can create as complicated patterns as you would using a so-called "matrix sequencer." Simply create a subtrack for each note that will be part of your arpeggio and then insert notes using the *pencil tool* to generate the new pattern (audio CD Example 3.23). As you can see the applications of this editor are several and it is really up to you to experiment with new ways to create music.

3.6 Alternative MIDI controllers

Keyboard controllers and MIDI keyboard synthesizers are the most common choice for inputting MIDI data into a sequencer. In fact we are so used to sequencing with such devices that often the modern musician/producer associates the composition process with a keyboard instrument. As a matter of fact this can sometimes be limiting, and it can narrow the overall spectrum of your projects. This is mainly due to the fact that many musicians are not keyboard players, and their keyboard performing skills can pose a big obstacle for their sequencing projects. Fortunately, keyboard controllers are not the only way to input MIDI data into a sequencer. Here's a description, analysis, and few practical applications of some of the other MIDI controllers available on the market.

3.6.1 Guitar/Bass-to-MIDI converters

One of the most widely used and widespread alternative MIDI controllers is the guitar-to-MIDI converter. This technology allows a regular acoustic or electric guitar to be connected to a MIDI system and to output MIDI messages and notes to any MIDI device, including a sequencer. This technology has been around for many years and constantly perfected by companies like Roland and Yamaha. Even though the models vary from manufacturer to manufacturer, the principle on which this type of controller is based is simple: a pickup (divided into six segments, one for each string) is mounted next to the bridge of the guitar (Figure 3.10). The pickup detects the frequency of the notes played on each single string by analyzing their cycles. This information is passed to a breakout unit that converts the frequencies in MIDI Note On and Note Off messages. From the unit, a regular MIDI OUT port sends the messages to the MIDI network. The pickup can also detect bending of the strings, which is translated in Pitch Bend messages.

Figure 3.10 Guitar-to-MIDI converter pickup: Yamaha G50.

Even though it takes a little practice to adjust your playing style to this type of converter, you will find yourself entering a whole new world of sonorities and possibilities. Imagine sequencing exotic synth lead parts using the natural legato and bending features of your guitar. Once again, the possibilities are really endless. And after getting used to it, you start wondering how you could have sequenced without it. If you are an electric bass player, you are not out of luck. There are pickups for all sorts of string instruments: 4, 5, and 6 strings. Keep in mind, though, that because of the pitch tracking system of the device, the delay between the acoustic note and the MIDI note is inversely proportional to the frequency of the note. Therefore for very low notes the delay is noticeable, which might discourage some bass players. This is due to the fact the lower frequencies have a longer cycle and so the pickup has to wait longer before detecting a full cycle. A way around this is to use a "piccolo bass," meaning a bass with strings that sound one octave higher. This is an alternative controller that is worth having in your studio either as your main controller or as an occasional substitute for your keyboard. Listen to Examples 3.24 and 3.25 on the audio CD to compare between a part sequenced using a regular keyboard controller and sequenced via a MIDI guitar controller.

As I mentioned earlier, practical applications of the guitar/bass-to-MIDI technology vary from simple sequencing of melodic or harmonic parts using the voice leading of six-stringed instruments to more creative applications, such as sequencing of drum parts using the "feel" offered by the guitar or bass. Other applications are more targeted to live performance, where, by assigning different strings to different MIDI channels and patches, you can create a "one-man band" and generate interesting layers and ensemble combinations. The same concept can also be used in a studio/sequencing situation. You can record-enable multiple tracks (maximum of six, one per string), all receiving on different MIDI channels, and assign their outputs to separate devices and MIDI channels. By pressing the record button you will record separate parts on separate tracks. You might assign a low sustained pad to the low E and A strings, use the D and G strings for sound effects or tenor voice counterpoint, and use the B and high E for the melody. The advantage of this approach is that you can capture a much better groove and "live feel" than from sequencing the tracks separately. Experiment with these techniques, and let your imagination run free. You will be surprised by the results.

3.6.2 MIDI drums and pads

MIDI drums and MIDI percussion pads were the precursors of a series of alternative controllers. They have been more successful and widespread than others, mainly because they started as mere "triggers" of Note On/Off messages without having to convert a definite pitch. This factor makes the MIDI drums the fastest alternative controller. The percussive MIDI controllers can vary, depending on their size, style, and features. Nowadays you can find three main types of percussive controllers: drums, pads, and triggers. Drum controllers recreate the entire drum kit, including bass drum and cymbals (Figure 3.11a). Each pad of the kit has several velocity-sensitive zones that can be assigned to different notes and MIDI channels. Roland has mastered this technique with their V-Drum series. There are several configurations available, ranging from a full studio kit to a portable "live gig" setup. MIDI pads are smaller controllers meant to complement and not substitute for an entire drum kit (Figure 3.11b). They usually feature four, six, or eight pads that can be played with either sticks or hands (to simulate percussive instruments such as bongos and congas). The MIDI channel and note assignment can be set for each pad individually. This

a

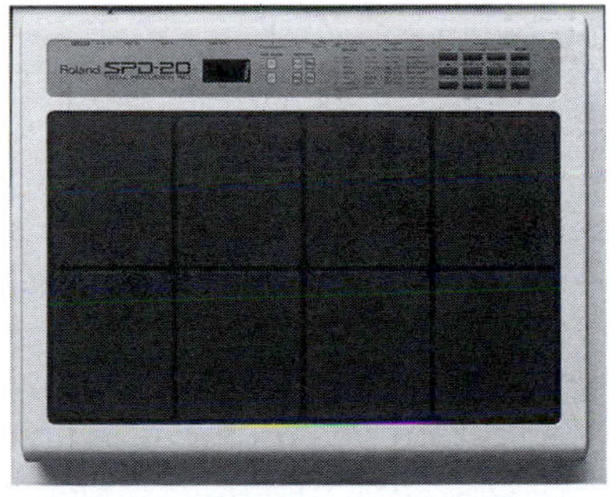

b

Figure 3.11 (a) MIDI drums and (b) MIDI percussion pads (Courtesy of Roland Corporation U.S.).

type of controller is ideal for a small studio situation and for infrequent use on certain projects. Another creative way to program a percussion/drum controller is to use it as a MIDI "pitched" instrument by assigning different pads to different notes of a pitched instrument or synth patch. You can also assign certain pads to trigger loops or samples and use them as background layers for a one-man-band type of performance.

A third type of device that takes advantage of audio-to-MIDI conversion is an audio-to-MIDI trigger. This is a simpler device that doesn't include the actual pads to trigger MIDI messages but instead allows you to connect, to a regular audio input, any audio source (it works better with percussive sounds with fast transients). The audio signal is then used to trigger MIDI notes. The unit (usually a one-rack-space unit) has several audio inputs on the back panel. You can connect a live instrument (such as an acoustic drum kit with individual microphones for each piece of the kit) or use prerecorded material output from a multi-track recorder. The audio signal on each channel will trigger different MIDI notes, depending on the settings you programmed. These devices (such as the old Alesis D4 and the newer DM5) are pretty inexpensive and fairly accurate.

The use of drum, percussion, and MIDI triggers in your studio will increase dramatically the way you sequence percussive parts. Listen to Examples 3.26 and 3.27 to hear the difference between a drum part sequenced using a keyboard controller and a similar part sequenced using a drum MIDI converter.

3.6.3 MIDI wind controllers

MIDI devices can be controlled not only by keyboard, guitar/bass, and percussive instruments but also by special electronic devices called *wind controllers* that emulate the shape, playing style, and features of saxophones and flutes. While this type of controller was popular until a few years ago, recently they seem to have faded from the active scene in the MIDI sequencing world. Nevertheless a wind controller is an incredible tool for improving your MIDI projects and, as in the case of the other alternative controllers, for firing up your imagination and creativity. Two of the most used wind controllers are the one made by Yamaha (the WX5) and the one made by AKAI (the EW3020). They both look very similar in recreating the look and feel of a soprano saxophone. These types of controller are appealing to both the seasoned saxophonist or professional flutist and the beginner. You can choose between several fingering position and mouthpiece settings that simulate the ones for standard saxophones and flutes. The main unit can be connected directly to any MIDI sound module through a regular MIDI cable, or it can be connected to an expander specifically designed to enhance the controls and the tone quality of the instrument. Among the most obvious applications is the sequencing of wind instrument parts that usually are particularly difficult to render from a keyboard controller. The problem of recording wind instrument parts from a keyboard is that most likely you will find yourself forgetting that wind instrument performers need to breathe between phrases. This will result in very long lines that don't sound realistic, mainly because they wouldn't be playable in a live situation. By using a wind controller, the volume and the Note On/Off messages are directly linked to the amount of air you blow into its mouthpiece. This has two main effects: (1) You will be able to accurately control the dynamics of your performance (on a keyboard controller you would have to use a fader to send CC 7 volume to achieve a similar effect). (2) You won't be able to sequence phrases that are longer than your breath, which will result in well-balanced and convincing lines. In addition to the regular keys and mouthpiece, these wind controllers have a series of extra controllers, such as Pitch Bend wheel, program-change buttons, and quick-transpose keys.

3.7 Complex tempo tracks: tempo and meter changes

In Chapter 2 we learned how to create a basic click track, where the tempo would stay steady all the way through the project. While this could work for some basic sequences it is definitely restrictive for more advanced and complicated compositions. Modern sequencers are extremely flexible when it comes to tempo and meter changes. In the following paragraphs we learn how to program complex tempo changes, *rallentandos, accelerandos*, and meter changes.

The tempo and meter changes in a sequencer can be inserted and edited in three different ways: through an edit list window, graphically through a graphic editor, and through *tap-tempo,* meaning a live input of MIDI messages that can control the tempo of the project in real time. No matter which method or methods you use, they all have the same result on the programmed tempo and meter changed. Keep in mind that not only is a tempo map used to follow and accommodate the changes according to the score of your music, but it can also be used in several creative ways to improve the quality of your production, as we are going to learn later.

Tempo changes are usually recorded in a separate track that acts as the "conductor" (sometimes called *tempo track*) of your virtual orchestra. In fact in DP this track is called the *conductor track*. These changes affect the tempo and the meter of your composition. You can program and insert any tempo/meter changes you want. The way your sequencer follows these changes can vary according to your needs. If you are using a steady tempo all the way through your composition, then you will disengage the tempo track and therefore your sequencer will follow and react to the tempo specified in the tempo slider in the transport window. This is the simplest use of the tempo control (and also the most limited in terms of features). You basically can set only one tempo for your entire project, and you can change it only by altering the tempo parameter in the transport window of your sequencer. This action can be done at any time, even during playback. This method, though, doesn't give you any control over programmed tempo changes. By switching to the "conductor track" mode (in DP and PT it is called "Conductor," while in CSX it is called "master track" and in LP "Tempo Track"), the sequencer will follow the tempo and meter changes you programmed or recorded in advance. Let's take a look at how to program and edit tempo/meter changes in all four sequencers.

In DP the tempo and meter changes are stored and edited from the "Conductor Track," which by default is the first track listed in the track list window. If you select it and then open the list editor window (*Shift-E*) you will be prompted with the familiar list editor environment we used for editing MIDI data. In the case of tempo and meter changes, the techniques are similar to the ones used to edit regular MIDI messages. You can insert tempo and meter data by selecting the "Input" (uppercase "I" icon) menu from the editor title bar and choose either "tempo" or "meter." Select the location where you want to insert the change (in the case of tempo changes select the basic pulse: quarter note, 8th note, etc.) and press Return. The new inserted value will appear in both the list editor and track list windows. You can repeat the same procedure to insert other tempos and other meter changes at any measure of your project. In order to have the sequencer follow these changes, you must enable the "conductor track" mode from the control window. To do so, click on the "Open Drawer" arrow located in the top right corner of the Transport window until you see the Tempo Control page. Click on the Tempo Slider drop-down menu and select "Conductor Track." After doing so you won't be able to use the tempo slider to change tempos anymore, since the sequencer will be locked into the conduct track. This technique allows you to create drastic tempo changes but is not very convenient if your goal is to create smooth *accelerandos* and *rallentandos*, since you would have to insert many tempo change events one after the other. To effectively create a smooth transition between tempos, you have two techniques. The first involves the insertion of a continuous series of tempo change events using the "Modify Conductor Track" function found in the Project menu (Figure 3.12).

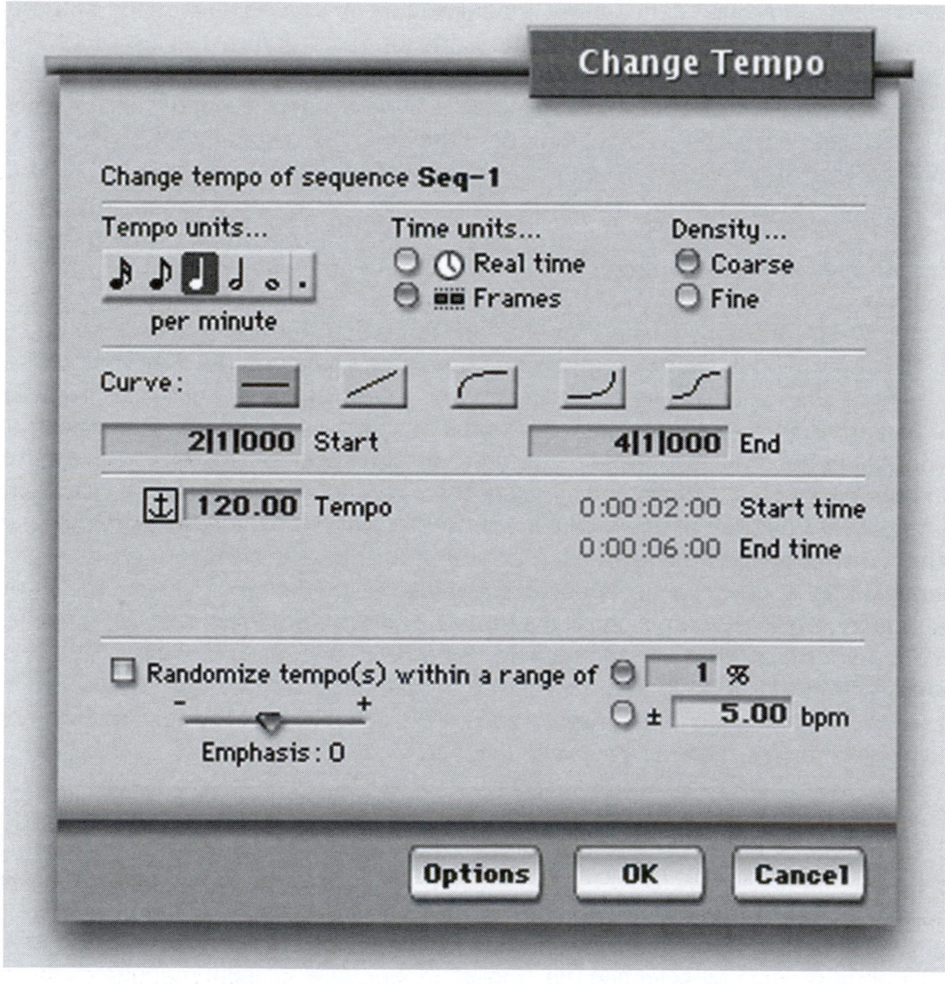

Figure 3.12 The "Change Tempo" window in DP.

After selecting this option, you will have to punch in the start and end positions where you want the tempo change to take place, using the Start and End fields. Choose the type of curve you wish the change to have (linear, exponential, or polynomial) using the Curve parameters, and select the beginning and end tempos (if you select the steady curve, you will see only one tempo field marked Start; if you choose any other curve, a tempo field marked End will appear automatically). You can also randomize the change using either a percentage parameter or a tempo variable by clicking on the Randomize Tempo(s) button and by specifying the randomization range in either percentage or ±BPM. After clicking on the "OK" button you will see in the track list window that a series of data has been inserted in the conductor track. Make sure that the "conductor track" mode is enabled in the transport window, and press play; you will see the tempo of your project smoothly change according to the data and the type of curve you have chosen. The tempo change data you just inserted can be edited in the list editor window but can also be edited using the graphic window. In order to do so, select the conductor track and click on the graphic editor icon in the transport area (or press *Shift-G*). Instead of the usual piano roll layout, you will see a graphic representation of the changes in tempo over time (*x* axis) and the tempo value in BPM (*y* axis). By using the usual tools available from the tool palette, you can insert data (*pencil tool*) or reshape the tempo change data already inserted (*reshape tool*). You can also select different shapes to create smooth accelerandos or gradual rallentandos.

Similar results can be achieved in CSX using techniques comparable to the one used in DP. You can access the tempo and meter change data through the "Tempo Track" information panel located in the Project menu. This graphic editor is very similar to the one used in DP. By using the usual tools, you can insert, change, or delete tempo and meter data. In order to be able to edit the tempo track data you have to be sure to have activated the master track button either by clicking on the "Tempo" button in the transport window or by clicking on the blue button in the upper left corner of the tempo window. You are free to insert single tempo changes or a series of data to create smooth changes. Any inserted data can be moved and altered by simply selecting and moving their insertion points. To insert different time signatures, simply input the type of signature you want to insert in the relative field in the top area of the tempo track window and by using the *pencil tool* insert the meter changes in the line between the ruler and the tempo window. If you want to delete a time signature, simply select the *eraser tool* and click on the items you want to delete.

In PT the tempo and meter changes are inserted and edited using the Tempo and Meter rulers. If your session is not set up to show the tempo and meter rulers, choose the "Ruler View Show" in the Display menu and make sure that "Tempo" and "Meter" are checked. To insert a tempo or a meter change, simply select "Change Tempo" or "Change Meter" from the MIDI menu (or click, respectively, on the small icons to the left of the rulers area). In PT there is no graphical or list editor window for the tempo and meter data, which makes it a bit limited for serious MIDI sequencing. In addition there is no possibility to create automatic accelerandos and rallentandos; instead you will have to manually insert single tempo changes to create smooth variations. In order for PT to be able to follow the tempo and meter changes, you have to activate the conductor mode by clicking on the conductor button located in the lower right area (next to the meter indicator) of the transport window. If the conductor mode is not activated, then PT will follow the tempo programmed in the tempo slider in the transport window.

In LP the tempo changes can be inserted from either the tempo list editor or the graphic tempo editor. They are both accessible from the transport window by clicking on and holding the "Synchronization" button located between the "Metronome" and "Solo" buttons. If you select the "Open graphic tempo" option you will see a familiar window reminiscent of the Hyper Editor. In fact it is a Hyper Editor specially programmed to insert and edit tempo changes. As we learned previously, you can use the regular editing tools to insert (*pencil*), delete (*eraser*), and reshape (*arrow*) the tempo change data. One of the advantages of LP is that you can have up to nine so-called "Tempo alternatives." This function, accessible from the Options menu in the Graphic Tempo editor, allows you to select nine different tempo tracks and therefore to experiment with different tempo options before committing to the final choice. It is the same as having "alternate takes" for your tempo track. This is a very useful feature. The tempo change data can also be edited and inserted using a tempo list editor accessible from the transport window. Click and hold on the "Synchronization" button and select "Open Tempo List." You will be prompted with a familiar window showing you a list of tempo changes inserted in your project. Once again you can use the regular editing tools to insert, delete or edit tempo change data. Changes to the meter of the project can be made from a dedicated window that can be opened by selecting the option "Signature/Key Change List Editor" from the Options menu (or press *Command-O*). This editor looks very much alike the tempo list editor. You can insert both meter and key changes by pressing the *Command* key and clicking on an existing item in the window. After a new item is inserted, you can change its parameters by simply clicking and typing the new values or alternatively by clicking and moving the mouse up and down to increase or decrease the values. A method similar to the one learned for DP is also available here in order to create smooth tempo changes based on preconfigured curves. If you select "Tempo Operations" from the Options menu of either the meter change or tempo change window, you will be prompted with a window similar to the one seen in DP. In this window you can specify the start and end tempos, the region in which the tempo change has to occur (in both bars and real time), the type of curve, the curvature, and the density in rhythmic subdivisions from 1/1 through 1/32 (for subdivisions higher than 1/8, use the option key). Use the Operation drop-down menu at the top of the window to select preset tempo operations such as "Create Tempo Curve" and "Create Constant Tempo." Click on "Do it" to apply the changes.

3.7.1 Creative use of tempo changes

Not only is the insertion and editing of tempo changes a valuable tool to make your MIDI project conform to the score of your composition, but it is also an extremely flexible device that can transform your sequences into something less mechanical and stiff. In fact even if your project is based on a steady tempo from beginning to end, your sequence can greatly improve if you take advantage of the tempo change feature. One of the best examples is to vary slightly the tempo of your sequence in key points of your composition. If you are sequencing a pop tune with a steady tempo, try to insert few tempo changes along the way to give a more natural flow to the song and to mask the mechanical nature of the sequencer. One of the biggest problems related to MIDI sequencing and the use of loops is that they both sound fine for a short time but after a while become repetitive and monotonous. This is due mainly to the lack of variation in tempo and groove, two aspects that make a real drummer and real musicians irreplaceable. By inserting tiny tempo changes in your conductor track, though, you can simulate the fluctuations that a live ensemble would necessarily create. You don't have to insert continuous variation, just a few events in the right points of your production. In general do not slow down the tempo unless this is required by the arrangement. Most of the rhythm section tends to speed up and not to slow down. By increasing the tempo slightly and little by little you also help create more excitement and simulate the animation that a live rhythm section would engender. Another thing to keep in mind is that the tempo usually increases slightly in crucial passages, such as during the transitions from the verse to the chorus of a song or during an exciting solo of a lead instrument. These tempo increments should be minimal and barely noticeable—in fact they shouldn't be noticeable at all, they should be felt but not heard. Therefore try to increase the tempo from a minimum of 0.5 BPM to a maximum of 1 BPM. This will create the effect of a live ensemble that translates the excitement of execution into tiny tempo increments. Listen to Examples 3.28 and 3.29 on the audio CD to compare the same excerpts played, respectively without and with tempo changes.

One of the best and most effective practical applications of creative tempo changes is the function called "Tap Tempo." This option allows you to literally conduct your sequencer by inputting MIDI messages from a controller that acts like a conductor's baton. This way you have very natural tempo changes that the virtual orchestra will follow during the performance of your composition. To activate this feature in DP, select "Receive Sync" from the Setup menu, choose "Tap Tempo," and select the source of your sync. This will instruct DP to wait for incoming MIDI messages (such as notes and controllers) to receive the sync and tempo. Next select the type of message and device that will be used to "conduct" your virtual orchestra. Usually select a note and the main MIDI controller in your studio. After making these settings, click "OK" and set DP for slave to external sync by pressing *Command-7* or selecting "Slave To External Sync" from the "Setup" menu. As you will notice, the Tempo Control parameter in the transport window has changed; it now indicates "Tap Tempo." If you press Play, DP will wait for you to tap on the controller and conduct the sequence. You can also record the tempo changes to insert this way by record-enabling the Conductor track and pressing the Record button. You can then edit the tempo change data inserted with the Tap Tempo control with either the graphic or list tempo editor, as we saw earlier in this chapter. This technique performs magic for compositions that feature rubato passages, such as classical orchestral scores or solo sections. Remember that I highly advise to always record your parts with a click first and then to use Tap Tempo or regular tempo changes to recreate rubato, accelerandos, and rallentandos. This will give you much higher control over the editing and quantization options, since even during rubato passages your parts will still be following the bars and tempo laid out by the conductor track. Listen to Example 3.30 on the audio CD to experience the flexibility and power of the Tap Tempo technique.

A similar method, called *Tempo Interpreter*, can be used in LP. Via this technique, LP can receive sync from either an external MIDI controller or the keyboard of the computer. To activate this function, you first have to set up which key command will be used to "conduct" your sequence. To do so, select the option "Key Commands" in the "Preferences" submenu located under the Logic Pro menu. This window allows you to set up keyboard shortcuts for basically every function available in LP. As the default you should have the "Tempo" function controlled by the Enter key. If you don't want to browse through all the different shortcuts, simply type in "tempo" in the "Find" field. If you plan to use a MIDI device to control the tempo, you can assign any controller, channel, and note in the parameter window to the left side of the Key Commands window. Next activate the Tempo Interpreter by selecting "Manual Sync (Tempo Interpreter)" from the Sync button in the transport window. This will set LP to wait for synchronization coming in from the key command shortcuts you set earlier. You can change several parameters related to the Tempo Interpreter by selecting "Open Tempo Interpreter" from the Sync button in the transport window. In Table 3.1 you can find a short description of each parameter.

Table 3.1 Tempo Interpreter parameters in LP

Parameter	Description	Notes
Tap Step	Sets the basic rhythmic value the sequencer will assign to the manual taps	Usually 1/4 is a good starting point
Window	Determines the size of the region that will determine the tempo changes	Larger values create more drastic tempo changes, smaller values create smoother and more precise changes
Tempo Response	Control the sensitivity related to tempo changes	Larger values imply greater sensibility
Max Tempo Change	Controls the maximum tempo change allowed	
Tap Count-In	Sets the count off you have to input before LP detects the tempo	
Smoothing	Controls how responsive LP is to your "conducting"	
Tempo Recording	If selected, allows you to record the tempo changes in the tempo track	
Pre and Post	Visual indicators that give you feedback about your "conducting" style. *Pre* displays the taps you input; *post* displays the taps that have been accepted by LP	A yellow flash means that the tap was within the allowed range, a red flash indicates that the tap was out of the allowed range

After setting up all the parameters, you can start "conducting" your sequence by tapping on the key you selected as your "virtual baton." After four beats (or whatever number of beats you chose in the Tap Count-In option), the sequence will start playing according to the tempo you tap, and it will follow any variation.

CSX uses a slightly different approach to achieve similar results to those of DP and LP. In CSX you record the conducting pattern on a regular MIDI track by tapping on any key of your MIDI controller (the sustain pedal won't work). After recording the conducting pattern of tempo changes, you can "transfer" the MIDI events recorded to the Tempo Track by selecting the part you just recorded and choosing the option "Merge Tempo from Tapping" from the Functions submenu in the MIDI menu. You have to indicate the type of rhythmic subdivision you tapped (1/4, 1/8, etc.). The "Begins at Bar Start" option allows you to choose if the first MIDI message will automatically start at the beginning of a bar after the computer calculates the tempo changes. Make sure the Tempo parameter is set to "Track" in the transport window. If you press the play button, the sequence will now follow the tempo you "conducted" earlier from your MIDI controller. The tempo data can be edited in the tempo graphic editor, as seen previously. You can also record tempo changes in real time by using the "Tempo Recording" slider available in the tempo graphic window. When in playback mode, you can record tempo changes by moving the slider horizontally. This technique is particularly useful to quickly insert accelerandos and rallentandos on the fly, without using Tap Tempo.

As you can see, the tempo change feature can be an incredible tool for improving your sequencing techniques and skills. Used in the right way it can really bring your virtual orchestra to life and help you overcome the intrinsic mechanical nature of the sequencer.

3.8 Tempo changes and audio tracks

When inserting, editing, and programming tempo changes, you can really appreciate the flexibility of the MIDI system. Since all the MIDI data recorded in a sequencer are only descriptions of performance actions, it is fairly easy to have them follow tempo and meter changes. The same cannot be said, though, for audio tracks. This type of track is much less flexible and forgiving than its MIDI counterpart. As you have probably already noticed if you had some audio loops in the sequence you were using to experiment with tempo changes, audio tracks usually do not adapt automatically to the tempo changes programmed in your sequence (there are exceptions, such as audio files in Recycle format, but I will discuss this in the next chapter). While this used to be a major problem in early MIDI/audio sequencers, nowadays there are several techniques that can make audio tracks easily adjustable to tempo changes. In the following paragraphs we are going to learn how to overcome the static nature of audio files.

3.8.1 Time-stretching audio files

One way to have audio files adjust to tempo changes is to use the time-stretching technique we learned in Chapter 2. This approach involves changing the tempo of a loop or sound bite without changing its pitch. It is based on complex algorithms (such as *phase vocoder*, and *time domain*) that try to "guess," through a frequency analysis, which samples need to be eliminated (when you increase the tempo of an audio file) or added (when you decrease the tempo of an audio file). Once you know at which tempo the original audio file was recorded it is fairly easy to adjust its tempo to the new tempo track. Keep in mind that as a general rule it is always better to increase the tempo of an audio file than to decrease it, since in the first case the computer will have to eliminate some of the samples, while in the second it will have to generate new samples through a process of interpolation. Let's take a look at how the four sequencers allow you to adjust the tempo of audio files to a different tempo than the original one at which they were recorded.

In DP you can use the function called "Adjust Soundbites to Sequence Tempo." After selecting the sound bite you want to modify, simply select this function from the Audio menu. The sequencer will change the tempo of the audio file selected according to the actual tempo marker of where the bite is located. Another quick way to time-stretch an audio file is to open the graphic editor for the sound bite by double-clicking on the part in the track list window. From the graphic editor, simply move the cursor toward the top corners of the edges of the sound bite you want to time-stretch. The pointer will change into a *hand tool.* By moving the mouse horizontally, you will be able to adjust the length of the sound bite without changing its pitch, almost in real time. Keep in mind that with this technique, results may vary depending on the material you are working on. Usually more complex materials, such as full mixes or audio files with a high harmonics content, return much higher signal degradation than rhythmic or percussive parts. A set of independent tools called "Spectral Effects" (Figure 3.13) is also available to apply time-stretch and pitch-shift changes to audio sound bites. You can access it from the Audio menu. Choose the percentage (ratio) of the time-stretch that will be applied to the selected sound bites by either inserting the value in the field named "Tempo" (positive values will speed the audio up, while negative values will slow it down) or moving the ball in three-dimensional space with the mouse. Using the "Spectral Effects" tools you can also pitch-shift the selected audio material (change the pitch without changing the tempo). You can do so by specifying the transposition in the "Pitch" field.

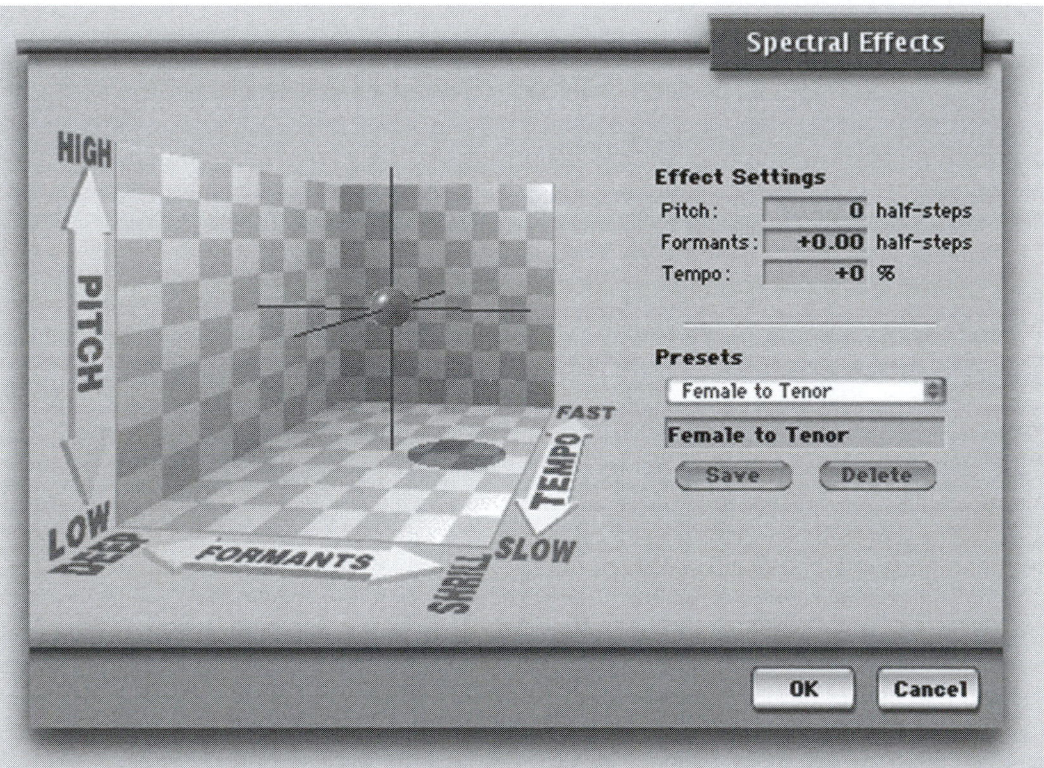

Figure 3.13 The "Spectral Effects" window in DP.

The same techniques can be applied in PT, where the time-stretching function can be accessed from the Audio Suite menu. By choosing "Time Compression Expansion," after having highlighted the region of audio you want to modify, you can alter its tempo using several parameters. In Chapter 2 we analyzed how to time-stretch a region by inserting the original tempo (*source*) and the new tempo (*destination*). In fact you can alter a region's tempo by using other parameters, such as length in samples, length of the region in real time, or length of the region in time code. These options are particularly useful when applying time-stretching to sound design or postproduction projects. In addition you can use a series of sliders that allow you to control parameters such as *ratio* (which allows you to set the length of the destination region in relation to the source length—by moving the slider to the right you increase the length of the destination region, and vice versa) and *cross-fade* (which allows you to control the amount of crossfade applied between samples that are either eliminated or added during the time-stretching process—a shorter setting is better for percussive parts with fast transients, while longer settings are more appropriate for long and sustained parts). The parameter *Min Pitch* has the function of limiting the frequency range of the time-stretching operation, resulting in a more accurate rendition of the original material. Use the *accuracy* slider to target the time-stretching algorithm to the material you are processing. A very useful way of quickly applying time-stretching to a region is to use the *TCE* (time compression expansion) *Trimmer tool.* This is a variation of the regular trimmer tool used to shorten or lengthen a region. To select the TCE, *click and hold* on the regular Trimmer Tool and select the TCE option (it looks like the trimmer tool but with the addition of a watch icon). With the TCE tool, instead of reducing or increasing the size of a region, you can stretch it to fit the tempo of your sequence. If used in conjunction with the grid mode (which allows you to drag elements in the edit window only by the grid specified in the *grid value fields*), you can easily reshape your audio material to fit any tempos.

To use the time-stretching function in CSX, select the part you want to alter and then choose the option "Time Stretch" from the Process submenu located in the Audio menu. In the window that appears on your screen you will see some of the familiar parameters we encountered in PT. The sections named "Input" (on the left side of the window) and "Output" (on the right side of the window) represent, respectively, the original tempo and the new tempo. You can alter the selected tempo of the part by using the length in samples as a reference, the length in seconds, the BPM, or the ratio between Input and Output (exactly as we saw in the case of PT). You can also choose among several different algorithms specifically targeted to different types of parts. The quality of the algorithms, along with the processing time, increases from Mode 1 to Advanced. The MPEX option is based on the algorithm developed by Prosoniq, and it usually guarantees best results for complex materials such as full stereo mixes. The "Drum Mode" option is targeted at percussion and drum parts. The *accuracy* parameter allows you to give priorities to the rhythmic aspect of the audio files you are processing (positive values) or to the harmonic content of the audio materials (negative values). Use the *preview* option to listen to the result of the process before actually creating a new sound file with the processed material. If you are satisfied with the quality, click "Process"; otherwise fine-tune the parameters to obtain the best result and then apply the changes to the audio file.

The time-stretching features of LP are comparable to the those of the other sequencers we have analyzed so far. For quick and precise adjustments of loop and in fact any other audio materials, you can use the two functions named "Adjust Objects Length to Locators" and "Adjust Objects Length to nearest Bar." The former will speed up or slow down your selection according to the area selected between the two locators; the latter will automatically adjust the speed of the audio material to fit into the closest bar. Both functions can be accessed from the "Objects" submenu located in the Functions menu of the main

Arrange window. For more complex time-stretching edits, you will need to use the powerful Time and Pitch Machine (Figure 3.14). This engine is similar to the one found in PT and DP. It can be accessed from the Sampler Editor window under the "Factory" menu (you can also use the shortcut *Control-T* from the Sampler Editor window). The Time and Pitch Machine allows you to control several time-stretching parameters, and you can decide to change the tempo of an audio track using criteria such as tempo change percentage, length in samples, length in time code (SMPTE), and length in bars. As we saw in CSX, you can choose among different algorithms targeted at different types of audio material in order to achieve the best quality. The algorithms are pretty much self-explanatory, targeting audio materials such as pads, monophonic, rhythmic, beats only, etc. Use "free" transposition if you want to have time and pitch independent of each other (this is the most likely situation in which you will work). Choose "classic" if instead you want to have both the tempo and the pitch affected simultaneously by the Time and Pitch Machine.

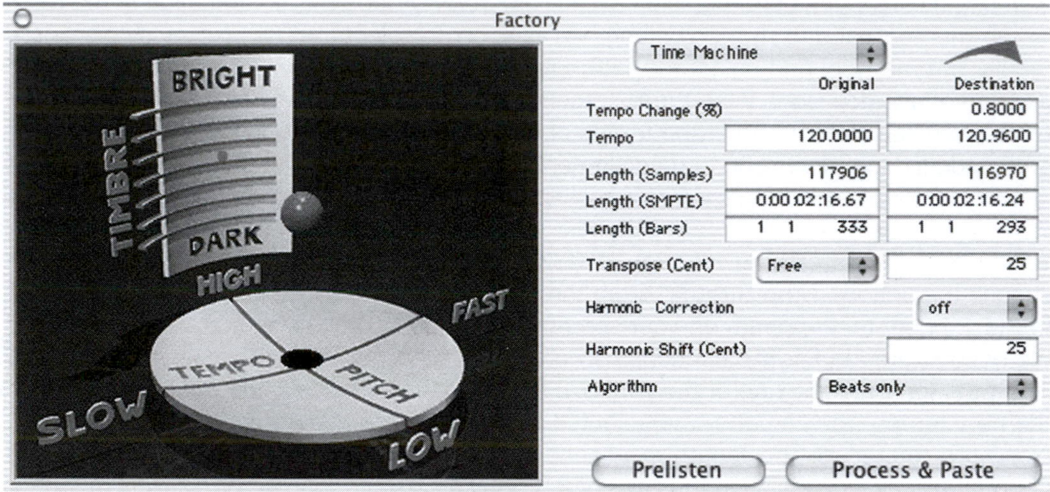

Figure 3.14 The "Time and Pitch Machine" window in LP.

You can see how powerful the time-stretching engines are and how useful they can be not only for matching loops recorded at different tempos but also to create sound effects and new sonorities. For example, you could overly slow down or speed up a vocal track or a solo instrument to come up with a completely new sound effect. Or you could time-stretch an entire track should you decide to slightly speed up your composition when it is too late to recall the musician for an additional recording session. Keep in mind, as I mentioned earlier, that it is better (if possible) to speed up the audio material than to slow it down. Also try to keep the tempo changes within a small range, usually up to ±10 BPM; for higher values, results may vary depending on the complexity of the audio materials.

3.9 Synchronization

Up to this point we built our studio around a computer sequencer that acted as the main central hub of the MIDI and audio network. While this is probably one of the most common situations, there are other scenarios in which other sequencers (hardware or software) and other multitrack recorders or video machines need to be included in your setup. If this is the case, then some new issues need to be analyzed in order to have a smooth integration of your main sequencer and the other devices. One of the main aspects we have to consider is the synchronization of time-based devices (such as drum machines, sequencers, video tapes, multitrack tape recorders, HD recorders, just to mention a few) to your main computer sequencer. All these devices store data either on tape or on random-access media (RAM and HD, for example). In order to be able to use them in the same studio situation, they all need to move at the same pace and to be always in sync with the others. When we set up our studio in Chapter 1 we didn't run into this issue because we only had one main sequencer that received and sent data to devices that didn't need any synchronization, such as sound modules and MIDI synthesizers.

There are two different categories of devices that can be synced to our sequencer: nonlinear and linear machines. The main difference between the two is based on the way they record and play back information. Nonlinear machines store data in a random-access way, meaning that data can be accessed immediately without waiting for the machine to actually reach a certain area of the storage media. A classic example of nonlinear machines are MIDI sequencers, HD recording systems, and drum machines. With a sequencer you are free to jump to any location almost instantly without having to wait to fast-forward or rewind to that location. This type of machine is the most flexible and easy to synchronize. Linear machines, on the other hand, store the information recorded one after the other, mostly on tape (in fact they are also referred as *tape-based* machines), which means that in order to reach a certain spot of your project you have to physically fast-forward or rewind the tape. This type of machine is definitely less flexible than the nonlinear type and presents a more complex synchronization setup. Next you are going to learn how to synchronize nonlinear machines to your sequencer. In the next chapter we will approach the synchronization of linear machines.

3.9.1 Synchronization of nonlinear machines

Let's analyze the synchronization process and techniques used for nonlinear devices. One of the advantages that nonlinear machines have over linear devices is that, in most cases, they feature a built-in MIDI interface, making it fairly easy to set them up for synchronization. The synchronization process for nonlinear machines involves the use of data that are received and transmitted using standard MIDI messages. There are mainly two types of sync protocol that can be used for nonlinear machines: MIDI Clock (MC) and MIDI Time Code (MTC). While they both allow you to synchronize multiple devices together, their nature and their features are very different.

MIDI Clock (sometimes also called *Beat Clock*) is a *tempo-based* sync protocol. This means that it is able to carry information about the tempo at which the sequence is programmed and therefore is ideal for MIDI studios that need to synchronize several nonlinear machines. To better understand the process involved, let's examine a practical example. Let's say we need to synchronize a drum machine to our main sequencer. Keep in mind that all the other devices will continue to function and are connected exactly as explained in Chapter 1. The only difference now is that have we added a drum machine that not only requires us to receive regular MIDI messages, but also needs to keep in sync with the main sequencer in order to play the programmed patterns in time with the other MIDI tracks. When trying to synchronize two or more devices, you have to set up one as the "master" and the others as the "slave." The master will distribute the sync to the slaves, while the slaves will wait to receive the sync from the master in order to play back. The master device is sometimes also referred to as *Internal Sync* and the slave as *External Sync*, depending on the software, model, or brand. Once you set up the master and slave devices (I recommend usually making the main sequencer the master), you should be ready to go. The slave devices will wait for incoming MIDI Clock messages; as soon as they detect the MIDI messages coming from the IN port, they will start playing in sync with the master. The MC data are System Real Time MIDI messages that are sent 24 times every quarter note. As long as no MC messages are lost in the connection between master and slaves, the devices will "move" at the same tempo and pace. Three other types of MIDI messages, called *start*, *stop*, and *continue*, complement MC by allowing the master device to instruct the slave machines to start the sequence, stop playback, or continue from the current position.

One of the problems of MC is that when a connected device receives a "start" message, it will start from the beginning of the sequence, making it fairly problematic to work on long sequences. Fortunately, an extra MIDI message, called *Song Position Pointer* (SPP), allows the master to control the current position of slave devices by describing its current position in 16th notes. When the slave devices receive an SPP message, they will set their counters to the position indicated and then wait for the incoming MC. As soon as the MC messages are sent and received from the master, the slave machines will start playing in sync. It is not uncommon for MIDI programmers and composers of loop-/groove-based sequences to favor groove boxes or hardware sequencers because of their particularly tight timing (e.g., the famous Akai MPC 2000). It is therefore fairly common to run into sessions that need MC in order to have all the devices in sync. The fairly basic setup is based mainly on the right assignment of the master and slave devices. In the next section we will analyze how the four main sequencers handle the master/slave assignment and how to set them up in the right way in order to have a problem-free session.

MIDI Time Code is a *real-time-based* synchronization protocol that contains no information on the tempo of the sequence but instead defines the passing of time. Its syncing format is divided into *hours:minutes:seconds:frames* (*hh:mm:ss:ff*). This format is derivative of the widely used protocol SMPTE (Society of Motion Picture and Television Engineers) and is based on the same format. While SMPTE is an analog-based signal used to synchronize linear to either linear or nonlinear machines, MTC is its digital translation, meaning a digital binary code sent over the MIDI network as MIDI messages. MTC falls in the System Common messages category. Whereas MC is directly related to the tempo of the sequence (since it is sent 24 times per quarter note), MTC is an absolute rendition of time and doesn't include any information about tempo or meter changes. If you plan to use this syncing protocol to sync two sequencers, you will have to have an identically programmed tempo/meter map in both devices; otherwise the bar numbers and tempos of the two sequences won't correspond. As we saw in the case of MC, you will have to set up one device as master and one or more devices as slave. The smallest time subdivisions on which MTC is based is the *quarter-frame message*. It takes eight quarter-frame messages to describe an entire MTC position (two messages for the frame, two for the seconds, two for the minutes, and two for the hours). As in the case of SMPTE, there are four main types of frame rates (in frames per second, fps): 30, 29.97, 25, and 24. In Table 3.2 you see a description of the different frame rates and their use in the industry.

Table 3.2 SMPTE frame rates

Frame rate	Use
24 fps	Film
25 fps	Color and black-and-white TV in Europe, Australia, and all countries using frequencies at 50 Hz
29.97 fps	Color TV in the United States, Japan, and all countries using frequencies at 60 Hz
30 fps	Black-and-white TV in the United States and Japan

As a standard the frame rates are assumed to be *nondrop formats*, meaning they express exactly the frame rate indicated. Another form of MTC (and SMPTE) is called *drop-frame*. This particular format drops frames number 00 and 01 at the beginning of each minute, except for the minutes falling on the tenth minute and its multiple. The drop-frame format is used to compensate the discrepancy between the 30-fps rate and the 29.97-fps rate, which is the one used in color TV in the United States and Japan. While for MC all the machine work with a standard resolution of 24 clicks per quarter note, with MTC the frame rate needs to be set up and needs to be the same for each synchronized machine. If you are scoring to a picture and you are synchronizing your sequencer to a video linear machine (more on this in the next chapter), you will have to set the frame rate of your devices to match the one used in the video sync track. Usually the video production company will tell you at which frame rate the tape was recorded. If you are not synchronizing your MIDI devices to any video source but instead are using MTC to synchronize audio devices and sequencers only, you can select the frame rate you want (usually 30 fps is a good choice since it gives the best resolution) as long as all the devices are set to the same frame rate. After choosing the frame rate for all the devices that need to be synchronized, you have to set the slave devices to "external sync" and put them in play. The slave devices will wait for incoming MTC messages, and they will start playback as soon as the first full message describing the position of the master, in *hh:mm:ss:ff* format, is received. While it is not the main intent of this book to go deeply into a discussion of the MIDI message format (for a full analysis I recommend the book *MIDI for the Professional*, by Paul Lehrman and Tim Tully, from Amsco Publishing), I would like to point out that MTC consists of four types of messages: *full*, *quarter-frame*, *cueing*, and *user bits*. The first two are for synchronization, while cueing messages are for automation and user bits messages, at the moment, are not used.

MC and MTC have advantages and disadvantages, and they have different practical applications. One of the advantages of MC is that it carries tempo information, minimizing the amount of programming required to insert tempo maps in all the devices that need to be synchronized. On the other hand, MTC has a much higher resolution than MC (at least for BPM rates lower than 300), meaning in general a tighter synchronization among devices. MTC also has the advantage of not requiring a "play" message, since slave devices will start playback as soon as MTC messages are received. MTC has the disadvantage of taking a fairly big chunk of the bandwidth reserved on the MIDI network (up to 8%) and therefore should be used only if synchronization is required and should be sent only to the devices that need it. Keep in mind that MC and MTC do not necessarily exclude each other in a studio situation. In fact it is quiet common to have several devices synchronized where some require MTC and others require MC. If this is the case it is recommended to have the same sequencer set as master for both MC and MTC. In Figure 3.15 you can see a variation of the studio setup learned in Chapter 1, with the addition of two devices, an HD recorder and a drum machine, slaved to the main computer sequencer through, respectively, MTC and MC.

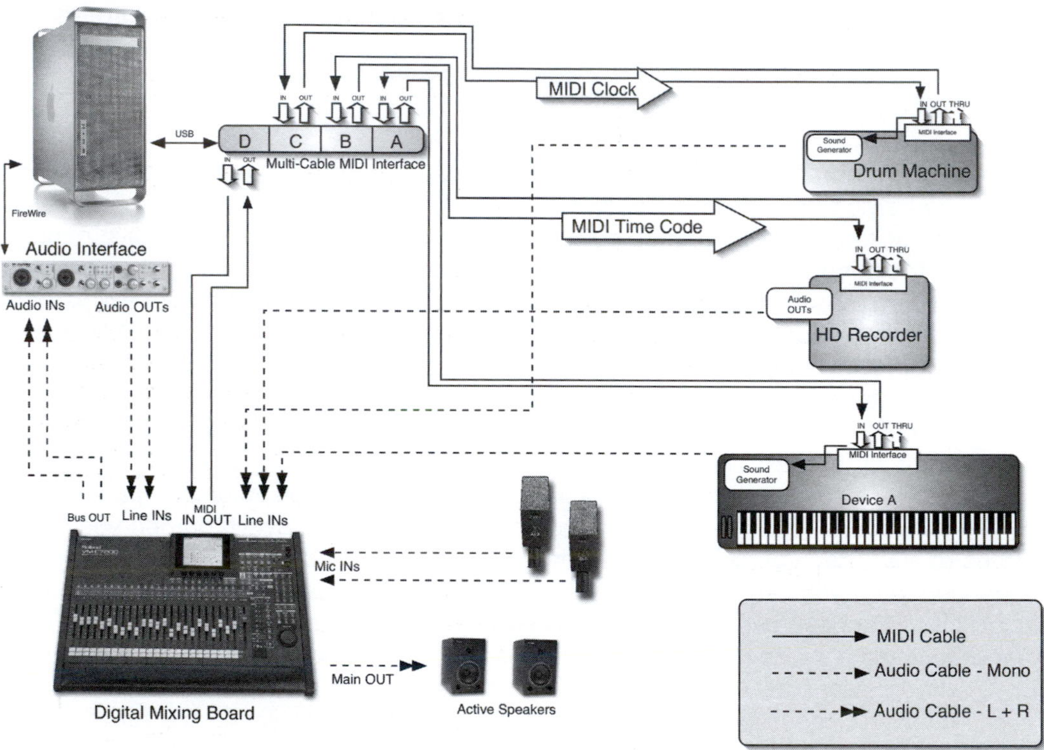

Figure 3.15 Studio setup with master/slave synchronization through MTC and MC (Courtesy of Apple Computer, Roland Corporation U.S., M-Audio, and AKG).

3.9.2 Sequencer setup for MC and MTC synchronization

As we just learned, one of the crucial steps to effectively synchronize nonlinear devices is to make sure the master–slave relationship is correctly established between the devices and that, in the case of MTC, the same frame rate is selected for all the devices. In Table 3.3 I sum up how to set up the four sequencers to be either master or slave and how to set up the synchronization parameters.

Even though the number of situations in which synchronization is required has been dropping in recent years because of the increasing computer power available, you will still find

Table 3.3 Internal (master) and External (slave) Synchronization settings for DP, PT, CSX, and LP

Sequencer	"Master" settings	"Slave" settings
DP	From the Setup menu select "Transmit Sync," choose the type of sync you want the sequencer to transmit (MTC or Beat Clock), and choose the MIDI cable on which to send the sync Select the frame rate from the "Frame Rate" submenu in the Setup menu	From the Setup menu select "Receive Sync" and choose the MIDI cable on which DP will receive the sync Select the type of sync you want the sequencer to receive Select the frame rate From the Setup menu select "Slave to External Sync" and press Play; DP will wait for incoming synchronization code to start playback
PT	Select "Show Session Setup" from the Windows menu In this window make sure the option "MTC to Port" is checked and that the device you want to slave is selected Set the frame rate If you want to transmit MC instead, simply select the "MIDI Beat Clock" option from the MIDI menu and check all the devices to which PT will send MC	The settings are similar to the ones explained in the case of PT as master, the only difference being that you will need to set PT to be slaved. To do so, click on the "Slave to External Sync" icon located in the transport window (the small watch icon) or by select the "Online" option from the Operations menu. From the "Show Session Setup" window located in the Window menu you can also select the error correction options: *none* = no error correction, *freewheel* = PT will keep playing for the number of frames specified in the field, even if MTC is interrupted (allows you to overcome short dropouts of the sync code), *Jam Sync* = PT can keep playing even when the time code is dropped
CSX	Select "Sync Setup" from the Transport menu Set the "Time Code Source" option to "None" Set the MIDI ports on which you want to send MTC and MC (select only the necessary ones)	Select "Sync Setup" from the Transport menu Set the "Time Code Source" option to "MIDI Timecode" Set the MIDI ports on which you want to receive MTC You can set up parameters such as *Drop out Time*, which has the same effect as *freewheel* in PT, and *Lock Out*, which allows you to control the number of frames that CSX needs to sync to MTC Make sure that CSX is in "external sync" mode by selecting "Sync Online" in the Transport menu
LP	Open the "Synchronization Setup Window" by clicking and holding on the synchronization button in the transport window In the "General" section set the Sync Mode parameter to "Internal" In the "MIDI" section select the ports on which you want LP to send MC and MTC	Open the "Synchronization Setup Window" by clicking and holding on the synchronization button in the transport window In the "General" section set the Sync Mode parameter to "MTC" or "MIDI Clock," depending on your needs Set the "Frame Rate" Make sure that LP is in slave mode by clicking on the Synchronization icon in the transport window

yourself, sooner or later, in a situation where you need to be able to synchronize some devices. It could be a situation where a client is bringing some old sequences recorded on a hardware sequencer or where you really like to record your drum patterns on your MPC 2000. The bottom line is that synchronization protocols such as MTC and even the older MC are going to be around for a while and therefore it is good to know how to deal with them. I suggest that you practice and make some tests with the equipment available in your studio and get comfortable setting up the devices and your sequencer in both master and slave modes.

3.10 Safeguard your work: back it up!

All the information you have acquired so far about sequencing and the creative use of your MIDI studio is important material with which you have practiced your production proficiency and with which you will improve your skills in the next chapter. We have been focusing mainly on learning setup procedures, studio arrangement, and software techniques from which your sequencing ability will benefit tremendously. One aspect, though, that is never stressed enough in the mainstream MIDI and sequencing literature is the importance of safeguarding your precious work during and after the composition and sequencing process. Usually, and unfortunately, you realize how important it is to regularly back up your data only after you have lost them. There is always a crucial moment in composers' careers when, the night before delivering the most important project of their life, the computer or HD decides to give up, crash, and put them through hell for a few hours. As you probably know by now, my working studio philosophy is based on the concept of a smooth and "well-oiled" technical environment that doesn't interfere with the creative process. It is for this reason that I want to devote the last part of this chapter to project backup issues and how to organize sessions, files, and documents on your HD.

As I just mentioned, disasters always strike when least needed, so it is essential to prevent them rather than try to fix them later. To have quick, easy, and efficient backup sessions it is crucial to have your files organized in a methodical way on your HD. This will speed up the process. Here are few tips about your HD's organization. In the best-case scenario you will have at least two separate HDs (let's call them 1 and 2). HD1 (usually your internal HD) will contain all the applications along with the operating system. HD2 (it could be either internal or external) is dedicated to projects and sessions and is used as the current working medium. As mentioned in Chapter 1, the two-HD setup is a maximal performance environment, since one HD's head will deal with all the operations related to the OS and to the running applications, while the second HD's head (in our examples HD2) will store and retrieve data related to the project/session you are currently working on. Besides being more efficient, this setup is recommended to avoid catastrophic crashes, usually more likely to occur on the OS and applications HD.

Now that your working data environment is set, it is time to think about how to organize your documents and sessions. Keep your projects very well organized on your HD. Do not allow two or more projects to share the same audio files folder. This could be dangerous during a backup session, since you might be intending to back up a certain project without realizing that you forgot to back up its precious and irreplaceable audio files. Keep your documents in general and your sessions in particular divided by client, sessions, and projects.

If you use different software to work on different projects, you might have separate folders dedicated to separate applications.

In Figure 3.16 you see the basic organization of a data storage system for a generic project studio. The main idea behind this organization is to have the documents separated from any other files or applications. Inside the "Documents" folder you will have other folders containing files categorized by generic types of applications (e.g., music, graphic, Internet, utilities). Inside each generic type you will organize your project first by clients and then for each client by session. This way you will always be sure of what you are backing up and where to find your sessions. Once you have your HD organized, you are ready to start backing up your data.

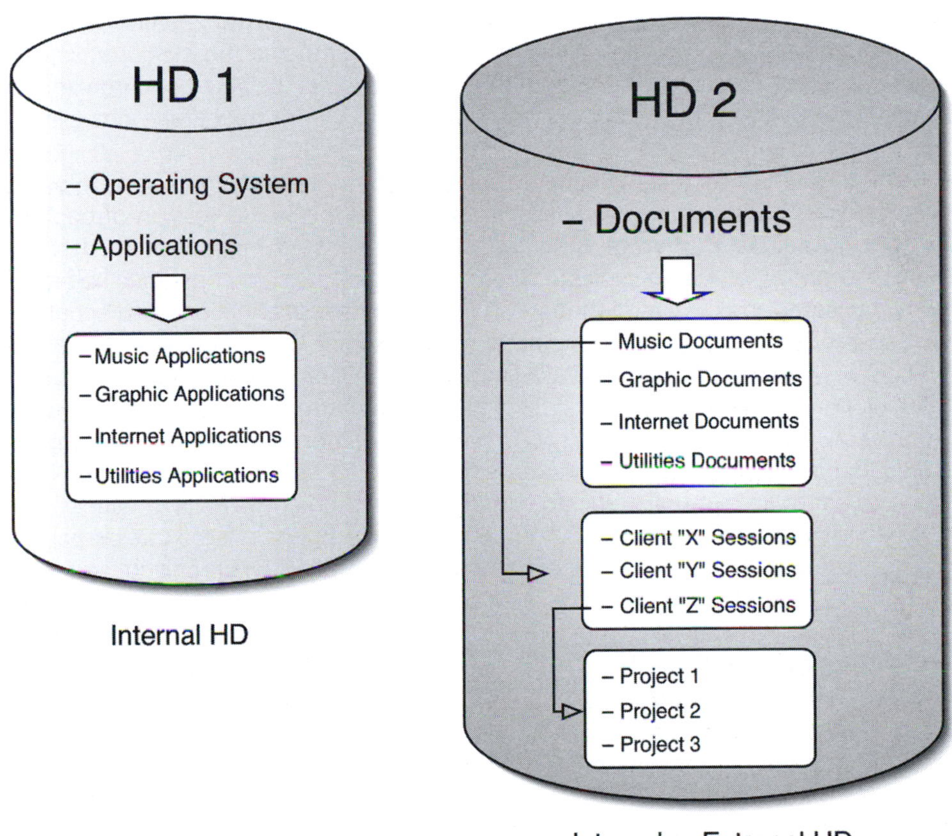

Figure 3.16 HD, folders, and file data organization.

3.10.1 Backup and archive

Even though the terms *backup* and *archive* are sometimes used to mean the same thing, their applications are fairly different. *Backup* refers to the copying of files related to a particular project and executed every so often during its realization. An *archive* session, on the other hand, is usually done at the end of the project, after all work is completed, and involves the copying of the files and the deletion of the original work project. In other words you back up your files at the end of every day of composition, sequencing, and recording, for the length of the entire production. At the end of the production, when the project is over and you have delivered the final material to your client, you can archive it, meaning copy the session to a different medium and delete it from the working HD in order to have free space to start a new project. Both actions can be done using different types of media. Most likely you will use a quick medium, such as a separate HD, for your fast, overnight backups, and CD-R/RW, DVD ±R/RW, or tape for safer backups and archive.

While the backup and archive session can be done manually by simply copying files from your working HD (HD2 in our example) to the backup media, there are software applications that can considerably facilitate this task. There are two main types of software used for backup: the so-called *backup utilities* (such as Retrospect by Danz, which has become the de facto standard backup application in many industries) and the *synchronization utilities* (such as Data Backup X by Prosoft Engineering Inc. or You Synchronize by You Software). Let's look at how the two systems work and what their differences are.

A synchronization utility is a simple application that compares the content of two folders, volumes, or directories and always keeps it updated to the latest version of each file by copying only the files that were modified since the last synchronization. It basically "mirrors" the content of the two selected locations so that they are constantly updated. This approach is great for quick, overnight backup sessions because it is fast and accurate and always leaves two versions of your projects, each of which can be accessed at any time. What I recommend is a third, external HD (let's call it HD3) as your overnight backup. Every day, after working on a project, spend 30 minutes synchronizing your working HD (HD2) with your backup HD (HD3). This way you know at any moment that you can count on a set of files you can access without problems. You can also use the mirrored copy (HD3) to bring the project to an external studio to work on it and then come back to your studio and resynchronize it with your working HD. You can save the settings related to a particular synchronization setup (such as the two locations that need to be synchronized, and eventual filters) in a file called "Set." You should create a separate set for each project so that when you want to synchronize the latest changes made to Project X, for example, the only thing you have to do is run the "Set Project X." The computer will compare the two folders and update each location with the latest file versions. This system is very flexible and it works great.

A backup utility works in a slightly different way. When you start a backup application, you have to create a backup set in which you specify the directory, folder, or volume you want to back up, the type of media where the files will be copied, whether you want the files to be compressed, and where you want to store the set itself. One of the main differences between this approach and the aforementioned synchronization system is that the backed-up files won't be directly accessible unless you restore them through the same application you used to back them up. While this system is efficient, you will sometimes be wondering where in fact your files are, and in case of a system failure you have to rely on the autorecovery capabilities of the backup software. The backup approach has some advantages, such as the capability of compressing the data, sometimes up to 50% (depending on the nature of the files), its reliability, and the ability to use several types of media, including tape, optical, and even remote servers. This type of application can also serve as an archiving system.

After your project is finished, mastered, and handed to the client it is time to archive it, meaning make a copy and delete it from the working HD and all the other places where you backed it up. Usually you archive projects that you are sure won't need any rework for a certain amount of time (this depends on your preferences and work cycle). I usually recommend archiving projects you are not going to work on for at least three months. To archive your project you can select either backup or synchronization software. I usually use backup software with the compression option to save some media space. Delete all the files from the working HDs after making your archive copy. When you next need to work on the project you archived, you will have to restore it to an available working HD.

Another important aspect of the backup/archive system is what I call the "3 rule." This involves always backing up and/or archiving to at least three different media and having one copy in a different physical location from the other two. For each major project, do quick, overnight backups on at least two separate HDs, and if possible use a third option, such as an optical medium (e.g., CD, DVD, or DVD-ROM). The same technique needs to be applied in the case of an archive session. For example, set your software to archive to tape and two DVDs simultaneously. Keep one of the two DVD copies in a location other than your studio. An option that has become more and more common in recent years is the *storage area network* solution, which can be applied to both Intranet and Internet servers. It is based on the idea of storing backup copies of your files to storage units connected on the same computer network. For a small session you can use fairly inexpensive hardware, such as an older computer with multiple HDs and an Ethernet connector. For a large session you will have to set up a more expensive, dedicated network based on a fast server. In any case, backup software and synchronization software are able to use the networked devices as viable storage units to set up your backup sessions.

3.10.2 How to calculate the size of a session

No matter which system and software you will be using, the bottom line is not to let time pass without protecting your data from fatal crashes. Take any steps and any necessary action to prevent problems that could stop your creative flow. When you are ready to back up a project, it is important to know how much HD space you will need and whether the session will fit on a single medium or require multiple media. While the MIDI data and the sequence file do not take much space (usually less than 700 KB), the project audio files take up the majority of HD space. Therefore to be able to estimate the size of a project is crucial.

In Table 3.4 you see a quick summary of the space it would take for 1 minute of mono audio at different sampling frequencies and bit resolutions. Knowing the number of tracks, the number of takes, the sampling frequency, the bit resolution, and the length of the project, you will be able to estimate the overall size of a session. For example, to calculate the size of the audio files related to a project that features 16 tracks recorded at 96 kHz and 24-bit resolution for a song that is 6 minutes long, you would use the following formula:

$$\text{(Number of tracks)} \times \text{(Length of project, in min)} \times \text{(Size, in MB, for 1 min of mono audio)}$$

In our example we would have

$$16 \times 6 \times 17.28 = 1659\,\text{MB}$$

Table 3.4 Size, in megabytes, for 1 minute of mono audio at different sample rates and bit resolutions

Sample rate/bit resolution	Size
192-kHz/24-bit	34.56 MB
96-kHz /24-bit	17.28 MB
44.1-kHz /24-bit	7.93 MB
44.1-kHz /16-bit	5.3 MB

Regarding the media you can use for your backup and archiving sessions, there are two issues that need to be considered: the reliability and duration of the medium and the maximum size of the medium. The reliability and duration depend largely on the type of storage technology used. Magnetic media, such as HDs, removable disks, and tapes, are generally pretty reliable (especially tape backup systems), and they offer large capacity. Optical media, such as CD and DVD, usually are good for smaller projects (especially CD), but they are best suited for short-term storage and data exchange. One of the best media for reliable backups and archiving sessions is the one based on magneto-optical (MO) technology, a hybrid of the two aforementioned systems, in which the data are written by a magnetic head only after the surface of the medium has been heated by a laser. Unfortunately this technology is not very popular anymore. I use MO disks, for example, for my sampler sound library, and it has been very reliable for more than 10 years. In Table 3.5 is a summary of the most used media and their specifications.

Table 3.5 Backup/archive media options and their main features

Type of medium	Technology	Size	Comments
CD R/RW	Optical	700 MB	Good for small and temporary backups
DVD R/RW	Optical	4.7–9.4 GB	Good for mid-sized backups
Tape	Magnetic	Up to 1000 GB	Good for big backups Reliable Not user friendly
HD	Magnetic	Up to 4000 GB	Fast Good for big backups Not very reliable for long-term storage
MO	Magneto/optical	Up to 9.1 GB	Good for mid-sized backups Very reliable Fairly slow

3.11 Summary and conclusion

In this chapter we covered some of the intermediate sequencing techniques that can improve your sequencing skills. While a basic quantization of your MIDI tracks can very often take the life and groove out of your parts, a more advanced quantization can fix the rhythmic errors and at the same time leave the original groove intact. Make use of quantization parameters such as sensitivity, strength, and swing to reduce the mechanical effect introduced by the quantization action. The "groove quantize" option enables you to apply different rhythmic styles and techniques to your parts. For example, you can have a straight 8th notes drum part converted into an 8th notes shuffle groove either by applying a preset groove or by creating your own styles.

Track layering can be applied to MIDI or audio tracks. This technique can be used creatively in several ways. By combining two or more MIDI tracks you can create new sonorities without having to play the same part several times or without having to copy the same part over and over. Simply assign the output of a track to several devices and MIDI channels at the same time. This technique is particularly effective in creating new patches and in complementing existing sounds. A similar layering technique can be used with MIDI and audio tracks in order to improve sampled acoustic sounds by adding one or more live instruments to the MIDI tracks.

Alternative MIDI track editors and MIDI controllers can improve your sequencing skills and, if used in the right way, inspire your creative process. The drum editor, for example, not only can be used to accurately edit rhythmic parts but can also become part of your sequence as a matrix editor and arpeggiator. Alternative controllers can be used to sequence parts in a more natural way. While it can be hard to render a woodwind solo from a keyboard controller, you can achieve a much more realistic result with a wind controller. The same principle applies in the case of MIDI drums/pads and guitar/bass-to-MIDI converters.

Complex tempo and meter changes can be automated and programmed through the use of a conductor track (or tempo track). Using either the graphic or list editors you can insert single events, such as tempo changes and meter changes, or continuous variations in order to create smooth accelerandos and rallentandos. You can also use the tempo map in creative ways. For example, you can insert subtle tempo changes to recreate the real feel of a live ensemble or slightly increase the tempo in certain sections of the composition in order to increase the overall rhythmic groove. With the "Tap-Tempo" function (also called "Tempo Interpreter"), you can conduct your sequencer using your controller as a virtual conducting "baton." Tempo changes can be a great tool for improving the groove of your productions, but they can also create problems when used in conjunction with audio tracks. While MIDI data are very flexible and can automatically adjust to the variation in tempo programmed in the conductor track, audio data need to be time-stretched in order to follow the tempo changes. Time-stretching requires the computer to calculate and adjust the samples of the audio files according to preprogrammed algorithms present in your sequencer. The sequencer will construct the new audio file according to the original tempo of the file and the "destination" tempo. In most sequencers this action can also be done in real time by simply clicking and dragging one side of the audio region (e.g., DP and PT).

By using different synchronization protocols we can have different devices and machines running together in perfect sync. While linear machines (meaning tape-based devices, such as multitrack tape recorders and video machines) can use a protocol called SMPTE to achieve synchronization, nonlinear machines (sequencers, drum machines, HD recorders, etc.) can be synchronized using either MIDI Clock or MIDI Time Code. These two protocols, which both use the MIDI network to send synchronization messages, require one device to be set up as master and the others as slaves. In the case of MC (which is a tempo-based protocol) the same message is sent 24 times for every quarter note. Through the use of an additional message, called Song Position Pointer (SPP), the master is also able to describe its position in the sequence by counting the number of 16th notes that passed since the beginning of the project. MTC is a real-time-based protocol that derives from the original SMPTE set. They both describe the passing of time in hours:minutes:seconds: frames. MIDI messages are sent each quarter-frame to describe the exact position of the master device. As soon as the slave devices receive the starting position, they will move to it and start playing, continuing as long as they receive MTC messages. Frame rates for both SMPTE and MTC include 24, 25, 29.97, and 30 fps. Two extra rates, called 29.97 drop and 30 drop, exist. While the latter is rarely used, the former is utilized to make up for the discrepancy between the video NTSC frame rate of 29.97 and the 30-fps rate used by broadcasters.

Backup and archiving sessions are crucial to the safeguarding of your precious creative work. Backup sessions are conducted during the realization of a project as quick, overnight copies after each working session. When a project is finished and you don't plan to work on it again for a long time, you will archive it, meaning make a copy on a separate medium (CD, DVD, etc.) and delete it from the original working HD. Software that can help in the process of backing up and archiving can be divided into "synchronization" applications and "backup/archiving" applications. The former keep a perfect copy of the current project on the main working HD and another medium (usually another HD), and they keep track of the latest modified files in both locations. The latter create a copy in a proprietary format (usually compressed up to 50%) that can be accessed only by restoring it onto an HD via the same software used to back it up. Media that can be used to back up and archive are HD, CD R/RW, DVD ±R/RW, tapes, and MO disks. No matter which system or media you use, remember to back up often and to keep several backup/archive copies of your most important projects.

3.12 Exercises

Exercise 3.1

Set up and record a sequence with the following features:

- Fourteen MIDI tracks
- Instrumentation and tempo that are free

Exercise 3.2

Using the sequence created in Exercise 3.1, quantize each track with a different groove and quantization setting. Take note of the parameters you chose for each track and compare them by listening first to each track individually and later to multiple tracks together.

Answer the following questions.

a. How does the "swing" parameter affect the quantized material?
b. How does the "sensitivity" parameter affect the quantized material?
c. How does the "strength" parameter affect the quantized material?
d. With which settings do you think your tracks improved the most?

Exercise 3.3

Using the sequence produced in Exercise 3.1, create MIDI layers for at least four tracks so that their outputs are assigned to at least two different MIDI channels and/or cables. Try to create some interesting combinations in terms of sound texture.

Exercise 3.4

Use the drum editor to create several patterns that will be added to your sequence.

Exercise 3.5

Add the following tempo change to your sequence:

- Bar 1, 120 BPM
- Bar 9, 100 BPM
- From bar 17 to bar 19, accelerando from 100 to 130 BPM
- Bar 24, 133 BPM
- From bar 28 to bar 32, rallentando from 133 to 90 BPM

Exercise 3.6

Create four audio tracks using some of the loops on the included CD, take four loops and insert them in your sequence at the following locations. Make sure to time-stretch the audio material so that it fits with the tempo changes programmed in the Exercise 3.6.

- First loop at bar 1
- Second loop at bar 9
- Third loop at bar 19
- Fourth loop at bar 24

Exercise 3.7

List and explain the main differences between the two synchronization protocols: MTC and MC.

Exercise 3.8

Calculate the space, in megabytes, required to store the audio files of the following two projects.

a. A 7-min project recorded at 196 kHz and 24-bit resolution with 24 mono tracks
b. A 3-min project recorded at 44.1 kHz and 24-bit resolution with 16 mono tracks

Exercise 3.9

Make a backup of the sequence you created in Exercise 3.1 and its audio files.

4 Advanced Sequencing Techniques

4.1 Introduction

In the first half of this book we learned information that ranged from how to set up a MIDI project studio to fairly complex synchronization settings, from basic session creation and MIDI track recording to sophisticated tempo changes and "groove" quantization techniques. While all this information is meant to improve sequencing skills and enhance your final productions, they focus mainly on predefined settings and options, leaving little space for personal customization. The goal of this chapter is to explore some of the advanced techniques offered by the main four sequencers analyzed in order to customize as much as we can their quantization, editing, effects, and automation parameters. As you know, what transforms a skilled composer into an original and creative one is a unique sound and unique melodies, harmonies, and orchestration style. The same can be said for the modern MIDI composer. The more we can bend the virtual orchestra to follow our own style (and not what the computer and the machine dictate), the more our production will sound fresh, original, and innovative. Here is where the customization of quantization parameters and the use of advanced tools play an important role in the creative process. In this chapter we will learn how to turn a good-sounding sequence into "our" great-sounding sequence! Let's move on to the advanced sequencing techniques.

4.2 Advanced quantization techniques

So far we have been using predefined templates to quantize the MIDI tracks we recorded. Although the implementation of parameters such as swing, sensitivity, and strength are a major improvement over the straight quantization we learned in Chapter 2, they are still fairly limited in terms of complete quantization freedom. What if we want certain MIDI parts to be able to adapt to the groove that we played on a different MIDI track, or what if we want to have a MIDI track follow the exact groove recorded on a separate audio track? Even better, wouldn't it be wonderful if we could control the quantization of audio tracks as we do on MIDI? All these options are actually available in several sequencers, and they represent one of the most exciting aspects of sequencing.

4.2.1 *Custom groove creation*

The grooves provided with some of the sequencers we have analyzed so far are basic starting points on which we can build a "groove" library that best fits our compositional style. We can create custom grooves from three different sources. First we can edit the existing ones using them as templates to make small or more drastic variations of the original. This allows us to tailor the quantization settings to a particular style, genre, playing style, or project. In addition we can create completely new grooves from scratch starting from prerecorded MIDI or audio tracks. This feature opens up an incredible number of possibilities. Imagine importing a drum track from your favorite drummer, extrapolating the intrinsic groove, and morphing it into any of your MIDI drums tracks. The same can be done from any MIDI track used as a source. The principle on which this technique is based is simple: By analyzing the MIDI notes on a MIDI track or the transients present on an audio file, the computer is able to extrapolate key points that the new groove will use as a reference for the new quantization algorithm. Let's look at how this technique works and how it can be used in specific applications.

4.2.2 *Editing a "groove"*

We already learned how to use the groove quantization feature in DP. Now let's analyze how to edit preexisting grooves first; this will make it easier to understand the process involved in creating custom grooves from scratch later. In order to edit one of the preexisting DNA grooves, simply select the "Groove Quantize" option found in the Region menu. Choose the category and groove you want to customize and click on the "Edit" button. This will open the Groove Editor window, where you can change the three main parameters of a groove: timing of the events, velocity, and length for each rhythmic subdivision (Figure 4.1). You can change each parameter for each grid value (e.g., 8th note, 16th note) specified in the original settings of the groove you are editing, independent of the others. Notice that this is different from the basic parameters available from the Groove Quantize window, which allows you to change timing, velocity, and duration only for the overall groove and not for each individual grid value.

The Groove Editor gives incredible flexibility in terms of timing, velocity, and duration of each event. For each grid point, you can change the timing by clicking and dragging left (meaning a rushing, or "ahead"-of-the-beat, feel) and right (a laying back, or "behind"-the-beat, feel). The velocity of each grid point can be set either via the mouse (click and drag on the velocity arrow on top of each point) or by inserting a numeric value in the respective field. The same techniques can be used for the duration parameter (click and drag left to reduce the duration value or right to increase it). Once you are done with the changes, click the "OK" button to save the changes or "Cancel" to discard them.

While using an already available groove to create a new one is a good starting point, you can go even farther by creating custom grooves based on preexisting MIDI parts in your sequence. This is a very flexible tool not only to increase the groove libraries and quantization options available for your productions but also, for example, to have several tracks to match perfectly the rhythmic feel that is embedded in the target MIDI track. In DP in order to create a groove from a preexisting part, first select the region you want to use as a template for the new groove from the track list window (usually two or four bars are a

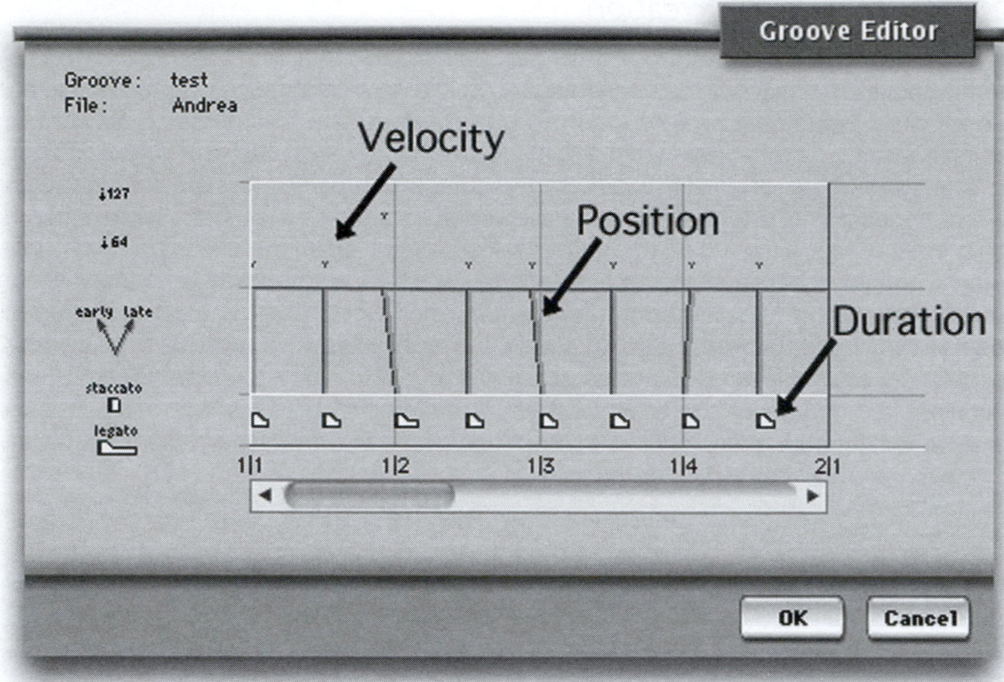

Figure 4.1 Groove Editor window in DP.

good starting point). Select the "Create Groove" option from the Region menu. You can create a new groove folder in order to keep your templates organized or select a preexisting one. Next select the basic grid values of the groove (8th notes, 16th notes, etc.) and the meter on which the grid is based, and insert a name for the new groove. When you click the "OK" button, a new groove will be inserted based on the rhythmic subdivision and MIDI events that were present in the region you selected. From now on that groove will be available from the regular Groove Quantize menu, and it can be used and applied to any region you want to quantize.

The same principle can be applied to your parts in CSX. In fact Cubase (and, as we will see later, LP and PT) is able to generate quantization grooves not only from MIDI parts but also from audio files, making it extremely interesting to create custom quantization settings based on your favorite audio material. Let's start with the procedure used to create a quantization groove from a MIDI part. First select one of the parts in a MIDI track on which you want to base your quantization setting. This could be a simple part on which you recorded a single-note rhythmic groove, or it could be a more complex part that you imported from another sequence or from a MIDI file created with another sequencer. Once you have selected the part, simply select the function "Part to Groove" from the Advanced Quantize submenu located in the MIDI main menu. The new groove will take the name of the part, so make sure to give it a representative name before creating the groove. To rename a part, you can use the "Event Infoline" at the top of the Arrange window. From now on when you select a part to quantize, you will have the option to apply the groove you just created.

4.2.3 *Audio to MIDI "groove" creation*

While the option I just described is interesting and definitely has practical applications, it seems a little bit limiting to be restricted to grooves based on MIDI parts, especially since the main point here is to "humanize" your MIDI parts as much as possible. In fact CSX offers an even better option that can greatly improve your MIDI tracks. It is possible to "clone" a groove from an audio file and transfer it as a quantization groove set that will be available for any quantization operation you need. While the procedure is a bit more elaborate than the one we just saw, it is definitely worth learning. The first step is to import the audio loop or audio file from which you want the groove "transplanted." Once you have the audio file in your sequence, double-click on it in order to open the sample editor. In this window there are two main modes: the regular edit mode, in which you can trim and cut the audio waveform, and the "Hitpoint" mode (Figure 4.2), in which the computer calculates virtual markers inside the audio file based on its transients and on its rhythmic features. The hitpoints inserted can be used for several operations, among them is the one that allows us to extrapolate the groove from the audio file. When you select the "Hitpoint" mode by clicking its button, CSX will ask you to insert some information about the audio file or loop you are using (Figure 4.2).

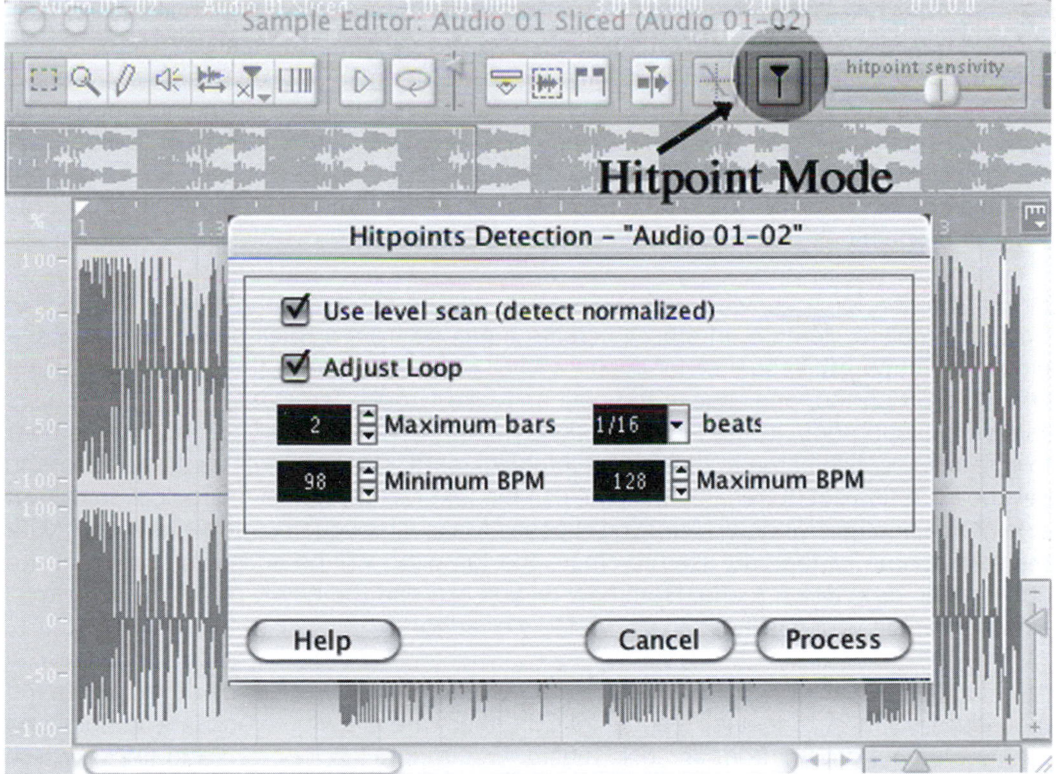

Figure 4.2 "Hitpoint" mode in CSX.

Usually you want to have the option "detect normalized" checked. This will instruct CSX to scan a temporary normalized version of the loop that guarantees a better result in finding the main transients (this option doesn't alter the original file). If the "Adjust Loop" selection is also "On," then CSX will try to guess the tempo of the loop based on the information provided in the bottom part of the window, such as number of bars ("Maximum bars"), basic beat subdivision ("beats"), and estimated tempo range ("Minimum" and "Maximum" BPM). After you click on the "Process" button, CSX will insert a number of hitpoints that coincide with the stronger transients of the audio loop. You can adjust the sensitivity of the "Hitpoints detection" process by moving the slider ("Hitpoint Sensitivity") at the top of the sample editor window. If you move it toward the left, the number of hitpoints will decrease; if you move it to the right, the number will increase. In order to effectively create a groove that captures the rhythmic essence of the loop, you should have hitpoints only on the major transients. Too many points will clutter the groove, making it too complicated; not enough points will leave you with a groove that won't be effective when used in the quantization process. If you are not satisfied with the automatic markers that the computer placed for you, it is always possible to move or delete some of them to better fit your needs. It is going to take few trials before you really start mastering this technique, but you will see how useful it can be.

After you have made the final decision on the number of hitpoints to keep, it is time to "transplant" the groove to a quantization set. Select the "Create Groove Quantize" option from the Advance submenu in the main Audio menu. The computer will build a new customized quantization set based on the audio loop and hitpoints you marked. The quantization grooves (including the one you just created) are going to be available from the "Quantize Setup" submenu in the MIDI menu. The groove will take the name of the audio part it was built on, so remember to rename the audio part in order to have a clear description of the groove using the "Event Infoline" at the top of the Arrange window. As you can see this opens up an incredible number of options and creative opportunities for improving your productions.

Similar results, for both MIDI and audio parts, can be achieved in LP using the "Make Groove Template" feature. In order to create a groove from a MIDI part, first record the groove on an empty sequence (or import a MIDI part). Next select the sequence you just created or imported, and choose the option "Make Groove Template" from the "Groove Templates" submenu, located in the Options menu. As we saw for CSX, in LP remember to give an indicative name to the sequence/object (part), since this will be the name of the new quantization set (to rename the part, use the Text Tool located in the Toolbox to the left side of the Arrange window). After you choose "Make Groove Template," a new item with the same name of the part used to create the template will be added at the top of the quantization drop-down menu. From now on you can use the newly created quantization setup based on your MIDI part in any window or editor.

Also, in LP you can apply to a MIDI part a groove extrapolated from an audio file. While the procedure is similar to the one learned for CSX, there are a few differences worth pointing out. To create a groove from an audio region, first import the loop or audio file on which you want your new groove to be based (see Chapter 2 to review this procedure if you need to). Trim the loop as necessary, and make sure the beginning of the sequence/object (part) corresponds to the beginning of the audio region. If you want to trim the audio region, select an even number of bars—usually two measures work best for this kind of operation. After your audio is ready, select it and choose "Make Groove Template" from the

"Groove Template" submenu located in the Options menu. LP will open the audio editor along with a special "Quantization" window for audio tracks (Figure 4.3).

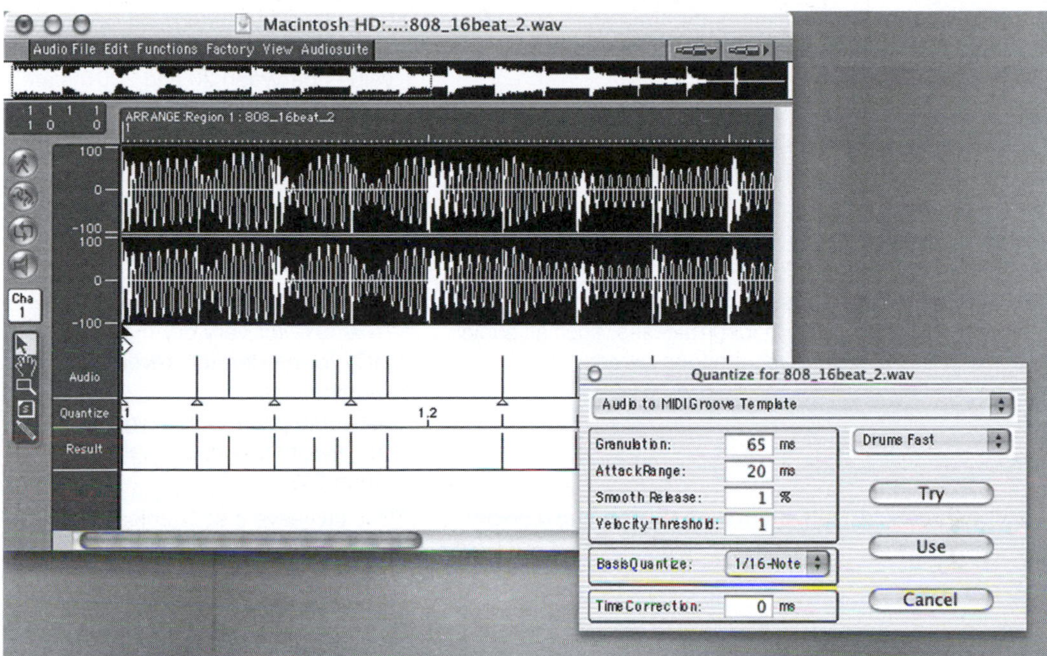

Figure 4.3 Extrapolating a groove from an audio track in LP.

To help the computer analyze and extrapolate the groove embedded in the audio loop, we have to provide a series of parameters regarding its rhythmic features, basic quantization, etc. A brief description of the parameters available in the Audio Quantization window is shown in Table 4.1.

In the drop-down menu (in the Quantize floating window) you can specify the type of audio material you are processing by choosing among a series of presets, such as drums, guitar, pop, and classical. Even though these categories are pretty generic, try to choose the one that best describes the type of audio material you are working on. In the main audio editor window you can visually check how the changes applied to the parameters in the Quantization window affect the final quantization groove. The three separate lines named "Audio," "Quantize," and "Result" respectively describe the hitpoints identified by the Logic engine, the hitpoints specified by you in the Basis Quantization field, and the final result, meaning the actual points that will be used by LP in the new quantization set. You can preview the result by clicking on the "Try" button. This will temporarily apply the groove to all the selected parts in the sequence. If you are satisfied, click on the "Use" button; otherwise you can keep changing the parameters or cancel the entire operation. From now on the new quantization set will be available in every window through the quantization list.

While a "groove finder" is not directly available in PT LE (version 6.4 and earlier), it is available in PT TDM and PT LE version 6.7 through an included audio engine called *Beat Detective* (BD).

Table 4.1 Audio-to-groove parameters in LP

Parameter	Description	Comments
Granulation	Indicates the time span of louder events of the audio material	Used by LP to determine velocity points Values between 20 and 200 ms are usually a good starting point
Attack Range	Determines the attack time of the phrasing of the audio material	For a drum part, choose a short attack time (e.g., less than 20 ms) For instruments with long attack times, such as pads or strings, use higher values (e.g., 40 ms)
Smooth Release	Used to process audio materials with long releases, such as audio files with long reverb	Usually use values between 0% and 6% (0% for very dry material and 6% for moderately reverberated audio parts) For audio parts with a high amount of reverberation, use values higher than 6%
Velocity Threshold	Determines the threshold below which the audio material is ignored	Usually leave it at 1, unless you are processing a very busy rhythmic part Increase the value of this parameter if the audio part has noticeable background noise
Basis Quantization	Allows you to insert "fake" hitpoints that are not detected by the computer based on loud transients	Used mainly to have more quantization points than the actual audio material features
Time Correction	Allows you to compensate for delays embedded in your MIDI setup (the time between a MIDI message sent and the actual sound produced by the sound generator of a MIDI device)	Use values between 0 and −15 ms

This engine works similarly to the ones found in the other applications I described earlier in this chapter. Let's take a look at its main features.

With BD you can achieve several tasks, all related to a selected audio region. You can, for example, find out the tempo of a loop (this feature is actually included in PT LE in the "Identify Beat" option under the Edit menu), or extrapolate the groove of an audio loop, and save it as a groove template to use on any MIDI or audio region. In order to extrapolate the groove from an audio loop, first you have to select the region of the loop you want the computer to analyze. Make sure the region selected is a complete set of bars, such as two or four complete measures. Next you have to define the selection in the BD window. To do so, open the BD window by selecting "Show Beat Detective" from the Windows menu. Change the Start/End and the Time signature parameters according to the audio material you selected, and then click on the "Capture Selection" option. Once BD has all the information about the loop, you have to let the computer analyze and detect the transient of the audio material. In the BD window, choose the "GrooveTemplate Extraction" mode, and select the right type of algorithm used to detect the transient: *High Emphasis* works best with high-frequency material, such as hi-hat and cymbals, while *Low Emphasis* works best with low-frequency material, such as toms and bass drums. When you click on the "Analyze" button, PT will insert so-called *beat triggers* (same as CSX hitpoints) according to the transients of the waveform analyzed. You can change the sensitivity of BD by using the "Sensitivity" slider. To extrapolate the groove based on the beat triggers generated by BD, simply click on the "Extract" button and insert a comment about the groove if you need one. The newly created groove template can be used directly in the current project ("Save to Groove Clipboard" option) or saved to disk ("Save to Disk" option). From now on you can use the extracted groove template to quantize any other MIDI or audio region.

As you can see, with the advanced quantization techniques we just learned, there are almost no limitations to what can be achieved in terms of rhythmic flexibility and precision. It is very important that you experiment with creating grooves of your own and use them as quantization templates for your projects. Start with generating short MIDI parts using a one-note rhythmic pattern, such as a swing feel you particularly like (MIDI drums kits and MIDI pads for this option are the ideal solution). Then use the MIDI part just created to generate a quantization groove and try to apply it to other MIDI parts/tracks. If your sequencer allows it, do the same starting from an audio loop. Start with very simple rhythmic grooves and audio material in order to get a better grasp of all the parameters involved in the process. When you feel comfortable, move to more complex grooves and patterns. The goal is to create a personal library of quantization grooves that you will use in your projects and that will contribute to give a unique signature and sonority to your productions.

4.2.4 Audio quantization

If you read carefully and practiced the exercises suggested at the end of each chapter, I am sure you will start thinking about how nice it would be if you could quantize not only MIDI parts but also audio regions. In fact by now you might be so "addicted" to quantization that you would use it for every single aspect of your production—so why not for audio parts? Well, you will be happy to learn that in fact there are possibilities for quantizing audio regions of your sequences, and that is exactly what we are going to learn in the following section.

Before learning the exact techniques related to each sequencer, we have to understand what the quantization of an audio region involves and how it differs from the procedure we learned in the case of MIDI parts. While the quantization of MIDI events is fairly simple for the sequencer since MIDI tracks hold only a digital description of the performance you executed on a controller, digital audio tracks contain a continuous flow of data that constitute the waveform. From a regular waveform it is impossible for the computer to distinguish between discrete events, such as a between a single snare and a single bass drum hit. While in MIDI those hits would be represented as two separate events, in an audio file they would be part of the same region, and therefore the computer wouldn't be able to treat them as separate entities. Things get even more complicated when the audio material we are trying to quantize is a complex mix of several instruments. Fortunately there are several ways to help the sequencer itemize an audio region. The main goal in order to quantize an audio part is literally to divide it in so-called "slices" that can be treated, and therefore later quantized, as if they were separate and independent events (Figure 4.4). This technique works best with drum parts or rhythmic loops in general. Since the procedure is based on the detection of fast transients (more typical of percussive parts) to determine where the slices need to be placed, it doesn't give the best results with slow-attack parts, such as strings or pads.

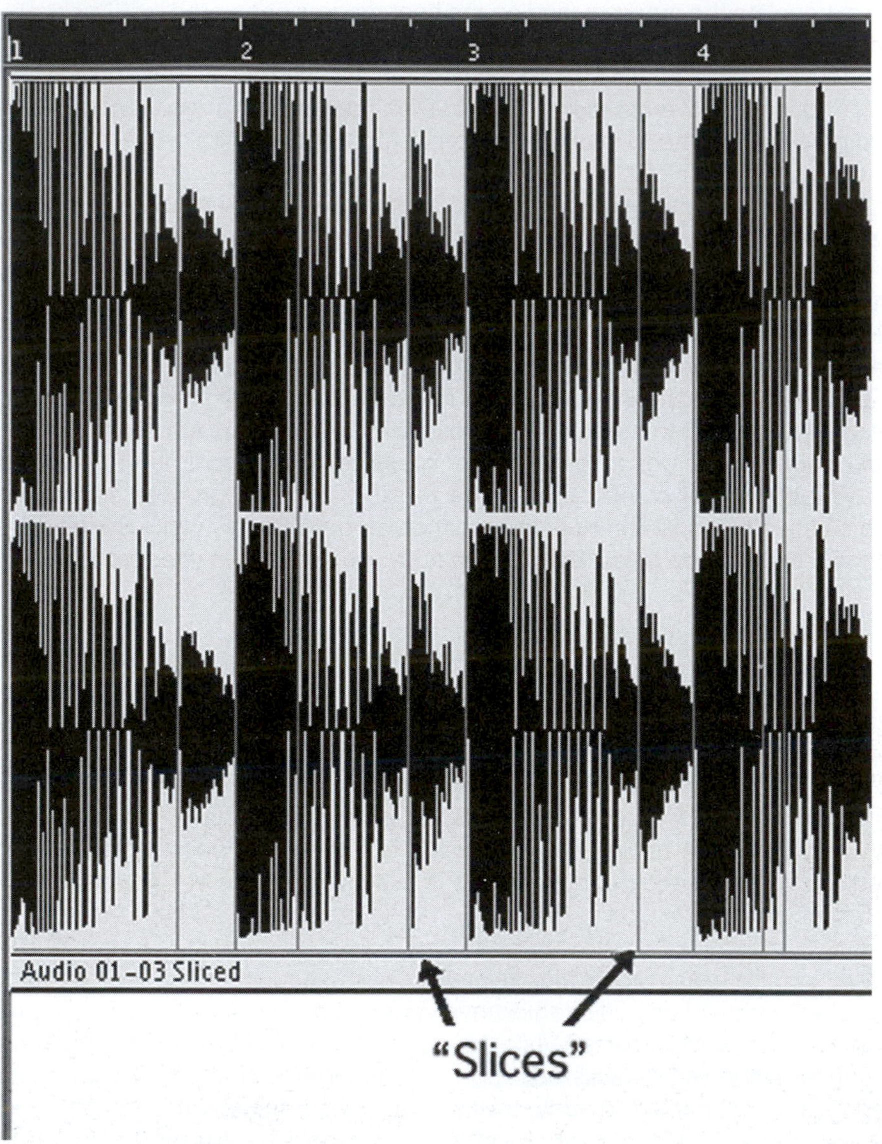

Figure 4.4 Example of a "sliced" audio loop in CSX.

There are specific applications that allow you to edit an audio file and automatically detect and create slices based on transients present in a waveform. One of the most used applications for this purpose is "Recycle," by Propellerheads. This application can "slice" a loop and save it in a proprietary format (.rex2) that can be imported by the majority of audio sequencers. What makes this format incredibly versatile is that each created slice can be treated as an independent event inside the original loop, making it easy not only to quantize the audio region but also to "anchor" the slices to bars and beats (as in the case of a MIDI track). This will allow you to have the audio region follow the tempo of the sequence in real time and without having to use the time-stretching feature of your sequencer. The "Recycle" software will not be covered here because it is outside the scope of this book. But there are in fact built-in tools in each sequencer that resemble the "Recycle" engine and allow us to achieve almost identical results.

In DP there is a manual technique that is as effective as the procedure found in "Recycle." In order to divide an audio region into slices, you have to import the audio loop first and find its tempo, as explained in Chapter 2. Also make sure the tempo of the sequence matches the tempo of the loop, which is crucial in order to divide the loop in rhythmically accurate slices. Open the graphic audio editor and select the grid value (called "Unit" in DP) according to the rhythmic resolution you want the slice to have. You find the "Unit" parameter in the upper right corner of the graphic editor window. Make sure the grid mode is enabled by clicking the Unit checkbox. Usually you want to choose the smallest rhythmic subdivision the loop is based on, but the way you handle this parameter depends on a number of factors, such as which rhythmic elements and positions are more important inside the loop. In order to slice the loop, select the *scissor tool* from the tool palette (if the tool palette is not open, press *Shift-O* to open it). With this tool you can click on any area of the loop and insert a cut that will be inserted only according to the grid value you selected earlier. You can even click and drag across the loop to insert slices for every grid value. For example, if you selected a 16th note grid, by clicking and dragging you will insert slices every 16th note (Figure 4.5a and b). Keep in mind that this "slicing" technique is grid based and not transient based.

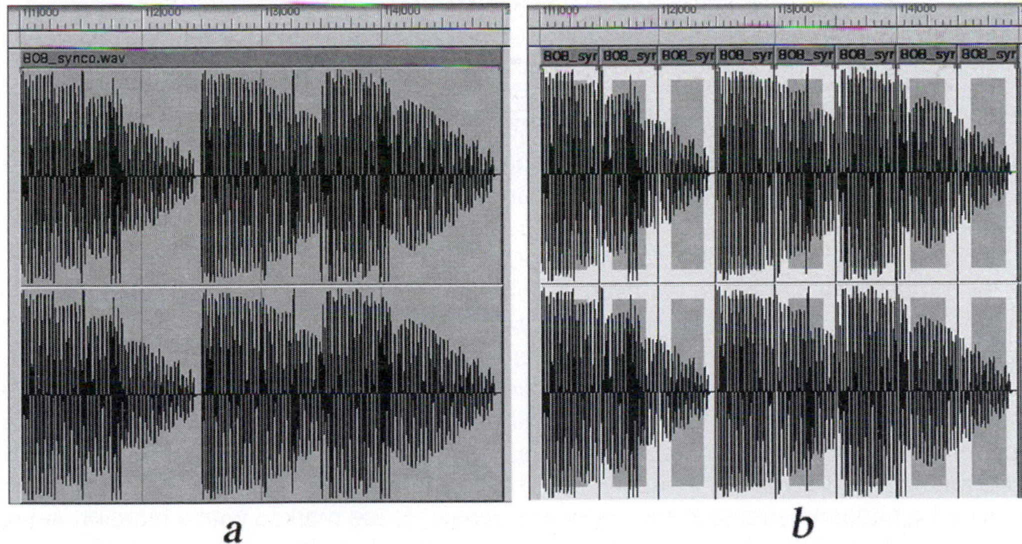

Figure 4.5 Example of (a) an original audio loop and (b) a "16th note sliced" audio loop in DP.

The most interesting aspect of generating several events from audio material is that the created slices (or events) can now be treated as individual entities and therefore quantized separately. You may remember that when we discussed the quantization filters for DP in Chapter 3, we learned that in the options available in the "Quantization" window we can decide which elements of a selected region we want to quantize. So far we have always chosen to quantize the attack of MIDI data. In fact, as we just learned, now we can expand the quantization feature to audio data as well and to each single slice of an audio region. The technique of slicing an audio file has two major applications. The first is to quantize audio files that need some rhythmic "makeover." For example, let's say you recorded a rhythmic guitar part on an audio track and that you are not satisfied with the rhythmic accuracy of your performance. You can slice the audio file recorded (this time you will have to do it manually with the grid set to "Off"; otherwise your slices will be inserted in the wrong place) and select the slices you want to quantize. Then simply select the "Quantize" option from the Region menu, and make sure that from the quantization window you select to quantize "Soundbites." This will tell the sequencer to move the selected audio slices to the closest grid point specified by the quantization value parameter.

Another extremely useful application of this technique involves being able to anchor the created slices to the measures and beats, exactly as a Recycle file would do. After you split an audio file in slices, the slices will be linked to their position, not in real time (as any regular audio file would be) but in bars and beats (as MIDI data are). Therefore when you change the tempo of a sequence, the computer will adjust the tempo of the audio files, not through the time-stretching algorithm but instead by simply moving the slices closer to each other (faster tempo) or farther from each other (slower tempo), allowing you to execute tempo changes in real time. This second application has some limitations though. First, the final result and quality of the tempo change greatly depend on the accuracy with which the slices were created. For simple rhythmic parts this technique works great, but for more complicated and rhythmically busy parts it might not work as well. Second, as we saw in the case of the time-stretching technique, usually you get better results with tempo changes that occur in a limited range (usually between +15 and −15 BPM). The same technique with some slight variations is also available in the other sequencers. Let's take a look at how the other applications handle the "slicing" technique.

In order to prepare an audio file to be quantized in CSX you use a similar procedure to the one we learned for creating customized grooves from audio loops. In fact CSX can use the same slices we created when analyzing an audio file in order to capture its groove. The first step, of course, is to import or record an audio file. Once the audio part is inserted on an audio track, you double-click on it to open the waveform editor. Use the "Hitpoints Mode" button (the same one we used in Section 4.1.2) to open the Hitpoint parameters window. The parameters are the same as in section 4.1.2. Once the computer has set the slices according to the transients of the waveform and the information provided about meter, tempo, and bars, you can freely move or delete slices that you believe were not placed in the right position. To delete a hitpoint, simply click on its arrow marker and drag it up into the ruler section of the window. To move a hitpoint, just click and drag it left or right, and to manually insert a hitpoint use the *pencil tool*. After you have made the right edits to the hitpoints so that all the important transients are marked with a hitpoint, select the "Create Audio Slices" option from the "Advanced" submenu found in the Audio menu. The newly created audio part will look very similar to the original, except for the slices marked by a thin line (Figure 4.4). If you change the tempo now, the slices will remain anchored to their bar/beat location, meaning the loop will change its tempo according to the tempo tracks or to the current tempo setting of the sequence. Notice that the name of the new audio part has changed into the original name of the file plus the extension "Sliced." This is because the new audio part is a special type of audio event that doesn't follow all the same rules of regular audio files.

CSX gives incredible control over the parameters of each slice. To edit a sliced audio part, double-click on it. The new audio editor (which is different from the waveform editor we use for "nonsliced" audio parts) allows you to control the start/end positions, volume, fades, mute, and name of each individual slice of a loop. In order to change these parameters, simply click on a slice, select the parameter you need to alter in the ruler above the waveform, and change its value.

By once again using the Beat Detective function available in PT, we can create independent regions from a single audio loop based on transients. These regions will then be "conformed" to the current tempo map in order to adapt the tempo of the original loop to the changes inserted in the tempo map. The first step is to create the "slices" according to the transients of the audio loop. This procedure is identical to the one involved in the groove extraction method (Section 4.2.3). Here's a brief summary.

Select the region of the loop you want to "slice" (make sure you select an exact number of bars and beats), open the BD window (found in the Windows menu), set the start/end positions and meter, select the algorithm you want to use (High or Low Emphasis), and click on "Analyze" to generate the slices. When the beat triggers are set (you can edit them using the Sensitivity slider), click on the "Separate" button to create individual audio regions based on the triggers. The "Trigger Pad" option allows you to insert a very short time delay between the beginning of each region and its "Sync Point" (meaning the actual marker that will be anchored to a bar/beat position) in order to guarantee that the attack of the transient is preserved. If you want to adjust a region to match the current tempo marker stored in the tempo track, simply choose the new separated audio regions in the Edit window, recapture the selection by inserting the start/end locations and the time signature of the regions selected, and click the "Conform" button. You can also choose among several parameters that allow you to decide which regions are going to be affected ("Exclude Within" parameter) and how much they are going to be affected ("Strength" parameter) by the "Conform" command. The "Swing" option lets you decide how much swing feel you want to apply to the selected region before conforming it.

In LP you can generate audio regions manually by using the *scissor tool*, just as we learned to do in DP. You have to import or record an audio track first, and then simply select the *scissor tool* from the tool palette and create the new regions by clicking on the audio file. The new regions will automatically stay linked to their position in bars and beats, so when you change the tempo they will automatically update their position to adapt to the new tempo. You can select the grid value at which the regions will be created from the Transport window by setting the "Display Format" parameter (the value visible under the "Time Signature" field in the Transport window). To quickly create multiple regions, hold the *option* key while clicking; this will create multiple regions for the entire length of the sequence you are working on based on the first rhythmic subdivision you choose.

Whereas the procedure I just explained is "grid based," another way to have LP separate regions automatically based on the transient of a waveform is to use the "Strip Silence" function. This procedure can be used for several purposes in fact, but in this section we are going to use it mainly to separate a sequence in several regions. The "Strip Silence" feature works like an offline gate processor, where audio material below a certain dynamic threshold will be cut. In fact the final result will be the original continuous audio file split into several regions, with the splits inserted where the signal was below the set threshold. You can take advantage of this function by opening the Audio Window from the Audio menu, selecting the audio file you want to strip silence on, and choosing the "Strip Silence" function in the Options menu. In the window that appears you can change parameters that allow you to fine-tune the gate function (Figure 4.6).

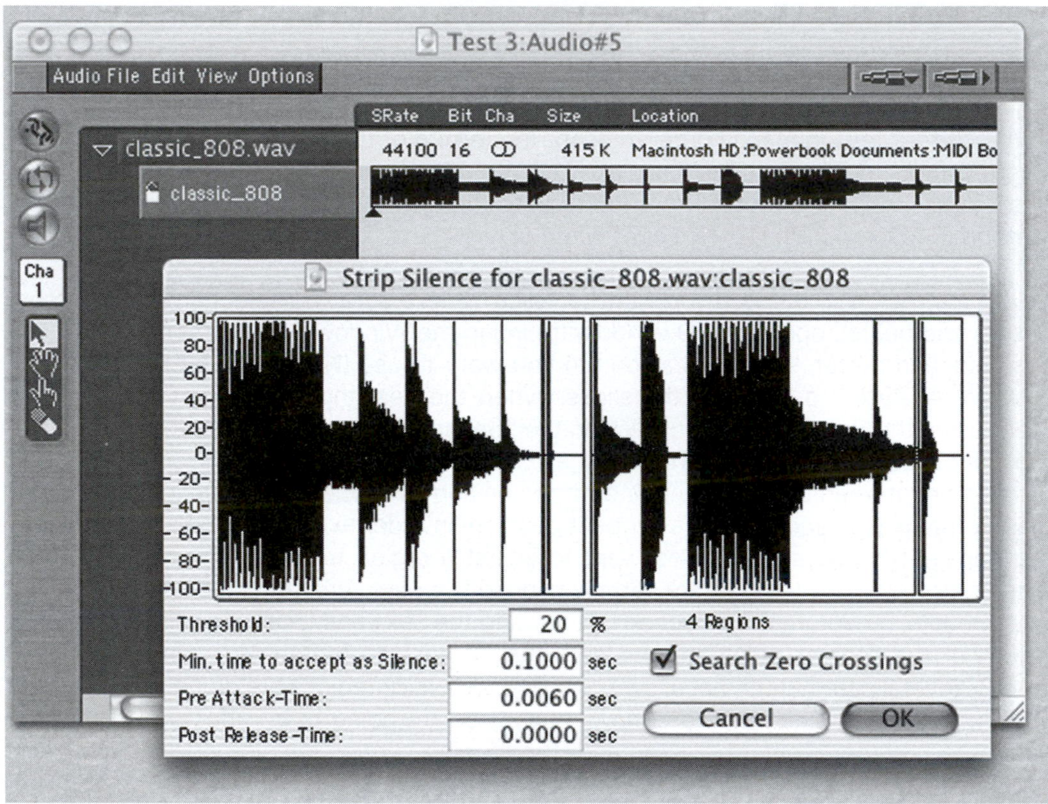

Figure 4.6 The "Strip Silence" function in LP.

Let's take a look at how these parameters affect the way the audio file is gated. The threshold sets the level (in percentage) that the amplitude of the signal has to pass in order to be kept as part of the region. This parameter has the largest impact on the actual separation of the regions. Low values will consider as "silence" only the quietest passages. On the other hand, if you select a high percentage value, you will gate most of the audio material. Usually I recommend values between 4% and 10%, depending on the dynamic characteristics of the audio file you are working with. The parameter called "Min. time to accept as silence" allows you to control how long a section has to be under the threshold in order to be considered silence. The Pre-Attack and Pre-Release times give you control, respectively, over how fast the silence will be stripped after the signal goes below the threshold and how fast the audio will be resumed after the signal goes back above the threshold. The "Search Zero Crossing" instructs the computer to look always for the samples on the Zero Crossing line to determine where to create the new regions, in order to avoid unpleasant digital clicks and pops. While the "Strip Silence" function can do the trick in separating several regions from a continuous audio file, I find it a bit more complicated and less effective than the functions we saw in CSX and PT TDM. The "Strip Silence" option works best with simple rhythmic loops, where the separation between the various percussive instruments is very clear. For more complex loops I recommend using a specialized application such as Recycle.

4.3 Advanced editing techniques

In Chapter 2 we learned basic editing techniques for both MIDI and audio tracks, such as straight quantization, basic graphic data editing, loop import, and basic track automation. In Chapter 3 we moved to more complex editing techniques, such as the drum editor, complex tempo changes, track layering, and groove quantization. In the next section of this chapter we will learn advanced editing techniques that can greatly speed up your work and also improve the quality of your productions. Such techniques vary from application to application, since each sequencer usually covers certain areas better than others. In this section I present each tool or technique with a general description and then go into details regarding how to use it and its practical application for each sequencer.

4.3.1 Advanced MIDI editors

Advanced MIDI editors can be used in several ways to either speed up your work or make it more flexible. I like to distinguish between two types of advanced MIDI editors, based on how they are applied and used. The first category includes editors that are not applied in real time but that need an "offline" processing time. The second category includes editing tools that can be applied in real time as "MIDI inserts" on a MIDI track or as a filter applied to the MIDI data before they are sent out to the devices. Let's take a look at the tools available in each category and how to take advantage of their features.

4.3.2 "Offline" global MIDI data transformers

While so far we learned how to edit MIDI data with a "micro" approach (meaning targeting a small number of MIDI messages or restricted regions), there are often situations (especially when you are under a tight deadline) where you need to process large sections of MIDI data or large regions at the same time. For this type of situation we need to be able to process the data with a "macro" editing approach; that is we need to be able to "batch-process" a series of data based on filters and conditions set in advance. Depending on the application, these types of "global" editors are sometimes called *logic editors* or *transformers,* or they simply assume the name of the specific functions associated with them, such as "Change Velocity" or "Change Duration." Behind these editors is the assumption that you want to quickly affect with a single command all the data or an entire category of messages contained in a selected region. Depending on the application and the type of editor you use, you can set filters and conditions to establish which data will be affected by the action and then choose the type of action you want to apply to the screened data. For example, let's say that we want to limit the velocity of all Note On velocity data recorded on a MIDI track in a region that includes measures 1 through 25 and that the boundaries of the velocities have to be between 10 and 100. With a so-called "logical" editor we can translate our goal into logical conditions that, if met, will achieve the wanted result. In our example we would have to set the first condition to filter only the Note On velocity data, the second condition would set the boundaries between bars 1 and 25, while the third condition would move all velocities lower than 10 to exactly 10 and, vice versa, all velocities higher than 110 to exactly 110. The practical applications of these types of editors are endless. Their real-life sequencing functions range from a simple velocity change filter to a sophisticated data transformer or from a basic quantization tool to an advanced "reverse pitch and position" function. Let's take a look at how each sequencer implements and integrates the logic editor capabilities.

In LP the logic editor, called the "Transform window," can be accessed from the "Window" menu by choosing the "Open Transform" option. In order to program it, after selecting a track or tracks, open the "Transform" editor (Figure 4.7). The window on which this editor is based is divided into two horizontal sections. The upper section (named "Select by Conditions") is where we set the conditions under which the computer selects the data to edit. The lower section (called "Operations on Selected Events") establishes the action that will be applied to the data that met the previous conditions. As you can see, it is very logical, and it seems more a mathematical tool than one for a musician's environment. Nevertheless this editor represents an extremely valuable resource for your sequencing projects.

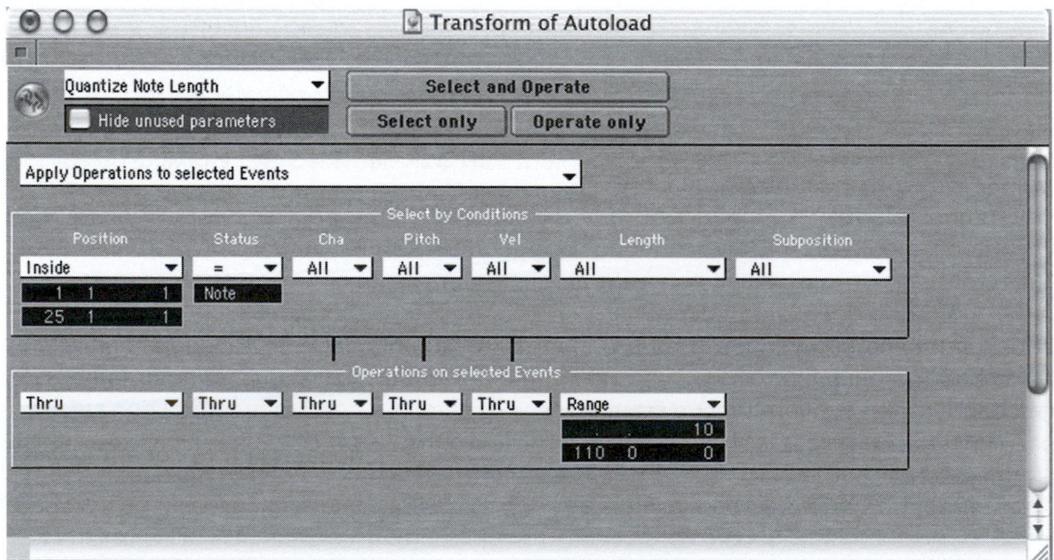

Figure 4.7 The "Transform" editor in LP.

LP's "Transform" editor is preprogrammed with useful presets in order to speed up some of the most common tasks. In the drop-down menu in the top left corner of the window you see the list of such functions (e.g., quantize note length, crescendo, half speed, double speed). If you want to program your own filters and actions choose the "Create User Set" option instead of selecting a preprogrammed function. This will clear all the fields in both sections of the window. The next step consists in setting the conditions under which the computer will select the events to be edited. Change each parameter (position, status, channel, etc.) according to the type of edit you want to achieve. In Table 4.2 you can find a quick explanation of each parameter.

Table 4.2 The "Transform" editor parameters in LP

Parameter	Function	Comments
Position	Main position of the event	
Status	Type of MIDI message	
Cha	MIDI channel	
1	First data byte value	In the case of a Note message it indicates the note number In the case of a CC it indicates the controller number
2	Second data byte value	In the case of a Note message it indicates the velocity In the case of a CC it indicates the controller value
Length	Length of the event	
Subposition	Positioning of the event within a measure	

The conditions of each parameter can be changed according to logical expressions such as equal, greater, smaller, inside, etc. After you select the type of data you want to be affected by the operation, move to the lower part of the window and choose the action you want to apply to the filtered data. Here you can choose among several operations, such as min, max, randomize, and fix. Refer to Table 4.3 for a more detailed explanation of some of the most common operations. Keep in mind that each operation affects the type of data located directly above. If you want to leave a certain type of data unaffected, simply set its operation to "Thru."

Once you have selected both the conditions and the operations, you are ready to apply the setting to the data. If you click on "Select and Operate," both actions will be taken. If instead you prefer to use the "Transformer" as a simple filter tool to choose certain types of data, you can click on the "Select Only" button. On the other hand, if you want the operation to be carried out on data and events selected manually, choose "Operate Only." A final option allows you to choose what to do with the edited data after the selection/operation process is executed. Use the drop-down menu to apply the operation to the selected events, apply and delete the unselected events, delete the selected events, or apply and copy the selected events to the clipboard. Keep in mind that after you create a new set in the "Transformer" it will be available from any editor window through their "Functions" menus.

In LP the logic editor is extremely powerful, I highly recommend using it regularly to speed up your work. Here is a practical application of the logic editor. Let's say we want to

Table 4.3 List of operations available in the "Transform" editor in LP

Operation	Function	Comments
Thru	Leaves the event unaltered	
Fix	Sets the event to the value indicated	If applied to the "Status" event, you can change the type of event and choose among a list of MIDI messages
Add/Sub	Adds the indicated value to or subtracts it from the original value of the event	
Min/Max	Replaces the events that are smaller or bigger than the set value	
Mul/Div	Multiplies or divides the events by the indicated value	
Range	Same as a combination of Min/Max	
Random	Creates a random value between the limits indicated	
+/− Rand	Adds a positive or negative value between zero and the indicated value	
Quantize	Quantizes the events to a multiple of the indicated value	Qua&Min is similar but with a combination of the Min operation
Exponent	The events are changed according to an exponential curve whose shape depends on the indicated value	Positive value = exponential curve Negative value = logarithmic curve

smoothly program the velocity data of all the close hi-hat notes in our drum track from a value of 1 to a value of 127 from bars 33 to 48. If we had to do it manually it would take a long time to make sure that every close hi-hat note (F#1) is processed and to calculate perfectly the ramp for each note. Using the logic editor instead, we can do it in a matter of minutes. The settings for this particular example are reproduced in Figure 4.8.

A similar editor to the one described in LP can be found in CSX. The so-called "Logical Editor" shares many of the concepts we learned about the "Transformer" in LP. In order to use this editor you have to first select a MIDI part in the main arrange window and then choose "Logical Editor" from the MIDI menu. The window that appears is very similar to the one we used in LP's "Transformer." The drop-down menu in the top left corner allows you to select which generic category of action the editor will apply to the data—for example, you can choose to delete the filtered data or copy them or transform them. The "Logical Editor" window has two different sections. The upper section allows you to create the filters and conditions that select which data will be passed to the operation; the lower part of the window hosts the operation section, where you can decide the action to be applied to the data that met the conditions. For each of the two sections you can create multiple lines (using the "Add Line" button), all containing different parameters. This allows you to generate extremely complex filters. In Figure 4.9 you can see the "Logical Filter" set up to add a value of 127 (operation) to all the notes that have a velocity lower than 100 (condition). The idea again is to speed up bulk editing operations by avoiding individual selection of data and instead globally choosing to apply a certain action only to the data that need to be altered.

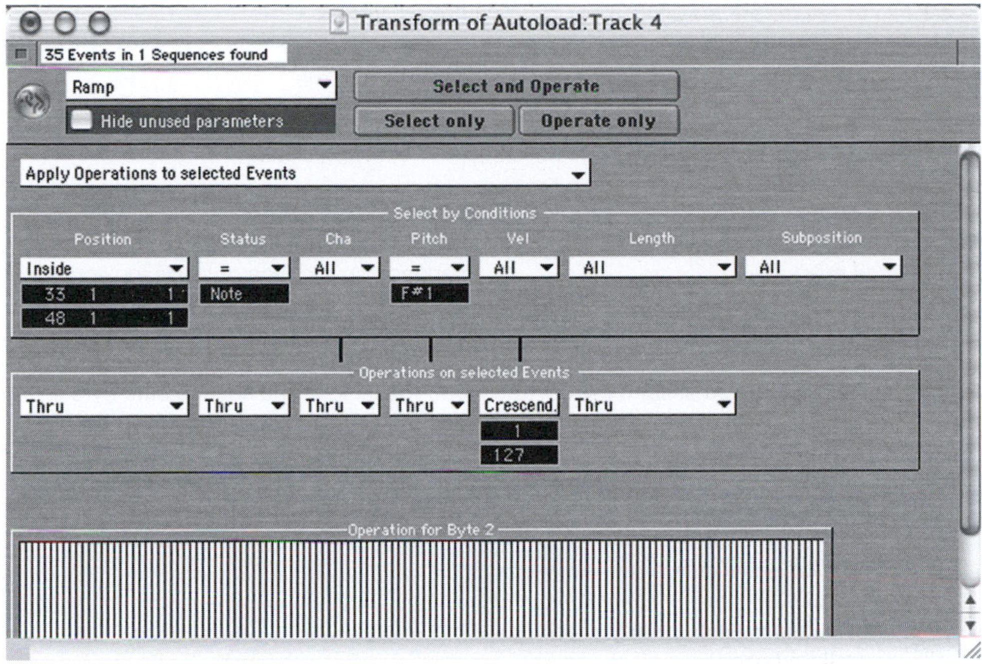

Figure 4.8 Practical application of the "Transform" editor in LP.

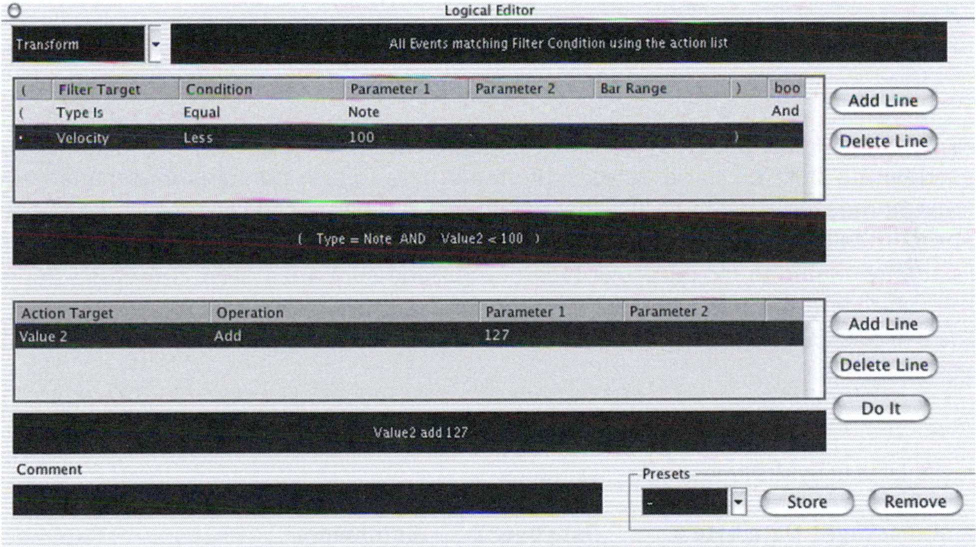

Figure 4.9 The "Logical Editor" in CSX.

For each line in the "condition" field you can set several parameters in order to define which data will be affected by the action. In Table 4.4 you find a short description of these parameters.

As you can see in Table 4.4, the MIDI messages in general can be filtered according to their so-called data bytes, meaning two of the major building blocks of a MIDI message

Table 4.4 List of operations available in the "Logical Editor" in CSX

Operation	Function	Comments
Position	Defines the range of the data in term of bars and beats	
Length	Selects the events based on their length	
Value 1	Represents the first data byte of a message	The meaning of the first data byte change depends on the type of message you are dealing with (refer to Table 4.5 for a list of Channel Voice messages and their data byte structure)
Value 2	Represents the second data byte of a message	The meaning of the second data byte change depends on the type of message you are dealing with (refer to Table 4.5 for a list of Channel Voice messages and their data byte structure)
Channel	Filters the data according to the MIDI channel on which they were recorded	
Type	The data will be filtered according to their type	For example, note, CC, aftertouch
Property	Filters the data based on their properties	For example, whether the data are muted or not
Value 3	Represents the third data byte of a message	This byte is seldom used

(along with the initial status byte). Most of the messages (especially the Channel Voice messages) are built on three bytes: the status byte (which contains information about the type of message and the MIDI channel on which the message is sent) and one or two (or sometimes more, depending on the type of message) data bytes, where only parameters strictly related to the particular type of message sent are stored. Therefore, depending on which message is sent, the data bytes assume different functions. In Table 4.5 you see a quick summary of the functions of bytes 1 and 2 for each Channel Voice MIDI message.

Table 4.5 Data byte/MIDI message assignments

Type of message	Data byte 1	Data byte 2	Comments
Note On	Note number	Velocity	
Note Off	Note number	Velocity	
Program Change	Program number	***	
Pitch Bend	LSB	MSB	
CC	ID number	Value	The ID number represents the CC number
Mono Aftertouch	Pressure value	***	
Poly Aftertouch	Note number	Pressure value	

Use the "condition" field to set the logical expression that will filter the data. The fields named Parameter 1 and Parameter 2 change according to the "Filter Target" choice, while the "Bar Range" option lets you choose the position range in which the filter and condition will work. If you use multiple lines to filter your data, you can choose between the two Boolean conditions "And" and "Or" to implement more complex conditions. Once the filters and conditions are set, you can move on and program the actions that will be applied to the filtered messages. To do so, simply select the "Action Target" (same options as the "Filter Target"), the mathematical operation in the "Operations" field, and the values that will be used by the operation (Parameters 1 and 2). If you prefer you can also program the "Logical Editor" by entering text-string instructions in the two fields located right below the two sections. Even though this option might look appealing to some MIDI programmers, I don't find it particularly interesting in terms of smoothness of workflow. Once you have set all the conditions and actions, you can apply them by simply clicking on the "Do It" button, or you can store them for later use by selecting the "Store" button. As we learned in LP, the "Logical Editor" is extremely powerful and can really be a lifesaver for quick and extensive editing tasks.

Some people might find the approach adopted by LP's "Transformer" and CSX's "Logical Editor" a bit intimidating, especially if they are new to the MIDI and sequencing environment. In fact, producers and composers sometimes try to avoid using such tools because they fear their unfamiliar interface will interfere with the creative process. If this is the case, there are other applications that give you some of the powerful features available in logical editors without their intimidating interface. In DP and PT, for example, we find not a real logical editor but instead a series of premade filters and tools that can achieve fairly complex results without any "programming" time. These tools are based on the principle of selecting a region of data (from any editor available) and then selecting one of the functions you want to apply. Among the most useful features available in DP that take advantage of this technology are "Change Velocity," "Change Duration," "Split Notes," and "Transpose." All four options can be found in DP's "Region" menu. Let's take a look at how to use these DP features.

The "Change Velocity" option allows you to alter the On and Off velocity of MIDI notes according to several criteria, which can be all edited and customized through a series of parameters. The velocities of the selected MIDI events can be changed using one of the following options: Set, Add, Scale, Limit, Compress/Expand, and Smooth. As you can see, most of these options are similar to the ones we analyzed in the logical editors of LP and CSX. The difference is that here the parameters are more easily accessible, since they are clearly laid out and have a specific function already assigned. In Table 4.6 you will see a detailed description of each function.

The "Change Velocity" function can be used in several ways to improve your projects. You can set it to quickly add or subtract absolute values ("Add") or relative values ("Scale") in order to lower a part without using volume automation—remember though that some patches are programmed to change their sonic features depending on the velocities of the notes, and therefore creating crescendos or decrescendos through velocity changes might affect the way your part sounds. I find the "Limit" function particularly useful. You will often find that the parts recorded will have a few notes with a higher velocity than the rest. While you could use the graphic editor to manually change their velocities, I recommend applying

Table 4.6 The parameters of the "Change Velocity" function in DP

Function	Description	Comments
Set	Sets the velocities of all selected MIDI notes to the value specified	Use this option if you want to flatten out a performance to increase its mechanical and drum-machine-like effect
Add	Adds the specified value to all the velocities of the selected MIDI events	Use this function to intensify (positive values) or diminish (negative values) how loud the selected notes will play
Scale	Similar to the previous function, except here the velocities can be reduced or increased by a percentage value	The velocities will be increased for values above 100% and decreased for values lower than 100%
Limit	Sets a low threshold and a high threshold; all velocities higher or lower than the thresholds will be moved to the values specified	This option is particularly useful if you have a region with some extremely high or low velocities that stick out compared to the majority of the other velocities
Compress/ Expand (Figure 4.10)	Allows you to limit the dynamic range of your region (compress) or to increase it (expand)	It works in a similar way to an audio signal compressor. If the velocities go over the threshold, then their values will be reduced according to the Ratio parameter (1:1, no compression; 8:1, maximum compression). If the Ratio is set to a value between 1:2 and 1:8, the effect works as an expander, meaning that if the velocities go over the set threshold, their values will be increased according to the Ratio. The Gain allows you to raise the velocity values of all the selected MIDI events in order to reestablish the original volume of the track
Smooth	Smoothly sets and changes the velocities of the selected events between two values; you can choose absolute values (0–127) or relative percentage (1–999%)	This function can be effectively used to quickly create crescendos and decrescendos of velocities. The "curvature" parameter allows you to control the shape of the curve applied to the velocities: 0 = linear, positive values = exponential, negative values = logarithmic

the "Limit" function to reduce the dynamic range of the MIDI part. This will create a much more cohesive and solid MIDI performance. Use the "Compress" option to gently remove uneven velocities from your parts. The parameters will vary according to the specifics of the parts you are working on, but I recommend starting with mild settings (low ratio and high threshold) to gently limit the highest velocities and then slowly to increase the ratio and lower the threshold until you obtain the perfect balance.

By using the "Change Duration" function in DP you can automatically alter the "Duration" parameter of the MIDI events included in a selected region. The parameters used in this function are similar to the ones listed in Table 4.6. I recommend the "Change Duration"

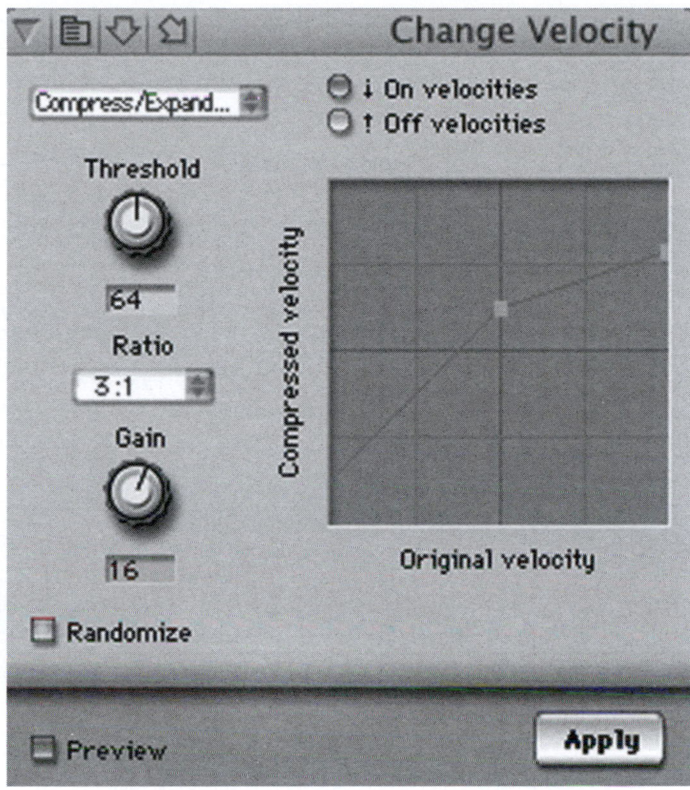

Figure 4.10 The Compress/Expand option in the "Change Velocity" function in DP.

option to quickly create legato or staccato passages. I find, for example, that when sequencing string parts it works well to apply a bit of legato articulation by selecting a region and changing the duration of the notes by using the "Extend Releases" option first (this will prolong the release of each note to the next attack) and then by using the "Add" option (use values between 10 and 20 ticks)—this will make each note overlap the following by exactly the amount of ticks you specified in the "Add" parameter. To create staccato parts, you can apply the same technique, but this time choose the "Subtract" option instead of "Add."

When I discussed the logical editors, you learned about functions that allow you to cut defined notes or events inside a selected region. A similar function, called "Split Notes," is available in DP. This editor, available from the Region menu, allows you to copy or cut selected notes from a track to another track (an already existing one or one created anew ad hoc). The targeted notes are selected according to several criteria that can be configured from the Split Notes window (Figure 4.11).

I recommend using the "Split Notes" function when, for example, you want to extract a single-note percussion instrument from a multi-instrument drum track. Simply select the note you want to cut, paste it to another track from the keyboard in the "Split Notes" window, select the operation you want to perform (cut), and select the target track from the "Send the note to" option. It also works great for cutting or copying outer voices in a polyphonic part: Use the "Top notes of each chord" or the "Bottom notes of each chord"

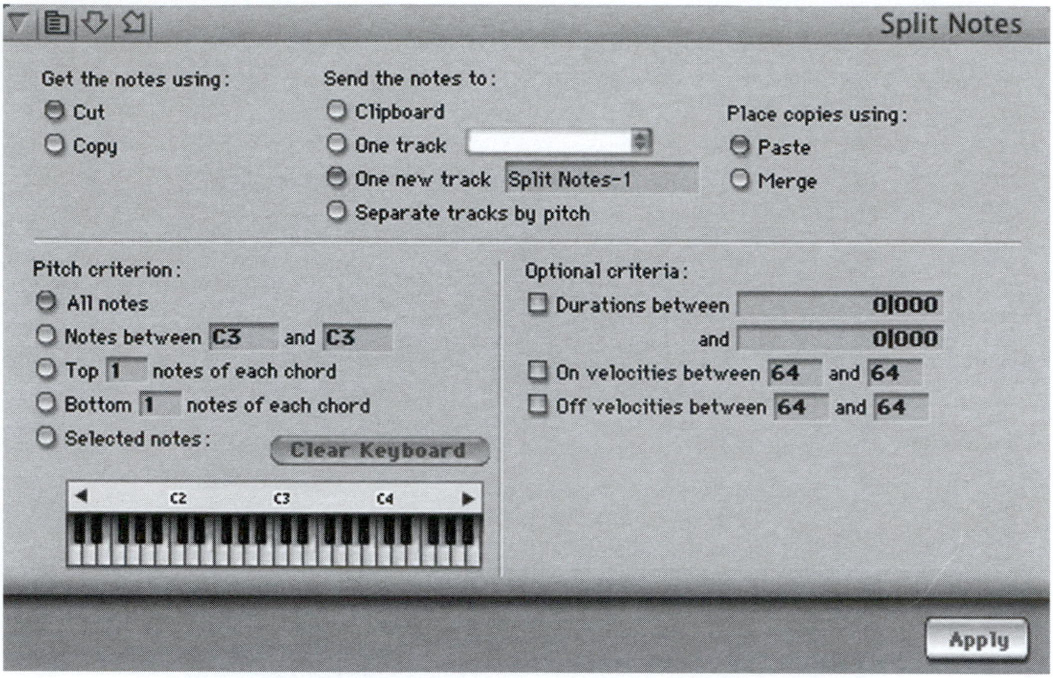

Figure 4.11 The "Split Notes" window in DP.

option to quickly select, respectively, the highest or lowest part in polyphonic tracks. Another very useful feature of the predefined logical functions in DP is the "Transpose" option (accessible from the Region menu). By selecting this function you can not only transpose the MIDI notes up or down in the selected region, but also harmonize them, meaning transpose the selected notes and keep the original pitches at the same time. The transposition can be made following a specific interval, diatonically according to a specific scale, from a specific scale to a target scale ("Key/Scale" option), and according to a custom map. I recommend this tool to quickly transpose extended regions or to try harmonized parts inside or outside a key without having to replay the harmonized voices.

PT uses a similar approach to DP. From the MIDI menu you can access functions such as "Change Velocity," "Change Duration," "Transpose," and "Split Notes." While these options give basic access to advanced editing tools, their parameters are much more simplified and streamlined than in LP or CSX.

4.3.3 "Real-time" MIDI effects

All the advanced MIDI tools I presented in the previous section are applied "offline," meaning that the computer calculates the function first and then applies the results by changing and inserting the new data in the track. This is a valuable way to alter the MIDI data, but it can be limiting if you are not 100% sure you want to commit to the result.

There are several ways to overcome the limitation of such a procedure. You can of course use the "unlimited undo's" function featured in the most advanced applications until you find the settings that satisfy you completely. Another way is to create an alternate take that is an exact copy of the original track and to keep it as a safety copy to which you can always revert in case you are not completely satisfied with the edits. There is another solution though that I often find the most convenient when I want to quickly experiment with edits such as transposition, quantization, and velocity change. This option is based on the strategy that instead of applying the edits "offline," you insert a so-called "MIDI filter" directly into the channel of the MIDI track you want to process. The filter is actually inserted right before the MIDI data are sent out to the respective cable and MIDI channel. The computer therefore alters the data in real time right before they reach the MIDI device. This technique has a few advantages and disadvantages, as shown in Table 4.7.

Table 4.7 Comparison of the advantages and disadvantages of the "real-time" MIDI effects

Advantages	Disadvantages	Comments
Settings are easily changeable		The parameters can be changed at any time, even during playback
Quick revert to original		You can bypass or delete the edit/effect at any time
Fairly good control over effects, even on MIDI tracks		A decent selection of MIDI effects (echo, delay, harmonizer, and arpeggios) is available
The edits/effects can be "printed" to track after the settings are considered satisfying		The "print MIDI effects" option lets you achieve the same result as the "offline" method after having experimented with several settings and parameters
	If the MIDI effects are used as inserts on a MIDI channel from the mix window, then settings are applied to the entire track; therefore to apply different settings to different sections or regions of the same track you have to create a new MIDI track and move the data on the newly created track, where you can apply a new MIDI effect with different parameters	CSX and LP allow you to apply effects not only through their mix windows but also through the Inspector window and the Parameters Box, respectively; this technique allows you to select different effects/filters for each part/ region
	The use of effects such as echo and multiple delays could cause "out of polyphony" problems	Since MIDI effects such as echo and delays are generated by retriggering the same MIDI notes over and over but with decreasing velocities, a device will use more notes than expected since each repetition will be calculated as a new triggered note

While the types of effects you use can vary from application to application, the procedure is very similar for all sequencers (PT is an exception since at the moment it doesn't allow you to insert MIDI effects on MIDI channels). The system is in fact very similar to the one used in the regular signal flow of an analog mixing board. In some cases (DP and CSX), the MIDI effects are inserted on the channel strip of the MIDI tracks from the Mixing Board window. Since the effects are applied in real time and inserted right before the MIDI Out of each track, you can change their parameters, bypass them, or delete them at any time. The actual effects available (and their parameters) are usually the same ones you can access from the "offline" list. To insert these MIDI effects, simply open the Mixing Windows of your sequencer and use one of the empty "insert" slots available on the channel of the MIDI track on which you want to apply the effect/filter. In DP, for example, you can insert MIDI effects such as arpeggiator, change duration/velocity, and echo (Figure 4.12).

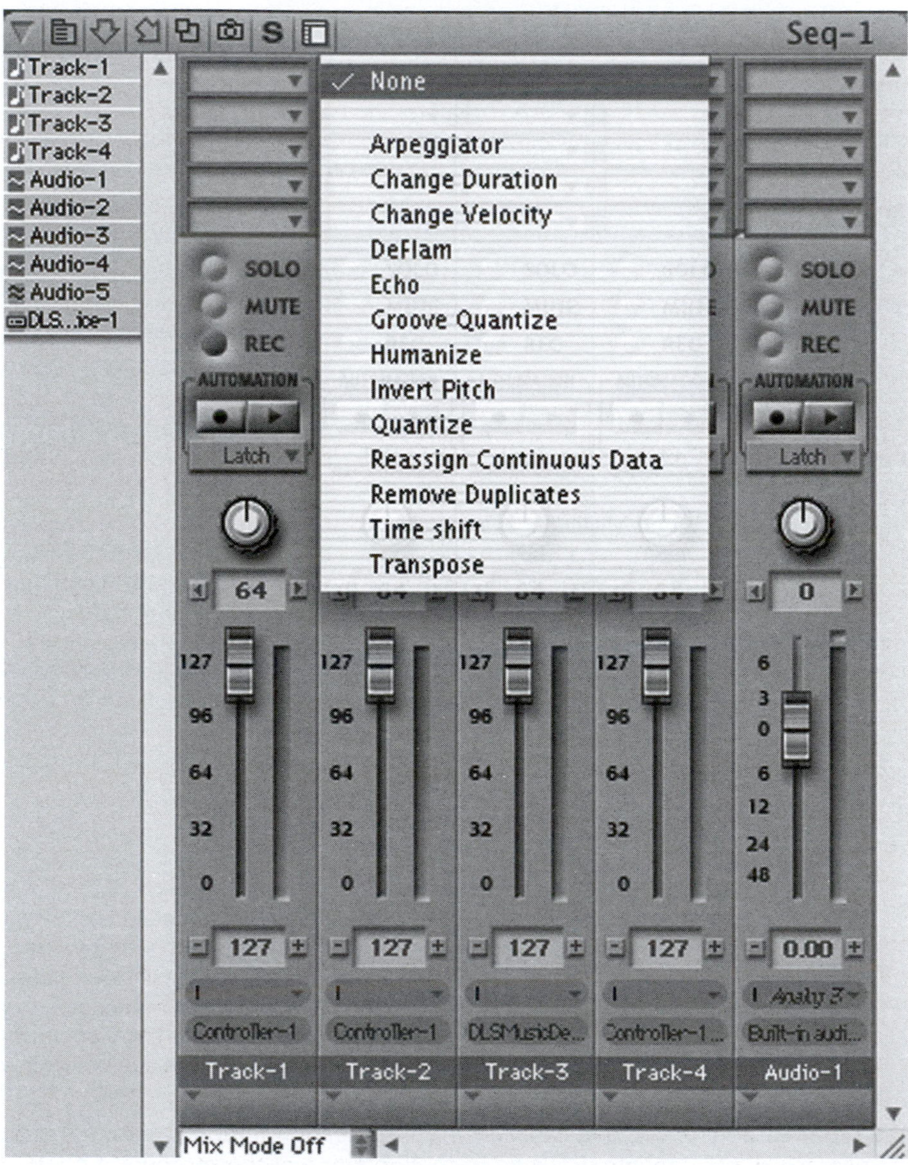

Figure 4.12 "Online" effects list in DP.

In the same way, in CSX you can insert a variety of very interesting real-time MIDI effects and also some more basic ones through the Insert slots available in the Mix window for each MIDI track. Among my favorite CSX real-time MIDI effects is the "Step Designer." This very interesting filter transforms the current MIDI track in a small step sequencer. You can program a total of 200 patterns that can be triggered in real time from a controller while the sequencer is playing. The notes generated by the step sequencer are sent out to the channel and cable to which the current MIDI track is assigned. This is a particular useful tool for live performance or for a creative recording of live patterns.

Another useful MIDI filter in CSX is the so-called "Chorder," which is a single-note-to-chord trigger. You can program the "Chorder" to trigger a chord by playing a single note from your MIDI controller. CSX also offers the possibility of using the "logical editor" as an insert on a MIDI track. The parameters available are the same ones we learned for the "offline" version. As I mentioned earlier, more basic effects can be accessed by selecting the Track Parameters tab in the Inspector window of each track. From here you can transpose the notes, alter their velocity (Vel. Shift), compress/expand the velocity of the notes (Vel. Comp.), and alter by compression/expansion their lengths (Len. Comp.). The last two parameters mentioned (Vel. Comp. and Len. Comp.) work as compressors if the nominator is smaller than the denominator (e.g., 1/2) and as expanders if the nominator is bigger than the denominator (e.g., 2/1).

The capabilities of LP in terms of MIDI effect insertion are as sophisticated as (if not more sophisticated than) the ones of DP and CSX. The best way to quickly filter a MIDI track through MIDI effects is to use the Sequence Parameter box (located in the upper left corner of the Arrange window) to insert filters such as transposition, velocity change, and delay. Use the "Dynamics" filter either to expand the difference between loud and soft notes (values higher than 100%) or to decrease the difference between loud and soft notes (values lower than 100%). Via the "Gate Time" option we can automatically create staccato effects (values lower than 100%) and legato effects (values higher than 100%). If you choose the *Fix* setting you will obtain a complete staccato, while opting for *Leg* will generate a complete legato effect.

Although these filters are fairly basic functions, in LP we can use the Environment editor to create incredibly complicated filters comparable to the ones found in CSX. To insert a new element in the environment assigned to your current project, select "New" from the Environment window and choose the object you want to create. Once the object is added to the environment, you have to make the right connections to have it connected to one of the MIDI devices or virtual instruments. To do so, simply *option-click* on the output arrow on the right side of its icon and select a destination device from the list of available outputs. To access the new effect, assign a selected track to this new virtual device from the Arrange window in the same way you would assign any MIDI track to an output. Some of my favorites functions accessible from the Environment editor are the "Arpeggiator" (Figure 4.13a) and the "Chord Memorizer" (Figure 4.13b). The "Arpeggiator" allows you to create sophisticated MIDI arpeggios by controlling parameters such as the direction of the arpeggio, its velocity, its range, the resolution/length of the notes, and its octave span.

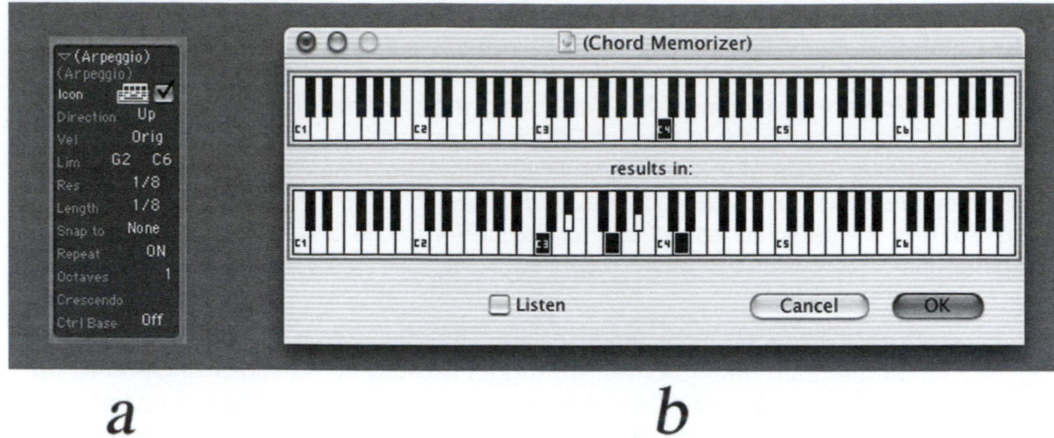

Figure 4.13 (a) "Arpeggiator" and (b) "Chord Memorizer" in LP.

The "Chord Memorizer," on the other hand, is designed to build complex chords that can be assigned and triggered by a single key or note. To edit the key/chord assignment, double-click on the "Chord Memorizer" icon and select the note that will trigger the chord (top keyboard) and the chord that will be triggered (bottom keyboard). Beside having some practical applications in a live environment, the "Chord Memorizer" can be very useful to sequence complex parts with complex chords, especially if piano is not your main instrument. Of course all the features available in the "Transform" editor can also be used to insert the same parameters we learned earlier.

4.4 Overview of audio track effects

Even though this book focuses on MIDI and its practical applications, I would like in this section to explain some important concepts about the use of effects applied to audio tracks. Since we just learned how we can use specifically designed effects with MIDI tracks, I believe it is also crucial to understand how effects such as reverb, delay, chorus, and compressor can be applied to audio tracks. While this section will cover the most important aspects of practical applications of audio plug-in effects in an audio sequencer, I encourage you to consult the excellent plug-ins guide by Mike Collins entitled *A Professional Guide to Audio Plug-ins and Virtual Instruments*, published by Focal Press.

The way effects are used and applied to audio tracks in a sequencer is very similar to how we use them in a more traditional analog-mixing environment. In effect the signal flow is the same. But instead of using external effect units, hardware patch bays, and cables to connect the devices and route the signal, we use only virtual connections created inside the application. In general there are two separate ways to add an effect to an audio source. The first is called *insert*; the second is called *send*. While they both have the result of adding some sort of effect to a *dry* (meaning without effect) signal, they differ substantially in their nature and applications. Let's learn more about these two techniques.

4.4.1 *"Insert" effects*

The simplest and often quickest way of adding an effect to an audio track is by directly using an "insert" on the audio track. This means to literally insert an effect in the virtual signal flow of the mixing board of your sequencer between the source (the actual audio material that is played back from the audio file stored on your HD) and the virtual "output" of the mixing board channel of the sequencer. By inserting the effect, you allow the entire signal of the track (meaning 100% of the signal) to be affected by the inserted effect (Figure 4.14). In a normal analog situation it would be the same as taking your guitar output and plugging it into the input of your effect unit and then taking the output of the effect unit and sending it to the mixing board. In this case the entire signal will be sent to the effect. Sometimes, depending on the effect you insert, you can still adjust the balance between the original dry signal and the effected wet signal.

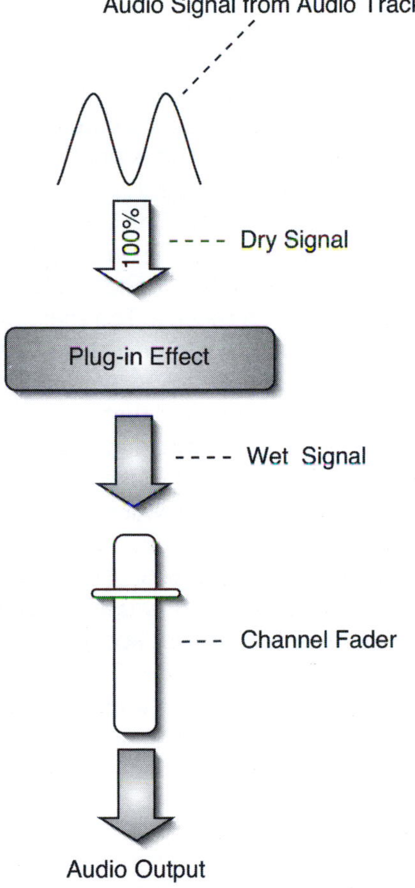

Figure 4.14 Effect of "Insert" on signal flow.

While this technique is quick and fairly easy to implement, it is used mostly for effects that require the entire audio signal to be processed. Such effects are mainly dynamic (such as compressors, limiters, gates, and expanders) and equalization effects. By using effects as inserts, you can speed up the process of adding effects without interrupting the creative flow. On the other hand, if you have to apply effects such as reverb, delay, and echo or other ambience effects, then the "insert" technique is not the most appropriate one, for a variety of reasons. The main one is exemplified by the high inefficiency of the "insert" technique. Imagine having a 32-track audio session where you need to apply reverb to all 32 tracks. By adding 32 reverb units as inserts you will choke the CPU of your computer, since reverbs are among the more CPU-hungry audio effects. In addition, if you want to change the overall ambience of the mix, you will have to modify the parameters of all 32 reverb effects, which is an incredible waste of creative time. Therefore for ambience effects (reverbs, delays, and echo), you should use a different technique to apply effects based on the "send" principle.

4.4.2 "Send" effects

The technique explained in the previous section is a valuable tool for effects that require the entire audio source to be affected, but there is a much more functional and efficient way to apply effects to audio tracks, based on the auxiliary "send" procedure. Keep in mind, though, that this technique is particularly indicated for effects you want to share among several tracks, such as reverbs, delays, and echo. The main idea here is that you will have a separate track (called *auxiliary track* or *effect track*, depending on the software you use) onto which you insert the effect (let's say a reverb). The auxiliary track receives the input from an internal channel path called a *bus*. On each audio track on which you want to apply the effect, you have to create a so-called "send" output and assign it to the same bus on which you set the auxiliary track to receive. The main difference from the "insert" technique we learned earlier is that here we can decide how much of the original dry signal is sent to the effect (and not only 100% like in the "insert" example); this gives us greater control over the balance between dry and wet signal (Figure 4.15). In addition we can share a single effect among several tracks by assigning each track's sends to the same bus. Each send will serve as an individual effect control for each track, while the fader of the auxiliary effect track (called the *effect return*) will serve as the general effect volume.

As I briefly mentioned before, the technique you use to apply effects to your project depends mainly on the type of effects you need to use. In Table 4.8 I list a series of common effects and the most frequent techniques for applying them to audio tracks.

On a practical level, the way the four sequencers handle the two techniques ("insert" and "send") are very similar. If you want to insert an effect on an audio track, all you have to do is open the mixing board window and click on the insert field of the channel strip of the audio track. After clicking on an empty "insert" slot, a list of all the plug-in effects available for your sequencer will appear.

If you want to use an effect as a send in DP and PT, you first have to create a so-called aux track. DP and PT use, respectively, the Project menu (select "Add Track" and then "Aux Track") and the File menu (select "New Track" and then "Mono" or "Stereo Aux Input") to create a new aux channel. Next assign the input of the aux channel to a bus (usually you want to choose the first available one) by clicking on the Input of the aux track and selecting a bus as Input. Insert on one of the available "Inserts" of the aux channel the effect you want to use and share among the tracks on the aux channel (e.g., Reverb). In DP and

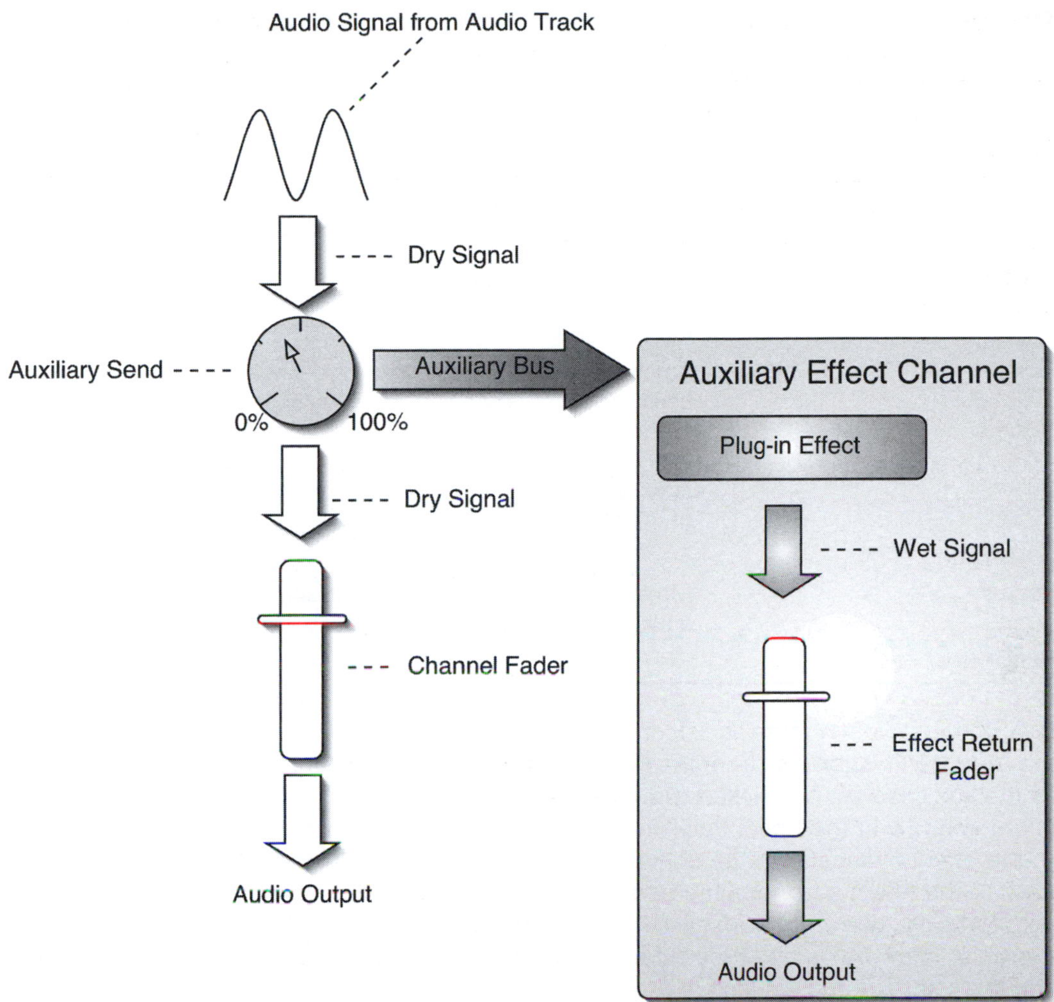

Figure 4.15 Effect of "Send" on signal flow.

PT the inserts are located at the top of the channel strip of each channel in the mix window. As a final step, assign the output of an available "send" of the tracks on which you want to apply the selected effect to the same bus number on which the aux channel is receiving (in DP and PT the sends are located below the inserts in the channel strip of each track from the mix window); now by raising the send level you can control the amount of dry signal that is sent through the bus to the aux channel, meaning to the effect. Use the fader of the aux channel to control the overall volume (return) of the effect.

In LP the auxiliary channels are usually present as default as part of your working "environment." Therefore choose an empty aux track, insert a defect in one of the "inserts" slots, and assign the input (I/O) to a bus. From the tracks on which you want to add the effect, simply click one of the "sends" and assign it to the same bus on which the aux channel just created is receiving. Next to the send you just assigned, a small rotary knob will appear, allowing you to determine the amount of signal sent to the bass/aux.

Table 4.8 Common application techniques for effects. "X" identifies the correct use

Type of effect	Insert	Send	Comments
Compressor	X		Any dynamic effect needs to be inserted in order to affect 100% of the original signal
Limiter	X		
Gate	X		
Expander	X		
Equalizer	X		The Eq. also needs to be inserted on the channel strip to filter the entire original signal
Reverb		X	Ambience effects, such as reverb, delay, and echo, are usually used as "send," since they are often shared among several channels at the same time. You can actually use them as "insert" (which I don't recommend) and control the dry/wet ratio by using the "Mix" control directly from the control menu of the effect
Delay/Echo	X	X	
Chorus	X	X	Modulation effects, such as chorus and flange, can be used equally well as "insert" or "send"
Flange	X	X	

In CSX you have to create an Effect track by selecting the option "FX Channel" located in the "Add Track" submenu from the Project menu. Once the channel is created, it will serve as the aux channel. Now insert the effect you want to use as "insert" through the mixing board window in the Effect track you just created. CSX automatically assigns the busses to the effect channels, so the only thing you have to do is assign the send of each audio track to the effect channel just created (as I mentioned CSX handles the bus assignments automatically; therefore from each "send" slot you will see a list of all the effect channels available, with the name of the effect inserted for each aux). In CSX you can alternatively see the inserts and the sends of the channels in the mixing board by selecting them from the information window located at the extreme left of the mix window.

A good and efficient management of the effects is crucial for a balanced and professional-sounding project, so I highly recommend you spend time getting familiar with the two techniques I have described ("insert" and "send"). Use your sequencer of choice as a starting point, and if possible try to experiment also with other sequencers in order to really master the art of effect handling. Also remember that effects in general are never the solution to badly orchestrated or poorly conceived music. Always try to see if your mix can be improved first by changing some aspects of the instrumentation. Effects are nice additions (and mostly final touches) that can help your projects to sparkle and shine.

4.5 Working with video: importing and exporting QuickTime movies

One of the many advantages of composing and orchestrating with an advanced audio/MIDI sequencer is that you can easily work with video material without resorting to fairly complicated synchronization techniques or "mysterious" synchronization protocols such as SMPTE (a technique I will describe in the next section). As I mentioned earlier, nonlinear devices are easier to synchronize because they store the data on "random-access" media. While video devices used to be tape based and therefore "linear," nowadays nonlinear video editing suites (computer based) are pretty much the standard in the professional (and even in the consumer) environment. This transition brought great flexibility not only to video editors, but also to composers and orchestrators who regularly work on projects involving scoring to video.

These days it is possible to import a scene, episode, or entire digitized movie to your sequencer and score directly to the picture without the need of external video devices, synchronizers, or complex synchronization protocols. Once you have a digitized movie (meaning a digital version of the original video source), you can easily import it to your sequencer and "anchor" the video tracks to the conductor track of your sequencer. The sequence and the video material will always stay in sync, since they are both played by the computer (which acts as virtual master for both sources) and "streamed" from the computer's HD. Another big advantage of a digitized video source to score to picture is that you will save a huge amount of time. Since the digitized video will behave as a nonlinear device, you won't experience delays related to fast-forwarding and rewinding of the tape that usually occur with linear video devices. One of the key issues, of course, is to be able to obtain a digitized version of the video (movie, documentary, commercial, etc.) you have to score. Because of the increasing use of nonlinear editing stations, you should be able to get a digital copy of the video directly from the video production company involved with the project. While composers working with audio and MIDI sequencers think that nowadays it should be standard procedure to obtain a digitized video, unfortunately not all video editing and production companies seem to share the same view. This means that more often than you wish you will end up with your video on a VHS tape without SMPTE sync (more on this later). In this case you will have to digitize it yourself. But let's take this one step at the time and see what options you should consider in case the video production company agrees to provide a digital copy of the video material.

Depending on the length of the project, you have a few options in terms of formats and compression settings of the video source. The best video format in which to have the source delivered to you is Apple's QuickTime (QT) movie. This format has really matured during the last 2 or 3 years; its video digital format features very good quality even at high compression rates. QT is preinstalled in every Macintosh computer, and, even better, all major sequencers support it, making it the de facto standard for composers and producers as the nonlinear temp format for scoring to picture. If the video production company agrees to deliver the video in QT format, then most of your problems are solved. The only thing you have to do is copy the video files onto your session HD (I recommend keeping them in the same directory as the project into which they will be imported) and select the import movie option in your sequencer (more details on this in a little bit).

Keep in mind that the size and resolution of the QT movie has an impact on the performance of the CPU of your computer. A full-screen movie with a low compression ratio will considerably slow down your computer (especially if the CPU is not one of the latest generation), and consequently the performance of your sequencer will suffer too. I therefore recommend reducing the size and increasing the compression of the movie you import in order to avoid having to sacrifice tracks or plug-in. Usually I like to have a small version of the movie with a resolution such as 320 × 240 and a compression such as Cinepak, Motion JPEG A (or B), or MPEG-4. This allows a quick reference for the scoring without taxing the computer's CPU too much.

Sometimes the video production company can give you the movie in DV (digital video) format. This option has advantages and disadvantages. The good news is that if you use the DV format to import the video to your sequencer you will actually be able to output the video to a separate video monitor through the FireWire port of your computer via a regular FireWire-to-RGB converter. The bad news is that DV uses a larger bandwidth than compressed QT; in addition, you will be using bandwidth on the FireWire port that should be left for the session HD and eventually for the audio interface. If you have a computer with a dual FireWire bus, make sure the DV output and the HD/audio interface output are connected to two separate busses. If the video production company is not going to provide a digital version of the video, you will have to digitize it yourself. This operation used to require expensive video cards and very powerful computers. Nowadays, fortunately, you can easily convert an analog video source into digital stream using a simple digital video converter that takes the analog input from your VCR and converts it into digital data by outputting the video in DV format to your computer through a FireWire port. You can use basic video software to run the conversion, such as iMovie or FinalCut Express. Don't forget, though, to always store the converted movie into the working HD/directory of the project you will be importing the movie into. Once the video material is ready, you can import it to your sequencer of choice and start scoring to picture without worrying about any synchronization issue. Let's take a look at how each sequencer handles the QT movies.

All four sequencers are able to import QT movies and automatically synchronize them to the sequence, and they all offer a series of parameters for customizing the start point and the synchronization features available. To import a QT movie in DP simply go to the Project menu, select the "Movie" option, and browse to the QT movie you want to import. Once the movie is loaded it is automatically locked to the current sequence. From the "mini menu" of the movie window you can control parameters such as the size of the window (half, normal, and double size), the starting point of the movie (in relation to SMPTE time), if you also want the movie locked to the sequence in edit mode (both graphical and numerical), and the video output option. Keep in mind that this last option can be changed only if you imported a movie in DV format, in which case you will be able to output the video to an external video monitor through the FireWire port and an FW-to-RGB converter. In order to improve the performance of your computer I recommend keeping the size of the movie to "normal" or "half." Also try to maintain the original aspect ratio of the movie instead of altering it. Once the movie is imported to the sequence, it is very easy to spot it, set markers and tempo changes, and start scoring to picture. If the "Lock icon" is selected in the menu bar of the movie window, the sequence and the movie are synchronized, meaning that whenever you press Play (or Fast-forward, Rewind, and Scroll) on the sequencer or on the movie, the other will follow. If you want the two items to be independent, simply deselect the "Lock icon." Also make sure that the frame rate of DP matches that of the imported movie by using the "Frame Rate" option available from the Setup menu. A thumbnail view is available through the "Sequence Editor" window.

In LP you can import a movie by choosing "Movie" or "Movie as Float" in the Options menu. The difference between the two options is that the former opens a regular window with the movie in it, while the latter opens a so-called "floating" movie window that will always stay on top of any other window. The setting regarding the SMPTE start time can be changed directly in the lower part of the movie window. Make sure the frame rate of the project matches that of the movie by using the Synchronization window accessible from the Transport window (click and hold on the synchronization button). By *Command-clicking* on the movie itself you can access more advanced settings (Figure 4.16), such as the movie size, continuous sync (which guarantees a closer and more accurate synchronization between the movie and the current sequence), show/hide the parameters of the movie, and video/movie settings.

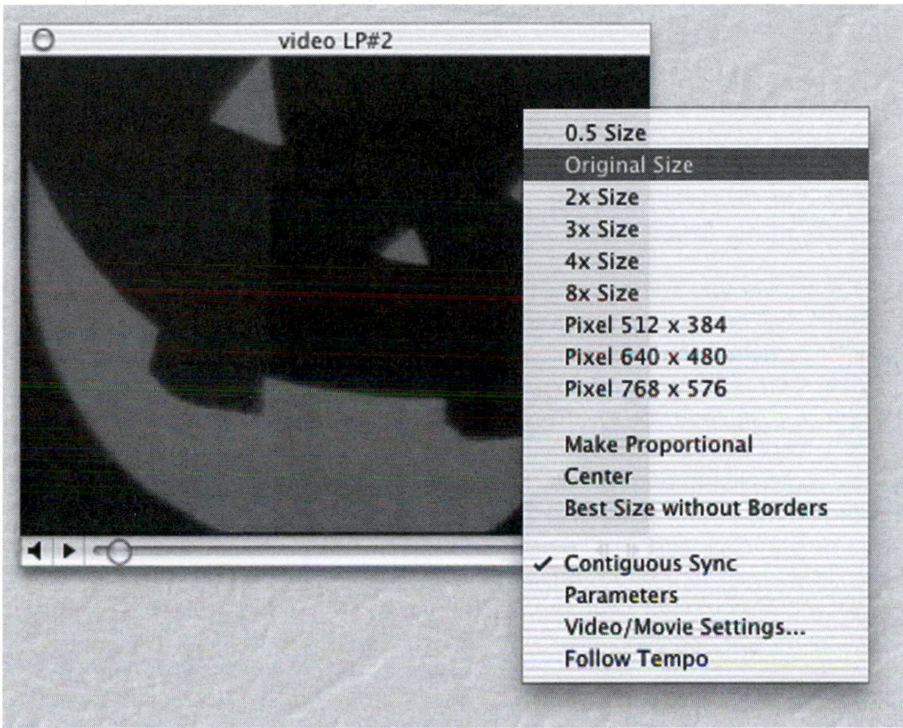

Figure 4.16 The video options in LP.

By selecting this last option you will be prompted with the video parameters window (you can also access this window by selecting the "Video" option under the "Song Settings" submenu found in the File menu). Here you can select several parameters related to the video output and the thumbnail settings. As we saw in DP, you can choose to output the movie to an external monitor through the FireWire port. The thumbnail function is very useful when you score to picture since it allows you to see a rendition of the video on a special track with poster-frames. This allows you always to have a constant visual representation of the relation between the sequence and the movie. To create a thumbnail video track, you can simply *click-and-hold* on an empty track's name and select the option "Thumbnail Video." The currently selected track will automatically be transformed into a thumbnail track. The latest version of Logic (Logic Pro 7) is also capable of exporting the movie along with the mix.

The same parameters and options are available in CSX, where, in order to import a video clip, you choose the option "Video File" found in the "Import" submenu in the File menu.

Once the video is imported, you can bring a floating window to the front by selecting "Video" under the Devices menu. In order to set more advanced parameters, such as video size, video output (including the FireWire output option), and the frame offset of the movie, you have to select the "Device Setup" option first (found in the Devices menu) and then select "Video Player." I really like how CSX automatically creates a thumbnail track and an audio track with the audio material of the video. This speeds up the workflow and is less distracting during the compositional process. As with the other sequencers, make sure the frame rate of the current sequence matches that of the imported movie. To do so, select the "Project Setup" option from the Project menu and change the frame rate parameter according to the movie you imported.

PT handles videos particularly well compared to the other applications, the main reason being that in PT not only can you import a video, but in fact you can also export the video with the audio track you mixed and created inside PT. This is an invaluable tool for the film scorer and composer, since most of the time you are required to deliver a temporary version and mix with the original picture in order for the producer to assess how well the music works with the video.

To import the movie you will be scoring, select "Import Movie" from the Movie menu. A thumbnail track will be inserted automatically in the edit window. Make sure the frame rate of the project matches that of the movie you imported by selecting the "Show Session Setup" option from the Window menu. This action will not automatically import the audio track of the movie; instead you will have to import the track separately by selecting "Import Audio from Current Movie" in the Movie menu. Keep in mind that the new audio material is added to the region list (the right side of the edit window) but not automatically in an audio track. You can create a new audio track and manually insert the movie's audio track from the region list. Pay particular attention to position the extracted audio track exactly at the beginning of the sequence, where the movie is located as default; otherwise you will run into synchronization problems between the audio track and the video. Parameters related to the movie, such as its size and sync offset, can be accessed from the Movie menu.

As I mentioned earlier, one of the most useful features of PT related to video tracks is that you can bounce to video directly from PT. Once you have your final project edited and mixed as you like, by selecting the option "Bounce to QuickTime Movie" from the Movie menu you can create a new video in QT format with the entire audio material recorded in your project. The parameters you have to set after selecting this option are the same as those of a regular bounce-to-disk operation (output source, sample rate, format, and resolution), with the exception of the file type, which is automatically set to QT video. Of course, the bounce function in general applies only to audio tracks; MIDI tracks need to be converted to audio to be included in the bounce. This process is explained in Chapter 6.

4.6 SMPTE: synchronization of linear to nonlinear machines

In Chapter 3 we learned how to synchronize nonlinear machines to nonlinear machines. This type of synchronization is fairly simple to set up since it relies on MIDI messages, such as Midi Time Code and MIDI Clock, to establish and maintain the synchronization among two or more machines. This is possible since the majority of nonlinear machines are equipped with a MIDI interface. Therefore by using the MIDI network we can exchange synchronization information between the devices in our studio. When we deal with linear machines, on the other hand, things get a little more complicated. The absence of a MIDI interface and the fact that we can't record or play back MIDI information directly onto and from a tape makes the synchronization of such devices less straightforward.

There are several practical applications that derive from the synchronization of linear machines to nonlinear devices. For example, if you really like the sound of analog tape and don't want to record your audio material directly to HD, you can synchronize a reel-to-reel analog tape machine with your digital sequencer—this setup allows you to run MIDI tracks (from the sequencer) along with analog audio tracks (from the reel-to-reel machine). Another application involves the synchronization of video machines such as VCRs with your audio/MIDI sequencer. As we saw in the previous section of this chapter, when we score to picture, in the ideal conditions we will be provided with a digitized version of the video. If this is not the case, however, and the only video format we have is a videotape, we will have to use "old-fashioned" synchronization tools, such as SMPTE code, in order to synchronize our VCR to our audio/MIDI sequencer. Let's take a look at this technique and at the steps involved in this procedure.

In the case of nonlinear machines, we learned that MTC allows us to synchronize several devices by describing the passing of time in the format *hh:mm:ss:ff*. Unfortunately it is not possible to record directly the MIDI data onto tape; therefore an adapted version of this format needs to be found to be able to record it to tape and subsequently synchronize the machines. SMPTE code (Society of Motion Picture and Television Engineers) does exactly this: It is an "analog" translation of the MTC, which can easily be recorded to tape. This code was in fact created in 1967 to make the editing of videotape easier, so to be precise we have to acknowledge that MTC is in fact a digital translation of SMPTE in MIDI messages. The bottom line, though, is that SMPTE is able to encode the digital format to tape by modulating (using a "biphase-marking" system) the 1 and 0 of the digital signal into a square wave (Figure 4.17). The frequencies of the wave depend on the frame rate, and they range approximately from 2400 Hz to 1000 Hz.

Figure 4.17 Linear graphic representation of SMPTE code.

The format and frame rates of SMPTE are the same as those we analyzed when we discussed the MTC protocol. In order to synchronize devices using SMPTE, we first have to record the SMPTE code onto an empty audio track of the linear machine. In the case of a multitrack tape recorder, choose one of the tracks placed at the edge of the tape, to reduce the "bleeding" of the SMPTE code onto adjacent tracks. SMPTE code has a very piercing sound and therefore can easily spill onto other tracks on tape (listen to Example 4.1 on the audio CD for an example of what SMPTE code sounds like). The operation of recording an

SMPTE signal onto an empty audio track is referred as *striping*. To stripe SMPTE code into an empty audio track you need an SMPTE generator, which is usually part of the most sophisticated multicable MIDI interfaces. On this type of interface, in addition to the regular MIDI In and Out ports, there are two extra audio connectors (normally labeled SMPTE In and SMPTE Out) used, respectively, to read and write SMPTE code. Before starting to record any audio track related to your project, you should stripe SMPTE for the entire length of the tape.

For example, let's say you have an analog 8-track reel-to-reel recorder you want to synchronize to your sequencer and MIDI studio (Figure 4.18). First you record the code onto track 8 of the analog machine for the entire length of the tape. Depending on which SMPTE generator/MIDI interface you have, you usually can decide the starting time of the code either through software or through parameters accessible from the front panel of the interface. It is a good idea to stripe starting not from 00:00:00:00 but from 00:59:00:00. This assumes you leave 1 minute "open" before the starting of the project, which will occur at approximately 01:00:00:00. This is recommended for two main reasons: (1) It is better to have a little bit of lead-in time (1 minute is enough), since it will take a few seconds for the machines to synchronize together. (2) Synchronizers (especially older ones) sometimes have a hard time recognizing a starting time of 00:00:00:00. After the SMPTE code is striped, you have basically "time-stamped" the tape so that you can now assign specific location times for every position of the tape.

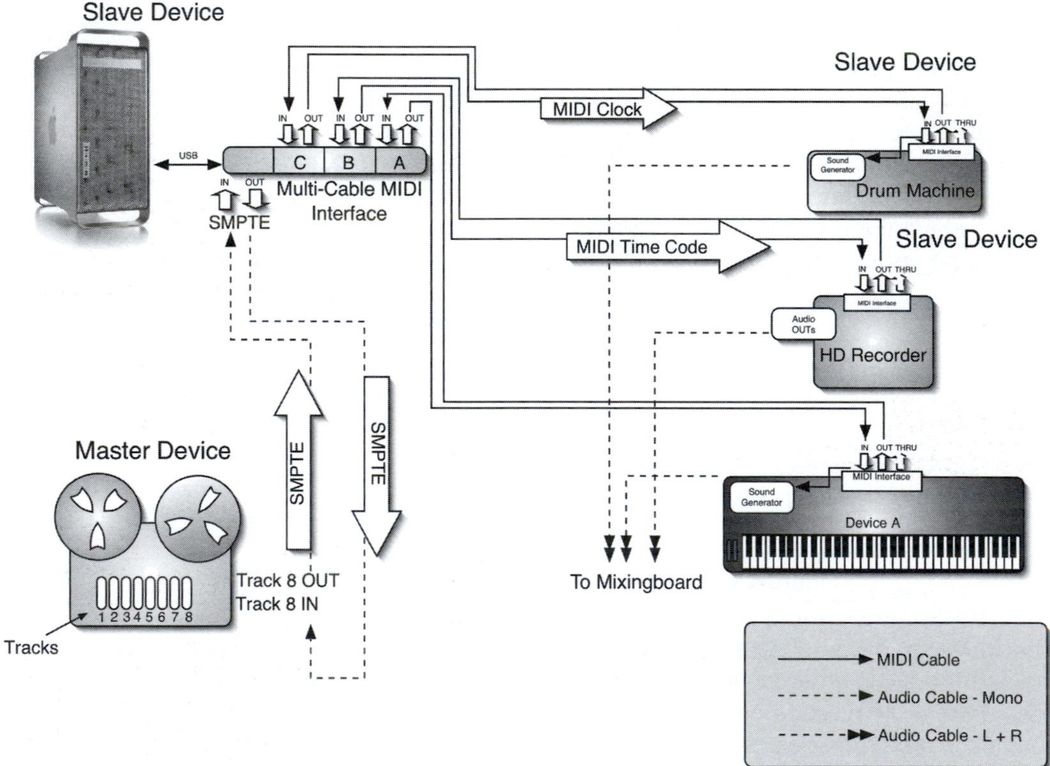

Figure 4.18 Studio setup with master/slave synchronization through SMPTE, MTC, and MC (Courtesy of Apple Computer).

One of the disadvantages of operating with linear machines is their relatively limited flexibility. In other words, whereas nonlinear devices can locate and move to any position in time inside a specific project/sequence without literally fast-forwarding (or rewinding) to the specified location, linear machines usually cannot. This means that when synchronizing linear to nonlinear machines, the former should be set as "master" while the latter (more flexible) need to be set as "slave." Make sure that on your sequencer (as in the case of any master/slave connection) the frame rate matches that with which the tape was striped. Once the tape is striped with the SMPTE code, every time you press the "Play" button on the multitrack tape recorder, the head will read the SMPTE code recorded on one of its tracks (in this example track 8). The code will be sent to the MIDI interface through the output of track 8 to the SMPTE input of the MIDI interface and here translated in MTC. The sequencer will start following the incoming MTC. As soon as you press the "Stop" button on the multitrack tape recorder, the flow of timing data will stop, and therefore the sequencer will stop until the SMPTE/MTC data start flowing in again. In this way the computer (and any other MIDI device connected) will follow the SMPTE code that was striped originally. Remember to set the MIDI devices, sequencer included, as "slave" (for a review on this operation, refer to Chapter 3).

A similar technique is used when you want to synchronize a VCR to your sequencer for scoring to picture. In this case the tape provided by the video production company usually already has SMPTE code that was striped by the video editor. For this type of application, a stereo VCR is indispensable, since on one channel you will have the voice-over or dialog of the video and on the other you will have the SMPTE code (Figure 4.19).

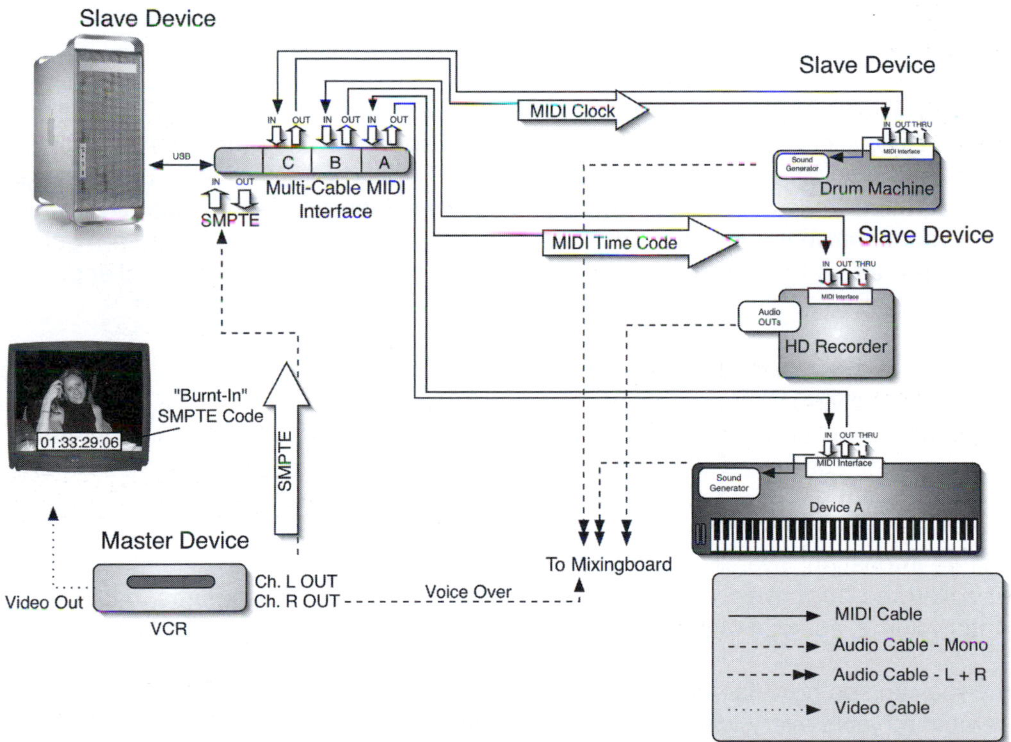

Figure 4.19 Studio setup with VCR synchronization through SMPTE (Courtesy of Apple Computer).

In the example in Figure 4.19, the left channel of the VCR outputs the prerecorded SMPTE code to the MIDI interface, which translates it into MTC, synchronizing the sequencer and all the other MIDI devices that need synchronization. The right channel of the VCR is sent to the mixing board with the voice-over signal. The video output of the VCR is sent to a video monitor with the video signal. Keep in mind that the video signal usually has a copy of the SMPTE code "burnt-in" to the picture so that the composer can have a visual reference of the code. In this way you can easily spot a movie and set the most important markers by looking at the picture and at the burnt-in SMPTE code.

To set up the sequencer to act as a slave, the directions are the same as in Chapter 3 when learning how to synchronize nonlinear to nonlinear devices. Whenever you press the "Play" button on the VCR, the SMPTE code is sent to the MIDI interface, translated in MTC, and sent to the sequencer and to all the MIDI devices that need synchronization. When you press the "Stop" button, the SMPTE and MTC will be interrupted, and consequentially all the synchronized MIDI devices will stop waiting for new incoming code. While for several years SMPTE has played a very important role in synchronizing linear devices to nonlinear devices, with the evolution of computers and nonlinear devices in both the audio and video fields, more and more the synchronization aspect of a studio will be left to nonlinear machines, mainly because of their flexibility and efficiency in dealing with such issues.

4.7 MIDI System Exclusive messages: the "MIDI dump" function

As part of the advanced sequencing techniques, I would like to discuss one aspect of the MIDI specification that is often overlooked or ignored by most books, especially regarding its practical implications. While we learned in the first chapter that System Exclusive messages (SysEx) are MIDI messages that are not standard and that they vary depending on the type of device, model, and manufacturer, we also learned that they are very powerful messages that allow us to get deep into the "brain" of a MIDI device. As a composer and producer, the chances that you will be spending time writing System Exclusive messages during your creative composition sessions are pretty slim, and therefore I am not going to talk in detail about these messages. Instead I want to discuss a very practical application related to SysEx.

One use of the SysEx messages has to do with the fact that through them a MIDI device is able to take a virtual "snapshot" of all or some (depending on the settings) of its internal parameters and send them on the MIDI network to a connected device through its MIDI OUT port. If the connected device is a sequencer, then we can record these messages and eventually play them back to restore the parameters and settings we recorded earlier. The parameters that can be sent out as SysEx include the general system settings of the device (such as intonation, overall volume, and effects assignments), the multitimbral settings, also called the *performance* or *mix* settings (such as channel assignments, patch assignments, volume, and pan of each part), and the parameters of all or individual patches.

Now you can see how useful this type of function can be in a studio where you may have to switch projects every week or even more often. While it is true that almost all devices nowadays can store all this information in their own internal memory locations, it is also true that eventually you will run out of memory locations to store new patches or new performances. I usually recommend avoiding storing the multitimbral settings on only the devices' internal memory, for a variety of reasons. First, unless you use expansion cards, it is not possible to have a backup copy of these data, and if you read Chapter 3 carefully you know how I feel about not having any backup of a project's data. Second, I like to keep all the data related to a certain project together in the same file for functionality, transportability, and organization reasons. The answer to these issues lies in the so-called SysEx MIDI Dump function. As I mentioned earlier a SysEx MIDI dump allows you to make a copy, through MIDI messages, of the internal parameters of a device. Depending on the device, you usually access this function from the "MIDI," "Utility," or "Global" menu (a specific device might have a different menu layout, but the procedure would be the same). Under these menus is a function called "MIDI Bulk Dump" that allows you to duplicate the entire internal memory (or part of it) of the device and send it through the MIDI OUT to a connected sequencer. When you prepare to make a MIDI dump transfer, you have to select which information and parameters you want to output. Usually you have several choices. I have summarized the most commons one in Table 4.9.

Table 4.9 "MIDI Bulk Dump" options

Type of parameter	Function	Comments
All	Sends a MIDI dump of the entire memory of the machine, saving everything	It is a good idea to make a global dump once in a while to have a fresh copy of all the patches, performances, and system settings. I also recommend performing such a MIDI dump right after buying the device so that you will always be able to restore the device to its original conditions if necessary
Performances	Sends a MIDI dump of the performances, settings and the multitimbral parameters	This function, depending on the manufacturer and model of the keyboard, can sometimes also be referred to as "Mix" or "Multi" You can also select a range of performances (or a single performance) to send as a MIDI dump
Patches	Sends a MIDI dump of the patches	You can also select a range of patches (or a single patch) to send as a MIDI dump
Temporary memory	Allows you to send as a MIDI dump only the current performance or patch that is being edited	This is particularly helpful if you adopt the MIDI dump technique as a regular way to store your devices' settings

Let's learn in more practical terms how to organize SysEx dump sessions and how to integrate them with your projects and sequences. After finishing your work on a project or after a temporary sequencing session, it is a good idea to store all the settings of your synthesizers, sound modules, and MIDI devices in general as SysEx MIDI dump. You can in fact choose among several ways of organizing your MIDI dump data. You should create a new MIDI track for each device and record each dump for each track/device separately. If you choose to do so, remember these few hints: Mute the other tracks before recording the SysEx, because these data take a big chunk of the bandwidth of the MIDI network, and the last thing you want to have is a MIDI device receiving performance data while at the same time sending SysEx. This may result in corrupted SysEx messages or MIDI system crashes. Also remember to mute the SysEx tracks when playing back the regular performance data, for the same reason. Another option you have (which I prefer) is to create a separate sequence, as part of your current project, where you record only the MIDI dump. This means that, depending on your sequencer of choice, you will have to either create a separate sequence just for the SysEx data (DP) or create a new "arrangement" that will be dedicated only to hold the SysEx data (CSX and LP). Another option is to create a new single track and to take advantage of multiple takes/playlists to accommodate several MIDI dumps at the same time (DP and PT). Make sure when you record the SysEx for a certain device that the other tracks are muted to avoid unnecessary MIDI data flow. Let's take a closer look at how each sequencer can be optimized to handle a SysEx MIDI dump session.

In DP I suggest creating a separate sequence to store the MIDI dump data. You can create a new sequence by using the drop-down menu called "Sequence Menu" in the top left corner of the Track List window. Once you select the new sequence, rename it "MIDI Dumps" and make sure to leave only the MIDI tracks you need for the SysEx dump of each device. Record-enable the first track (name the track with the device's model), and make sure you're ready to send the dump from the device. How a MIDI device sends MIDI SysEx dumps depends on the manufacturer and the model. In most situations you have to browse to the MIDI Dump menu of your device, select the data you want to include in the SysEx messages, start recording on your sequencer, and then trigger the transmission of the SysEx data by pressing the "Enter" key on the device's front panel. Once the data are recorded, you can see them as regular MIDI data, except they will appear in hexadecimal code (Figure 4.20) and you can't really edit them without damaging the dump irreparably, unless you are a seasoned MIDI programmer.

Since SysEx messages take a big chunk of bandwidth of the MIDI network and they are not so common in terms of basic sequencing techniques, sometimes sequencers are set by default to ignore such messages. It is good practice to double-check that there are no MIDI filters programmed to ignore SysEx messages before recording a MIDI dump. In DP you can check the status of the MIDI filters by opening the "Set Input Filters" in the Setup menu (Figure 4.21). Make sure the "System Exclusive" item is checked.

Remember that you have to rerecord the MIDI dump after every session if you made some changes to the current setup of a device. In order to "recall" the settings when you reopen the project, simply select the sequence with the MIDI dump data, and play back the tracks containing the SysEx messages. Make sure that each track is output to the correct device. As soon as the playback of the data begins, the devices will display a message that acknowledges the receiving of a MIDI dump. After a few seconds (or a few minutes, depending on the amount of data you originally transferred as SysEx), the device will go back to regular performance or patch mode, with all the settings you originally programmed.

Figure 4.20 Example of a SysEx MIDI dump in DP.

In LP and CSX I recommend creating a separate arrangement/project for the MIDI dumps. To create a new Arrangement in LP, choose the "New" option from the File menu, while in CSX select "New Project" from the File menu. As we learned for DP, create tracks for each device, and make sure they are output to each device for which you record a MIDI SysEx dump. In LP in order to check if the SysEx messages are filtered, open the "Song Settings" from the File menu and choose the "MIDI Options" parameter. Make sure the SysEx icon is not grayed out and that the "System Exclusive MIDI Thru" function is not active (this option is used only when hardware controller units are employed to program an external sound module or synthesizer). In CSX you can check the MIDI filter options by opening the main Preferences menu located in the Cubase SX menu and selecting the "MIDI Filter" submenu. Make sure the SysEx parameter is not checked in the "record" column. In PT you can either create a separate session or reserve some of the MIDI tracks for the SysEx MIDI dump. Once again, make sure the SysEx messages are not filtered by selecting the "MIDI Input Filter" option from the MIDI menu.

No matter which sequencer you use or which technique you prefer, I can't stress enough how important it is to take advantage of the SysEx MIDI dump features of modern MIDI devices. It is going to save you a lot of time that can be put into the creative flow, and it is also a good way of avoiding major disasters in terms of data loss, since the SysEx dumps will be saved and (hopefully) backed up with all the other project information.

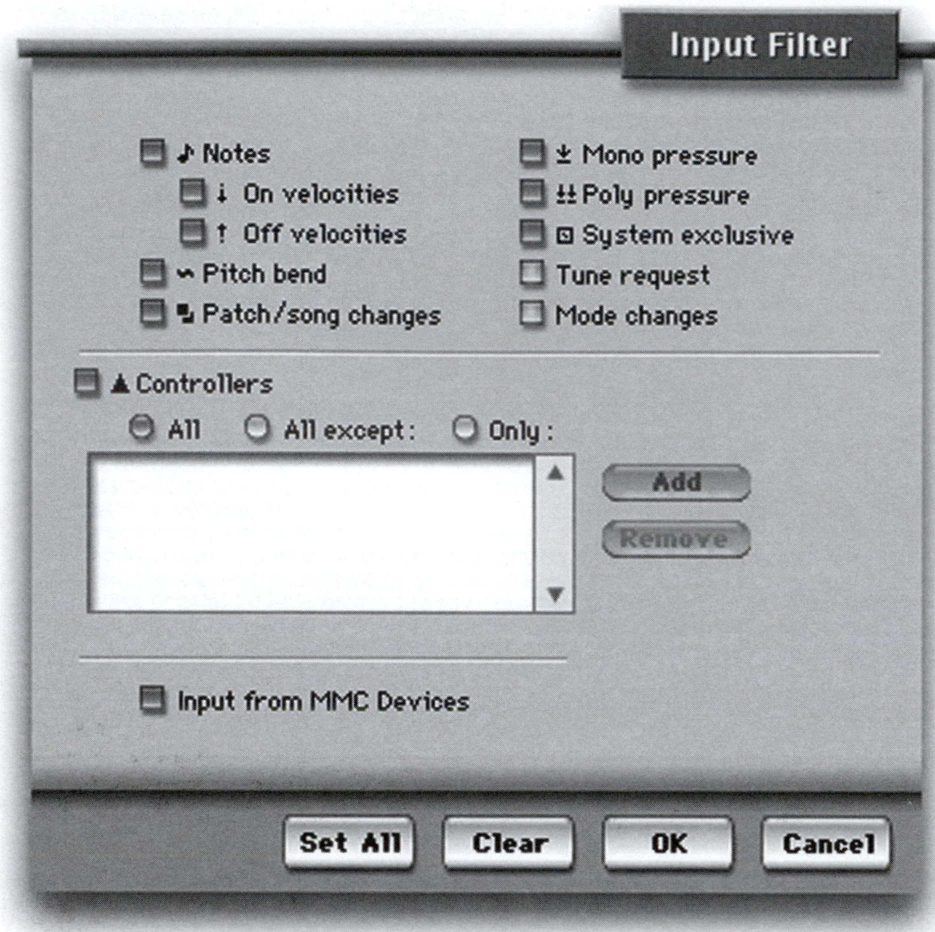

Figure 4.21 The "Input Filter" option in DP.

4.8 Summary and conclusion

In this chapter we focused on advanced sequencing techniques. Having learned the basics of MIDI and some intermediate features, it was time to make the big jump and master complex techniques, such as custom groove-quantize creation, quantization of audio material, use of MIDI and audio plug-ins.

The creation of customized grooves to use during the quantization session is one of the most intriguing features of modern sequencing. Custom grooves can be created from scratch or "morphed" from MIDI or audio parts. In DP we can customize the quantization grooves by manually creating and altering the parameters of the groove through the "Create Groove" option found in the Region menu. The groove can be based on a preexisting MIDI part already present in your sequence. CSX, LP, and PT allow you to create new quantization grooves not only from an existing MIDI part but also from audio material. This option greatly expands the boundaries of sequencing and helps reduce the "stiffness" often associated with traditional quantization techniques. While the technique used to create custom grooves in CSX, LP, and PT is similar to the one learned for DP, the procedure applied to create grooves based on audio is slightly more complex. In CSX you have to create, from the waveform editor, so-called "hitpoints" that correspond to the most prominent and rhythmically important waveform transients. The markers created will be used by CSX as reference to calculate the parameters for the new groove. In LP a similar result is achieved using the "Make Groove Template" from the "Groove Template" submenu that is located in the Options menu. A series of parameters, such as granulation, attack range, smooth release, and velocity threshold, then allows you to choose the best settings for the specific audio material you are working on. In PT TDM and PT LE 6.7 you can access similar features regarding groove quantization and audio files using the engine called "Beat Detective" which is available as a separate bundle for PT LE 6.4 and earlier owners. With BD you can achieve several tasks all related to a selected audio region. You can, for example, find out the tempo of a loop or extrapolate the groove of an audio loop and save it as a groove template usable on any MIDI or audio region.

All the grooves created through the aforementioned procedures can be used and accessed from any quantization window in your sequencer and utilized for both MIDI and audio parts. In order to be able to quantize an audio part, the part itself needs to be prepared. It has to be "sliced" according to either a fixed grid (8th, 16th notes, etc.) or by using its waveform's transients. Once the audio part is cut into several smaller subsections, each region can be moved and quantized as if it were an independent element, greatly increasing the quantization flexibility of audio tracks. If your sequencer is able to import and directly handle audio files in "Recycle's rex2" format (which means audio files or loops that are already processed and "sliced"), then your "recycled" audio files will follow automatically the tempo changes programmed in the sequence. If not, you will have to slice them manually from the waveform editor of your sequencer.

Even though the editing techniques we learned in Chapter 3 are extremely useful and constitute the most common procedures for editing MIDI data, there are more advanced options when it comes to creative use of MIDI. The so-called "offline" editors allow you to apply some predetermined editing "macros" according to parameters and conditions set in the editor. While certain sequencers, such as CSX and LP, provide dedicated offline editors that allow you to filter and transform MIDI data according to the nature, position, and type of the data, DP and PT have more structured and predefined filters that can achieve similar results with less programming involved. Such filters include, for example, "Change Velocity," "Change Duration," "Transpose," and "Split Notes." In LP and CSX you can use, respectively, the "Transformer" and the "Logical Editor" to create your own MIDI filters tailored to specific tasks and editing goals. By setting the conditions (meaning which events will be affected) and the functions (meaning which actions will be applied to the filtered data), you can edit large sections of your project with a single command. In DP and CSX, most of the offline editing techniques can also be applied in real time through the use of MIDI plug-ins that act as filters inserted in the mixing board channel of the track right before the MIDI data are sent out to a MIDI device. Such filters can include MIDI delays, quantization, arpeggiators, velocity change, duration change, etc. In LP and CSX you can access basic MIDI filters from, respectively, the Parameters Box and the Inspector window. Real-time MIDI plug-ins are handled by LP through the Environment editor.

Audio tracks allow for greater manipulation through effects and plug-ins than MIDI tracks. Since in the case of external MIDI devices the generation of the actual sound resides outside the computer and the sequencer (software synthesizers are an exception), we cannot use the internal sequencer plug-ins to affect MIDI tracks directly. The audio material recorded on audio tracks, on the other hand, is generated inside the computer. We can therefore take advantage of the plug-in system to treat the audio material. The way effects are used and applied to audio tracks in a sequencer is very similar to the more traditional analog mixing environment. By using an effect as "insert," we let 100% of the original dry audio signal be affected. This technique is used mainly for dynamic effects (such as compressor, limiter, gate, and expander) or equalizers. While the "insert" technique works well in situations where the effect doesn't need to be shared and we want to affect only the track on which it was inserted, it is not the most flexible and efficient way of using effects that need to be shared among several tracks. For this particular purpose we can use the "send" technique, which involves the use of an auxiliary channel on which we insert an effect (usually an ambience effect, such as reverb, delay, or echo). The aux channel is set to receive the input from a bus (an internal signal path that connects different internal components of the virtual mixing board in our sequencer). Each channel we want to be affected by the effect inserted on the aux channel will be sent to the aux through the use of a so-called "send," which is output to the same bus on which the aux is receiving. By controlling the amount of signal sent to the aux through the "send" fader, we can control how much effect will be applied to the tracks.

With the introduction of more powerful computers, modern sequencers can handle MIDI and audio as well as import video. This function is particularly useful when scoring to picture. By importing a video source inside the sequencer, we avoid having to deal with synchronization protocols used to synchronize the computer and an external video machine. If the video production company you are working for can provide you with a digitized version of the movie, then the only thing you have to do is to copy the video file on your working HD and import it to your sequence. All four sequencers handle this function in similar ways. They all can import video material in QuickTime format. PT can also export the video with the mix you created in the sequence through the "Bounce to QuickTime Movie" option. If you don't have a QuickTime version of the movie, you can fairly quickly and inexpensively digitize it on your own by using a simple analog-to-FireWire video converter.

In Chapter 3 we learned how to synchronize nonlinear to nonlinear machines through the use of specific MIDI messages such as MTC and MIDI Clock. If we want to synchronize linear machines to nonlinear machines instead, we need to use a variation of MTC that can be recorded on tape, since it is not possible, due to a limitation in bandwidth, to store the MTC directly on tape. SMPTE is a synchronization code used in this type of synchronization. It has the same format as MTC (*hh:mm:ss:ff*), and it is able to encode the digital format to tape by modulating (using a "biphase-marking" system) the 1s and 0s of the digital signal into a square wave. The first step to synchronize a linear machine to a nonlinear machine is to "stripe" SMPTE code onto an empty track of the linear device using an SMPTE generator (which usually can be found in the most sophisticated MIDI interfaces). By striping the code we basically time-stamp the tape. After the tape is striped, the output of the track with SMPTE code on it is then sent to the SMPTE input of the MIDI interface. Here it is translated in MTC by the MIDI interface, and it is sent to the sequencer and to the connected MIDI devices that if set in "slave" mode will follow the SMPTE/MTC code frame by frame. This technique can be use to synchronize reel-to-reel tape machines, VCR, or digital tape recorders that lack other synchronization protocols.

As part of advanced editing techniques you learned how to use the SysEx MIDI dump data to store and recall entire configurations and settings from and to the MIDI devices in your studio. This is a very useful technique not only to quickly reset your studio with all the configurations related to a certain project but also to have a backup copy of the settings, multiple channel assignments, and patches you created for a project. In order to do a MIDI dump, make sure your sequencer is not set to filter and therefore ignore System Exclusive messages. Create a separate sequence/arrangement for the SysEx data (this will guarantee that you won't try to play back the regular sequence data and the SysEx data at the same time). Create or reserve a single track for each device you wish to send a MIDI dump from. Next record-enable one track at a time, press Record, and, after waiting a few seconds, start the MIDI dump from the selected device. You can choose to dump all the settings stored inside the device ("All") or specific parameters, such as patches, multiple channel assignments, or even single entries, such as a single patch. To recall the setting, simply assign the output of each track containing the SysEx dump to its respective device and press Play; the data will be received by the devices, and every setting saved in the dump will be restored.

If you have mastered the techniques explained in this chapter, you should congratulate yourself! From Chapter 1 up to this point you explored, learned, and practiced several concepts and techniques that allowed you to improve your sequencing skills in order to bring your productions to the next level and to be able to focus on the creative process without having the technical aspect of sequencing interfering too much with your compositional attitude. In the next chapter we will learn how to specifically deal with different sections of an ensemble in terms of sequencing techniques, sound layering, and MIDI editing.

4.9 Exercises

Exercise 4.1

By using any of the standard MIDI file sequences on the data part of the audio CD, import two MIDI parts and create two groove-quantize presets you consider useful for your particular style of writing.

Exercise 4.2

By using any of the audio loops on the CD, import two audio loops and create two groove-quantize presets you consider useful for your particular style of writing.

Exercise 4.3

By using any of the audio loops on the CD, import two audio loops (different from the ones used in Exercise 4.2), and with the "slicing" technique alter the original groove via different quantization options (try, for example, to apply an 8th note swing feel to an 8th note straight feel, and vice versa).

Exercise 4.4

- Set up and record a sequence with the following features:
  - Ten MIDI tracks
  - Two audio tracks with loops of live instruments
  - Free instrumentation and tempo

Exercise 4.5

With the sequence created in Exercise 4.4, apply the following edits using a combination of offline and real-time MIDI filters.

a. Transpose at least two MIDI tracks one octave higher.
b. Set an arpeggiator with a 16th note pattern.
c. Apply a note velocity reduction of −10 to the first four bars of a MIDI track of your choice.
d. Apply a note duration reduction of −5 to the first 16 bars of a MIDI track of your choice.
e. Select all the notes of a percussion instrument (such as a hi-hat) from a composite drums track; cut them and paste them on an empty track.

Exercise 4.6

On the two audio tracks created for the sequence of Exercise 4.4, apply a reverb effect as a send and apply one equalizer for each track as an insert.

Exercise 4.7

Explain the main differences between using MTC and SMPTE in a studio to synchronize external devices to your sequencer.

Exercise 4.8

Use the SysEx MIDI dump technique illustrated in this chapter to store the settings of all the MIDI devices in your studio that you used to record the sequence of Exercise 4.4. After recording the data, try to restore them into the devices.

5 Elements of MIDI Orchestration

5.1 Introduction

In the first four chapters we learned several concepts and techniques related to the practical aspects of sequencing, such as MIDI messages, editing techniques, quantization options, audio and MIDI track use, and synchronization procedures. While these subjects represent a very important part of the knowledge required to achieve a higher standard in modern and contemporary production using MIDI/audio sequencers, the musical aspect of such productions also plays an extremely important role. The orchestration techniques involved in the realization and production of a project based on MIDI and audio are crucial to obtaining professional final results. I have referred to the equipment involved in the contemporary composition and production process as "the orchestra of the 21st century." This concept involves a learning process based not only on technical skills and procedures but also (and sometimes mainly) on orchestration techniques and rules that are part of a more traditional approach to writing, arranging, and orchestrating. This often means that no matter how well you know your sequencer, synthesizer, and sound modules, if you orchestrate a part for the wrong instrument or out of its range the results will be disappointing and the final production unprofessional.

As I mentioned earlier, the role of the modern composer has changed drastically in the last few years. Nowadays the composer is often also the producer, the orchestrator, the arranger, the MIDI and audio engineer, and the mixing engineer. Such multifaceted involvement in the production process requires a wide range of skills. This is why I firmly believe that to have an edge over the competition and to be a successful contemporary producer and composer you have to learn some traditional orchestration and arranging notions and at the same time modern sequencing techniques.

In this chapter you are going to learn some of the most important elements of orchestration for acoustic and electronic instruments and how to merge the technical concepts you mastered in the previous chapters with the more traditional aspects of orchestration. Each of this chapter's main parts analyzes a different section of the modern orchestra, starting with the rhythm section and moving to the string section, the wind section, and finally synthesizers. For each I will provide key features, sequencing techniques, the instruments' range, and sound layering techniques in order to achieve the best results in terms of MIDI orchestration. This chapter is useful both to seasoned orchestrators who need information on how to create convincing mock-up versions of their scores and to seasoned MIDI programmers who need to brush up on their orchestration skills.

5.2 The rhythm section

The rhythm section constitutes the core of most commercial and contemporary compositions and productions. Since most of the current writing (especially in the pop and commercial style) is groove driven, the rhythm section has achieved a predominant role in most sequences and projects. The characteristics of the instruments that form the rhythm section can highly influence the main style of the compositions, and, depending on their qualities and timbres, the sequencing techniques involved can change drastically.

The modern rhythm section usually comprises the following instruments: piano, guitar(s), bass, drums, and percussion. I intentionally did not include synthesizers in this category because they are able to produce such a large, versatile, and heterogeneous group of textures and sonorities that they deserve a separate section (at the end of the chapter). As I mentioned earlier in the book, the best way to achieve more convincing and professional productions is to use a combination of real acoustic instruments and MIDI instruments. Therefore where it is possible, I highly recommend taking advantage of the audio capabilities of your sequencer and recording as many live instruments as the budget allows. Nevertheless, in most situations you will find yourself having to use MIDI samplers and synthesizers to sequence your projects. Let's analyze in detail how to get the most out of your MIDI gear to achieve a realistic feel and texture for the rhythm section instrumentation.

5.2.1 The piano

By the term *piano* we usually mean the so-called pianoforte, acoustic piano, or grand piano. This instrument has the widest range of any other acoustic instruments (Figure 5.1), covering more than eight octaves, from A0 (27.5 Hz) to C8 (4186 Hz).

Figure 5.1 The range of the acoustic piano.

The piano can play several roles within the rhythm section. It provides the necessary harmonic support by outlining the chord progression of a composition, and it provides the rhythmic support and definition required by the rest of the rhythm section. Because of its versatility in terms of dynamic range it can be used to reinforce passages in which other sections can take advantage of its fast attack, and it can be used as a solo instrument during soft passages. The acoustic piano is one of the acoustic instruments that, for obvious reasons, is easier to reproduce in a MIDI environment, mainly because its acoustic features can be fairly easy to synthesize and reproduce through sampling techniques. In addition, the controller used to sequence piano parts is usually a keyboard.

Even though this may seem obvious, one of the first aspects of sequencing a piano to pay attention to is making sure you use a sustain pedal attached to your keyboard controller to simulate the realistic feel of an acoustic piano. In my career I have seen many MIDI producers trying to sequence piano parts without a sustain pedal, something you should definitely avoid. What really makes a difference when sequencing piano parts is the sound or patch you use. Most of the midrange synthesizers you can buy nowadays have pretty decent acoustic piano patches that can be applied in pop, rock, or contemporary commercial music. In most cases they can't emulate a realistic piano texture in more traditional and acoustic music styles, such as jazz and classical. Most of the piano voices available in synthesizers are thin and bright, and they don't capture the real feel and warmth of an acoustic instrument.

In general, synthesizers have a limited amount of ROM (read-only memory) in which to store the raw samples used to construct the built-in patches. This means that the original samples need to be either compressed in size or limited in length, contributing to a poor sonic fidelity of the final sound. If you are looking for a realistic and versatile acoustic piano sonority, you can look for a so-called *ROM expansion board* for your synthesizer. Most of the professional synthesizers allow you to expand their sound palette with dedicated and proprietary expansion boards that usually contain high-quality samples of a specific instrument category (e.g., bass, piano, vintage keyboards, strings, brass) or musical style (Latin, Asian, drum and bass, techno, hip-hop, etc.). A better option is to use a *sampler*, meaning a MIDI device that doesn't reproduce sounds through synthesis but instead plays back waveforms stored in its internal RAM. The waveforms can be sampled from real acoustic instruments or from any other available source. The advantage of this technique is that (depending on the memory installed) the sampler can hold multiple samples, each longer than the ones stored in a synthesizer, resulting in a more realistic reproduction of the original acoustic sonority. I highly recommend the use of a sampler to reproduce any acoustic instrument, especially complex timbres such as strings, brass, and percussion.

In recent years, due to the increased power of computers, several software samplers have emerged and become the standard source for the sequencing of acoustic instruments (though not only acoustic). Software samplers (as well as software synthesizers) differ from their hardware counterpart mainly in respect to how the sound is generated. In the case of a hardware synthesizer/sampler, the sound is produced by a dedicated sound generator (in most cases a dedicated sound chip) installed inside the machine. Software synthesizers/samplers, on the other hand, run on a computer where the CPU is responsible for the calculation and generation of the waveforms that produce the sound (I will discuss software synthesizers later in this chapter). The advantage of software samplers is mainly that computers usually can take advantage of a larger amount of memory and that, if the particular software allows, they are able to stream the various samples directly from the HD, virtually eliminating any sample-size restriction associated with traditional synthesizers or hardware samplers. Among the most used software samplers are the GigaStudio by Tascam, Kontakt by Native Instruments, MachFive by MOTU (Figure 5.2), and EXS 24 by Apple/Emagic (Figure 5.3).

Another advantage that software samplers and more advanced hardware samplers have is that to each key you can assign *multiple samples* (instead of a single sample). Multisamples allow you to reach a higher level of realism in terms of sound and performance by assigning several waveforms of the same note to each key. Each multisample is triggered by a different-velocity On value. This means that when playing soft passages, the sampler will play back waveforms that were sampled by striking the keys of the acoustic piano with a softer pressure, while when playing louder passages the sampler will trigger waveforms sampled by striking the keys of the acoustic piano with a harder pressure. This technique is applied to any other sampled acoustic sound, such as string and brass. The number of multisamples allowed by the sampler depends on the sophistication of the software or hardware used.

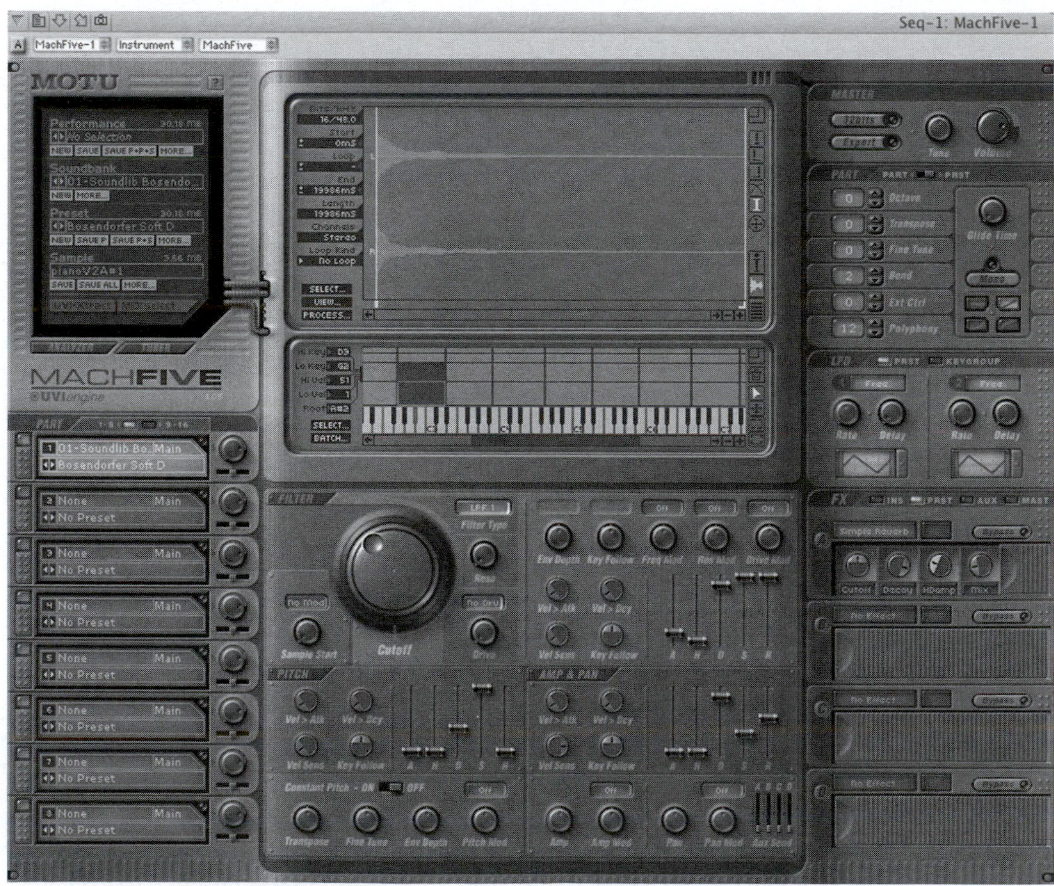

Figure 5.2 Software sampler MachFive by MOTU.

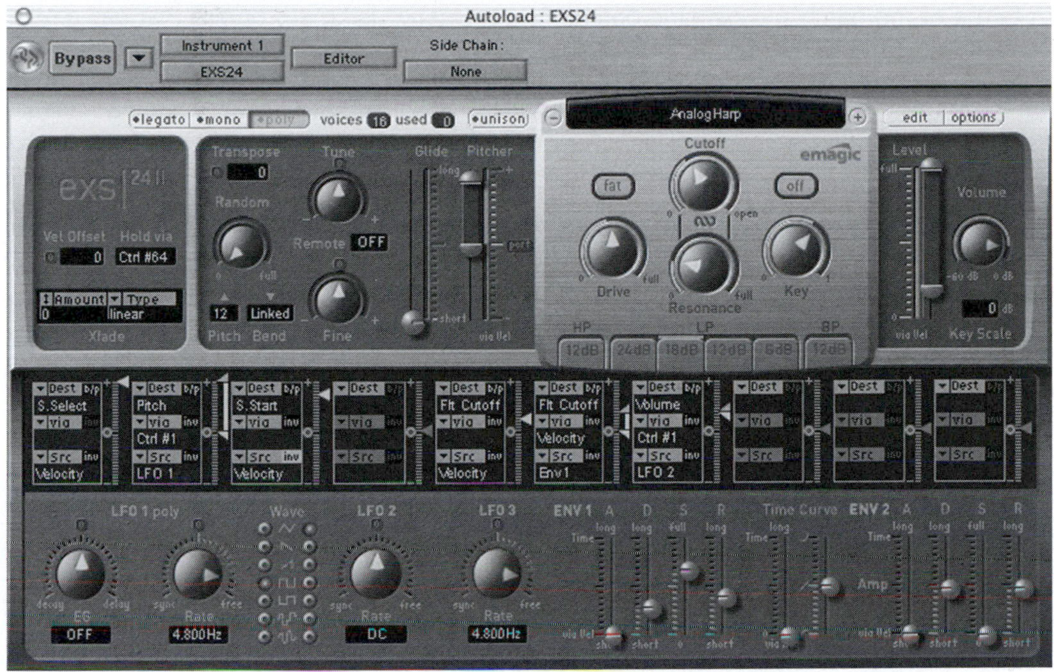

Figure 5.3 Software sampler EXS 24 by Apple/Emagic.

Another option that allows you to have the best of both worlds (MIDI and acoustic) when it comes to piano sequencing is based on the integration of an acoustic piano with a MIDI interface. I mentioned this type of device earlier in the book when discussing different types of keyboard controllers. Such devices involve the use of an acoustic piano to which hardware mechanisms are applied in order to translate the notes played on its keyboard in MIDI data, and vice versa, to translate incoming MIDI data into hammers and pedal actions. This allows you to sequence piano parts using a real piano keyboard action and, even more important, to play back the sequenced MIDI part through the acoustic mechanism of the piano. Among such devices the Yamaha Disklavier has been recognized as the most reliable and the most used in the industry. These types of devices are very flexible. You can use them to input the MIDI parts in a sequencer, edit the MIDI data as you wish, and then, when the final part is ready, simply play back the MIDI track and record the acoustic piano with microphones on a stereo audio track. The sound is that of a real acoustic instrument, while the data editing applied has the flexibility of a regular MIDI track.

In terms of sequencing techniques when laying down piano parts, I suggest avoiding as much as possible copying and pasting to repeat recurring passages. Instead, play the part live as much as you can. Small changes in velocity, duration, and position of the notes can add a much more realistic feel to the part you sequence. If your controller gives you the option, use a soft pedal (Controller 68) to attenuate the volume during soft passages. You can also achieve similar results via a generic expression pedal connected to your keyboard controller. This will allow you to control the volume of the MIDI devices connected to your MIDI network.

In terms of the patch and panning option, I recommend using a program that has a mild panning of the stereo image based on the range of the piano. This means the middle of the key range will be panned in the center, with the higher and lower ranges panned slightly left and right, respectively. Try to avoid extreme wide or narrow panning settings unless required by a particular style of music, for example solo piano pieces.

The category of pianos doesn't include only the pianoforte but, of course, other keyboard instruments, such as electric piano, Rhodes, Wurlitzer, and MIDI piano. The range of electric pianos is usually more limited than that of the grand piano, as shown in Figure 5.4a and b. For this type of sonority you can use either a sampler or a synthesizer, with very good results. The sampler will give you a sonority closer to the original "real" instrument, while a synthesizer will give a more creative approach in terms of texture and sonority. One thing you might consider experimenting with is patch layering with several piano sounds (pianoforte, electric piano, etc.) in order to create new sonorities or to correct timbre unbalances that might limit your production. I often find that layering two electric pianos with opposite acoustic features helps in achieving a much richer and appealing timbre. I always try to start with the best sound I can come up with before using effects such as EQ and dynamics to tweak it. Often a lack of bass frequencies in a patch can be fixed by simply layering a slightly different patch that features more of the low end of the spectrum. In the same way a lack of definition in the middle-high register can be resolved with the addition of an edgier patch.

Figure 5.4 The range of (a) the Rhodes Mark I and (b) the Wurlitzer 200 series electric pianos.

5.2.2 *The guitar*

The six string guitar is one of the most versatile instruments of the rhythm section. Its range is shown in Figure 5.5. Keep in mind that the guitar sounds an octave lower than written. The open strings (E, A, D, G, B, E) are common to all different types of guitars normally used in contemporary music production. Parts written for guitar usually are in keys in which the open strings are featured the most, such as the keys of C, G, D, A, and, to a certain extent, F. Special tuning techniques can be used though to facilitate the performance of parts not directly related to these keys.

When sequencing guitar parts there are few points that you have to keep in mind. As we will see with other acoustic instruments, with guitar sequencing it is very important to

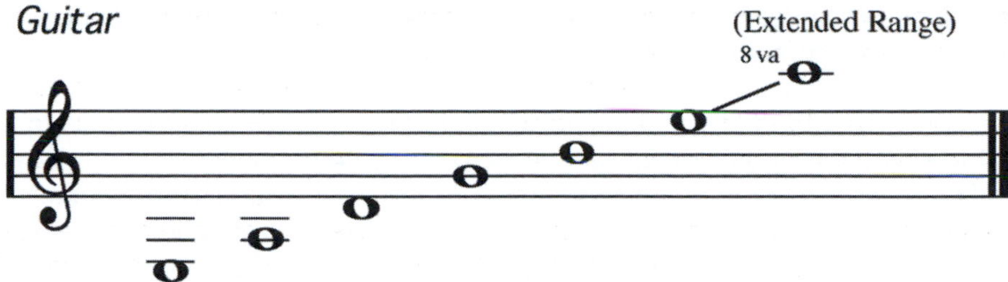

Figure 5.5 The range and open strings of the guitar.

start with a convincing and realistic part. If you try to sequence a guitar part that would be impossible to play on a real guitar, then your battle is lost right from the beginning. Always try to sequence parts that a guitarist would be able to play on a real instrument. This might seem obvious, but it is crucial always to keep this concept in mind for any acoustic part you are sequencing. In the case of the guitar it is important to use a convincing voicing of the notes you play. Definitely avoid using close "piano voicing" that would sound too clattered. Also avoid intervals smaller than a major second unless you use one of the open strings. If you are sequencing a melody, use the Pitch Bend and Modulation wheels to create, respectively, small sliding effects and light vibrato. This will add a nice realism and will cover up for some of the stiffness of the synthesized patch.

No matter how hard you try to sequence convincing guitar parts from a keyboard controller, the results will most likely be disappointing. For complex parts I definitely suggest using a guitar-to-MIDI controller like the one described in Chapter 3. The first advantage of using such a device is the correct voicing. A second, more subtle but no less important advantage is the timing with which each note of a chord would be sequenced. By using a keyboard most of the time you have the tendency to play all the notes forming a chord at the same time, as you would with a piano. When strumming the strings of a guitar, though, the notes forming a chord are separated by a few milliseconds, contributing to the peculiar guitar sonority. Even though through meticulous editing you could reach the same effect after sequencing the part with a keyboard controller, by using a guitar controller directly, not only can you save a lot of time, but you will achieve more convincing results. The same can be said for bending and vibrato effects. While the use of the Pitch Bend and Modulation wheels can be helpful in mimicking the expression techniques of a guitar, you will never get the same results as you would through a guitar-to-MIDI controller. Keep in mind that, depending on your playing skills and guitar controller, you might end up having to edit your parts even after sequencing them with such a device. Therefore I always recommend a little bit of "cleaning" after finishing recording such a part.

In terms of sounds and sonority, most of the concepts I pointed out when discussing the piano can be applied to the guitar. Samplers usually give you the best patches and the most realistic timbres when it comes to acoustic guitar. In particular, samples of nylon string guitar can be fairly accurate in reproducing the acoustic instrument. Steel string guitar can also be reproduced accurately, especially in arpeggio passages, while in strumming passages they are harder to replicate (almost impossible from a keyboard and a little bit easier from a guitar controller).

One technique I usually utilize in sequencing and recording MIDI guitar parts is to run the output of the synthesizer or sampler I am using to generate the guitar sound through a guitar amplifier in order to capture a warmer and more realistic sound. After sequencing the part and after making sure that all the notes are edited, quantized, and trimmed the way you want, send the audio signal of the module playing the guitar part through a guitar amplifier. Record the signal of the amp through a good microphone onto an audio track in the sequencer. The warmth and realism of the ambience captured by the microphone will add a great deal of smoothness and realistic texture to the synthesized guitar patch. This technique can be used with any type of guitar patch, from classical to folk, jazz, or rock. This approach works particularly well with guitars that need medium to heavy distortion or saturation applied to their sound. Even though it is possible to apply such effects through a plug-in inside the sequencer, I prefer using a real amp and the microphone technique in order to get as close as possible to a live situation. Listen to Examples 5.1 and 5.2 on the audio CD to compare a stiff guitar part sequenced using a keyboard controller played directly from the MIDI module and a more convincing part sequenced using a guitar-to-MIDI controller and the sound run through an amplifier to which I applied some mild saturation.

5.2.3 The bass

The bass is also a transposed instrument, like the guitar. It is written on staff paper one octave higher than it actually sounds. The two main types of basses are the electric (also called bass guitar) and the acoustic (also called double bass). The electric bass can have four, five, or six strings. The four-string bass features E, A, D, and G as open strings; the five-string bass usually has a lower B in order to add a deeper texture to the original setting; while the six-string version adds a top C that can be also tuned as a B, depending on the preferences of the player. Acoustic basses are somewhat more limited in range; they usually feature four strings (E, A, D, G), although some models have five. Classical double basses can have a so-called *extension* installed on the low E string that allows the player to go down to a low C. In Figure 5.6 you can see the range of the bass.

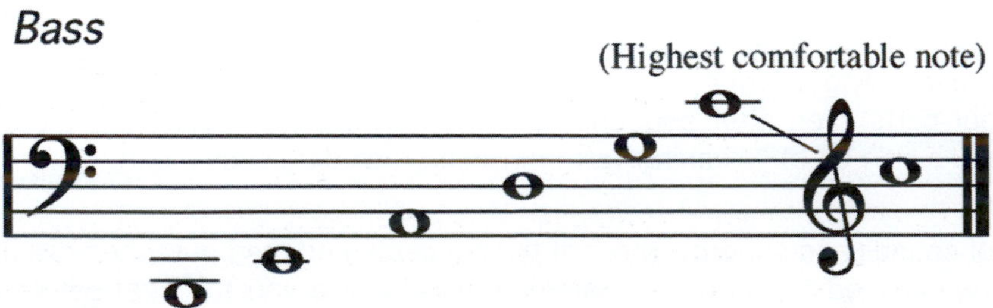

Figure 5.6 The range and open strings of the six-string bass.

What you learned when I discussed the guitar can be applied almost verbatim to the bass and electric bass. In sequencing acoustic and electric bass parts, try to avoid intervals lower than a perfect fourth unless you want to create a particular dissonant effect. If you have a bass part that requires a double-stop (meaning two notes played at the same time) with a major or minor 3rd interval, it is better and more realistic to use a major or minor 10th instead.

Most of the synthesizers available nowadays can reproduce convincing electric bass sounds, especially for certain musical styles, such as rock and pop. As in the case of guitar sounds, a sampler will give you better option when trying to reproduce the acoustic bass sonority. Use the amplifier technique illustrated earlier to capture a warmer and more realistic sound. If you don't have an amplifier at hand or your microphones are not of professional quality, you can also use a tube preamplifier to filter the output of your MIDI module. While this technique is not as effective, it is nevertheless more practical and less time consuming. While the sequencing of bass parts is somewhat easier than guitar parts using a keyboard controller (due to the fact that bass features mainly monophonic lines), I highly recommend sequencing bass parts with a bass-to-MIDI controller if possible. The feel and natural flow will increase dramatically, and this factor will contribute not only to the bass part itself but also to the overall groove of the project.

5.2.4 Drums and percussion

The results obtained when sequencing drums and percussion parts can vary substantially, depending on several factors, such as the style of music, the groove you are trying to reproduce, and the repetitiveness of the patterns. There are two key elements that play a very important role in drums and percussion sequencing: sounds and rhythmic feel. In terms of sounds and patches, both synthesizer and sampler offer a good palette of sonorities. Of course, while the latter are indicated more for recreating live drum sets or percussion ensembles (e.g., jazz ensembles, classical orchestra), the former are usually used more in "artificial" orchestration projects (e.g., techno, dance, pop), where the drums and percussion are targeted to a more creative use of timbres.

I recommend always having at hand a MIDI sound module entirely dedicated to drum sounds, in order to be able to quickly find the most appropriate sonorities in a short amount of time. These modules, such as the Alesis DM5, are a good starting point to sequence drum parts since they offer a comprehensive palette of sounds to start with. Later in the production you might want to replace some of the patches, but this type of module offers a fairly good starting point. By using a sampler (hardware or software), you can take advantage of the multisample technology, which in the case of drums and percussive sounds can greatly improve the accuracy and fidelity of your parts. This technology is particularly effective for drum sounds since they change depending on the strength at which a drummer hits a certain instrument. A snare, for example, features a totally different attack, release, and color if it is struck softly versus hard. The multisample technique allows you to experience a similar response by assigning several different samples to different MIDI velocities. When you play a drum part, the sound will change depending on the velocity at which you play the notes on your controller. Particularly interesting are new software synthesizers, such as BFD by Fxpansion (which I used for the drum parts of the examples CD), which allows you not only to control multisampled sounds but also to change the room and virtual microphone placement of the kit you are using in the sequence. The quality of the final result depends greatly on the style of music you are sequencing. Usually repetitive rhythms that involve little or no variation (such as dance and, to a certain extent, pop rhythms) can be sequenced, with very good results. Other styles, where variations and changing groove play a predominant role, are more problematic. A typical example is jazz. This is one of the hardest drum feels to reproduce with a MIDI sequencer. One of the main aspects that defines jazz is the swing feel that is outlined by the drummer, using a combination of patterns and accents that are almost impossible to recreate in a MIDI setting.

While the quantization techniques we learned are definitely a good starting point, there a few other approaches I want to suggest for getting the best possible rhythm tracks for your projects. First of all you should invest in a MIDI pad device in order to avoid playing drum parts on a keyboard controller. Using a regular keyboard can flatten out the groove considerably, thereby making your drum parts sound stiff and mechanical. Sequencing a swing feel pattern on a MIDI drum set or MIDI pads is going to improve the overall rhythmic feel of the production. Even if you are not familiar with drums and percussion techniques, it is a good idea to use such MIDI controllers to sequence your parts. Some of the most advanced MIDI pads allow you to use either sticks or hands to trigger the pads. This gives you a lot of flexibility when it comes to inputting percussion parts such as bongos and congas.

Another technique involves the use of a mix of MIDI and audio tracks in order to achieve a more realistic result. I always like to include at least one live percussive instrument, such as cymbals, shaker, or hi-hat (HH) along with the other MIDI drum tracks. As we learned at a macro level, the addition of one or more acoustic instruments can bring to life the other MIDI tracks, the same can be said at a micro level when sequencing drums and percussion. You will be surprised at how much your grove will improve by simply adding a live shaker or tambourine to your MIDI parts. These instruments are cheap and fairly easy to play but their impact on a groove can be crucial. Listen to Examples 5.3 and 5.4 on the audio CD and compare the same groove sequenced using only MIDI tracks and after replacing some of the MIDI tracks with live instruments. In the case of a swing feel, I recommend, if possible, using a live ride cymbal or hi-hat, since these two instruments are the most characteristic of this particular style.

As with any MIDI track, try to avoid as much as possible the use of the "copy and paste" command to extend your drum patterns. This definitely works for styles based on repetitive (drum-machine-like) grooves, but for music styles where a live drummer would be required it is better to play all parts at least for an entire section without using the paste command. This will help to take some of the stiffness of the MIDI sequencer out. Sequence each drum and percussion instrument separately and on a dedicated MIDI track. This makes the editing easier and quicker. The only exceptions to this are the bass drum (BD) and the snare drum (SD). I highly recommend playing these two voices together in order to get a tighter feel, similar to the one that a live drummer would get.

Another very important rule to keep in mind is that (as for any other instruments) to sound realistic, the parts you sequence need to be playable by a live musician. Just as you shouldn't sequence parts for instruments that are out of their playable range, in the same way you have to imagine a drummer performing the parts you are sequencing. One of the most common mistakes is to have more than four pieces of the drum set playing at the same time. Unless the drummer you are writing for has more than two arms and two legs, chances are very slim that he or she can play more than four percussive instruments simultaneously (one exception would be the use of the special BD or HH pedals sometimes found in custom drum kits). Always remember to keep it real; this is one of the most important rules. You can always double-check the number of notes triggered simultaneously for the drum parts in either the graphic editor or the drum editor.

When you combine MIDI and audio drums tracks in the same mix, the effect you achieve can be very satisfying in rhythmic feel. Keep in mind, though, that mixing together such different audio sources can sometimes be difficult, due to their substantially different sonic characteristics. The acoustic tracks (such as cymbals and shaker) can stand out too much compared to the more mellow and preprocessed MIDI tracks. In order to avoid problems during the mix and in order to give to the MIDI tracks a more natural ambience sonority, I usually rerecord through a couple of good condenser microphones, on separate audio tracks, the MIDI material played through the main speakers of the studio. This technique is similar to the one explained for the guitar and the bass. This will add a more natural reverberation to the MIDI parts and will allow you to avoid problems during the mix between MIDI and audio percussion tracks. If you want to achieve an even more "live" sound (perfect, for example, for jazz, funk, and orchestral percussion situations), you can record the entire drum mix played through the main speakers of your studio on a stereo track using a pair of good microphones. The farther you place the microphones from the speakers, the more you will be capturing the ambience of the room and therefore the more you will add "live" ambience. Vice versa, the closer to the speakers you get, the more punchy quality and direct sound you will get. Take your time finding the perfect spot for the material you are dealing with; make few attempts and choose the one that best fits the drum part you are sequencing. I usually find that a distance of 4 to 6 feet between speakers and microphones gives the best results, but this can change with the room, the microphones, and the speakers used.

For drums and percussion parts, it is becoming more common to bypass completely the MIDI sequencing techniques and use a series of loops to replace any attempt at live recording (MIDI or acoustic). I am not a big fan of this approach, for several reasons. While the use of loops can be interesting for embellishing your production and give a nice touch to a series of MIDI tracks, loops (for definition) are very repetitive and can really flatten your sequences and make it extremely boring and cold, especially if used as the primary source for the rhythmic tracks of your productions. While loops can be a good source of inspiration and can definitely be used for quick mock-up versions, they are uncreative and impersonal. I would rather listen to a simple MIDI drum track that is groovy and original than to a short sophisticated rhythm that keeps repeating over and over without variation. If a loop becomes an essential part of your project, then I highly suggest "spicing it up" with some variations using either MIDI parts, acoustic live recording of a few percussive elements (e.g., shaker, hi-hat, cymbals) or, even better, a combination of the two.

A good alternative to premixed stereo loops (which, unfortunately, constitute the majority of the loop-based rhythmic material available) are the multitrack versions of the prerecorded drum and percussion parts offered by companies like Discrete Drums. This type of drum track has the advantage of providing the original multitrack version of the loops, allowing a much higher degree of flexibility at both the editing and mixing stages. Having separate tracks for each groove allows you to replace single instruments of the drum set to create variety (every so often you can change the HH or the BD pattern, for example), plus during mixing you can apply different effects to each single track in order to achieve more realistic results. You can also use the multitrack sessions in a creative way by mixing and matching parts from different grooves (e.g., the HH part from a hip-hop groove with the BD and SD from a "funk" rhythm). No matter which technique or combination of techniques you use, always try to be creative, and do not let the sequencer or the loops dictate your music.

5.3 The string section: overview

The string section is one of the most discussed and controversial aspects of MIDI, especially with regard to the sonority and the sound library used to reproduce the acoustic instruments. The modern string section includes the violin, viola, cello, and bass. Each of these

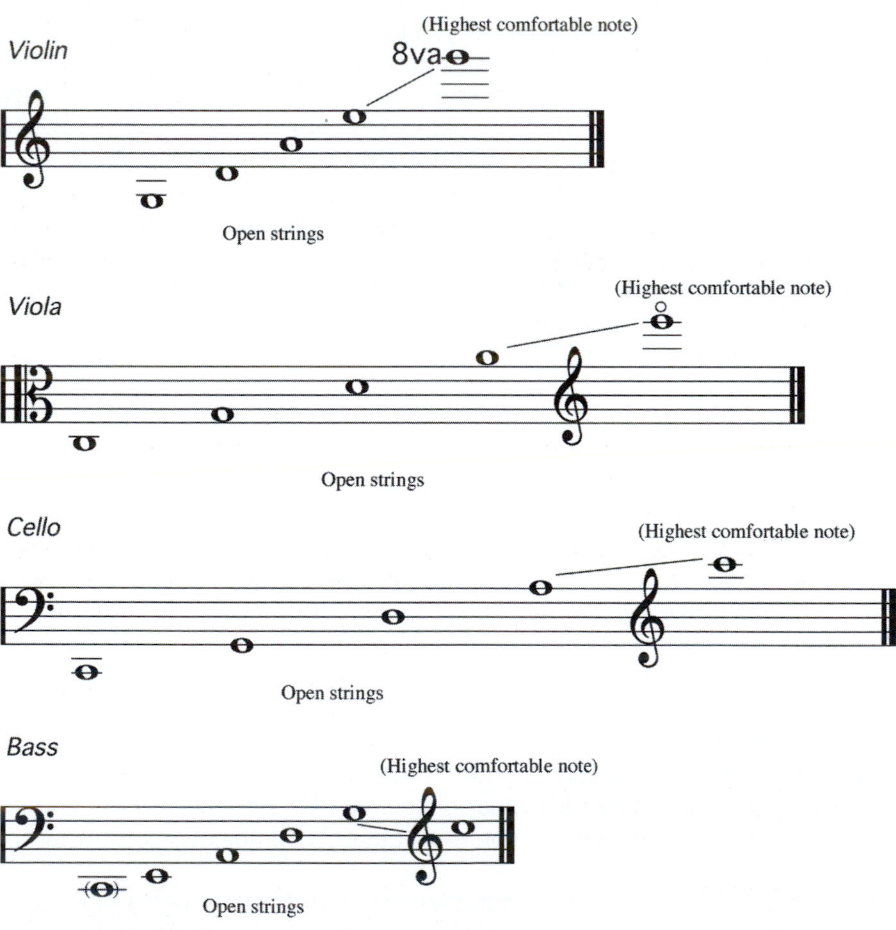

Figure 5.7 The range and open strings of the violin, viola, cello, and bass.

instruments presents different sonic characteristics, playing techniques, and applications. It is very limited to refer to such different sonorities with the generic term *strings,* as often found in many synthesizers' patch names. If you used only the synthesizer patch called *strings* to sequence string section parts, you greatly limited the real potential of your virtual orchestra, since these generic patches often do not take into consideration the very distinctive sonic character of each string instrument. This is why it is extremely important to understand the function and the characteristic of each instrument forming the string section in order to achieve better results when sequencing string parts. In Figure 5.7 you can see the ranges and the open-string subdivision of the violin, viola, cello, and acoustic bass.

Each of the four main instruments forming the string section has specific characteristics, so let's analyze each of them separately in order to outline their main features. Table 5.1 sums up the most important sonic elements and orchestration principles that apply to each one of them.

Table 5.1 Characteristics of the string family of instruments.

Instrument	Sonority	Range	Comments
Violin	Each string has a particular sonority: the E and G project the sound better, while the two inner strings (D and A) are more mellow Usually takes the lead part, such as the melody, in a section. If the section is big enough, the lines can be split in "divisi" style If you use the "divisi" option, try to keep the balance between sections by having more violins playing the top line (melody)	The higher range of the instrument has a thinner sound, so it is often a good idea to reinforce it with other violins or with violas	Usually works better in larger groups—the bigger the section, the sweeter and mellower the sonority Try to avoid sequencing violin lines for fewer than three violins in a section
Viola	Has a darker and warmer sound than the violin, mainly because of the lower range	Usually plays the inner notes of a chord when voiced with violins and cellos	Try to avoid sequencing viola lines for fewer than two violas in a section
Cello	Has the strongest sonority of the entire string family Can be used as a solo instrument even against a larger number of violins or violas	Has the same open string position of the viola but pitched one octave lower	The high register is loud and pleasantly edgy, while the lower register is warm and rich
Double bass	Has a deep, dark sonority In the higher register, can be melodic but not as agile as the cello	Features the lower register of the string family of instruments	Pizzicato style can be effective in both classical and jazz styles

5.3.1 Sequencing strings

When sequencing the string section, the first thing to keep in mind is that you must voice them (meaning choose the notes to apply to each section) in the same way you would voice the parts for the live acoustic instruments. Often by playing the parts on the keyboard you can be tempted to use piano voicing that would be perfect for a keyboard instrument but do not work as well on string instruments.

When voicing for strings there are a few rules to follow. Try to avoid intervals equal to or smaller than a major second in the higher range of the strings, as shown in Figure 5.8 (unless for some particular effects). This will guarantee a more powerful and stronger sonority. Try to avoid close voicing between sections. It is better to use open voicing techniques in order to reach maximum clarity and full sonority. By using mainly thirds you can achieve a very mellow and smooth texture, while the use of fourths helps to create the impression of a larger ensemble, even with a small string section. The fifth is used mainly in the lower register and in the bass and cello section in order to create a solid base on which the higher extensions of the chords can be built.

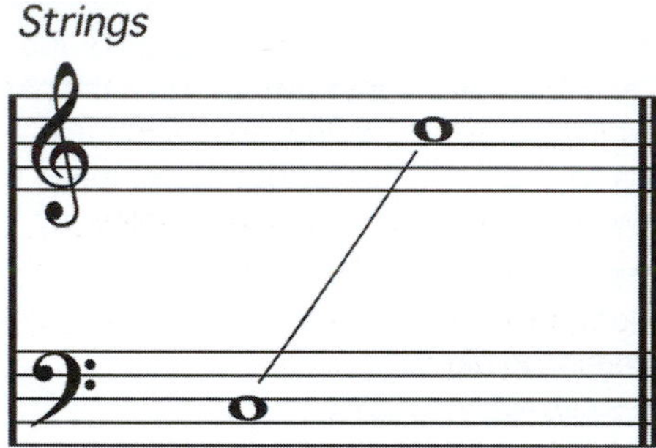

Figure 5.8 Full sonority range for strings.

Experiment with different types of voicing before committing to the final one. Try always to think horizontally and in terms of lines that each section would play, instead of vertically. By doing so you will achieve a much more coherent and realistic-sounding string section, as if your virtual orchestra was really composed by several independent players. When sequencing the different sections, play each line separately on a different track. For divisi create a separate new track and record the split section on independent tracks. If possible assign each section of the divisi to separate MIDI channels and slightly different patches in order to have more control over the individual volumes and pan of each part.

5.3.2 Sonorities and sound libraries

As with the other instruments analyzed so far, a sampler (instead of a synthesizer) gives you the best options in terms of sounds. This is particularly true for the string section, where the complexity, richness, and selection of tones are greater than for any other acoustic instrument. While synthesized strings can be effectively used for more creative productions in musical styles that do not necessarily need real acoustic sonorities, more acoustically oriented projects that demand a realistic sonority must employ sampler libraries. In terms of sounds, here are some suggestions for improving your "virtual" orchestra. What makes a MIDI string part "alive" is variation and the correct use of all the different colors that are peculiar to the string family. The use of different bowing techniques and color effects, such as *sul tasto, sul ponticello, col legno, tremolo, pizzicato,* and *glissando,* will improve the realism of the sequenced string parts. Table 5.2 gives you a quick description of these techniques and sonorities.

Table 5.2 Bowing techniques

Color/technique	Description	Comment
Sul tasto	Indicates that the players need to use the bow closer to the fingerboard	Produces a mellow and velvety sound
Sul ponticello	Indicates that the player needs to use the bow closer to the bridge	The opposite of sul tasto, produces a harsh and edgy sound
Col legno	Indicates that the player needs to use the reverse side of the bow (the wooden part) to tap or strike the strings	Used mainly for staccato or rhythmic effects
Tremolo	Achieved by bowing the string with rapid alternation between up and down bow	Creates a dramatic and suspended feel
Pizzicato	Achieved by plucking the strings with the fingers	Creates a gentle percussive effect
Glissando	Achieved by sliding the left hand between two notes on the same string (it can also be referred as *portamento*)	Can be used for a particular effect

You can achieve different timbers and colors either by using program changes when needed or by assigning different patches for each bowing technique to different MIDI channels, which is a better solution if your setup allows. When you need a different color, you will sequence that particular part on the most appropriate track. Keep in mind that this procedure will take several MIDI channels and therefore can be used only if your studio has enough sound modules or software samplers available. For smaller studios, the use of program changes to switch the patch/sonority is a valid alternative. Unfortunately, if the switch needs to be quick, sometimes this approach can be problematic due to the delay introduced by the sound generator during the patch change. There are few software samplers, such as the GigaStudio by Tascam, where different layers of patches can be switched in real time by using specially programmed keys of the keyboard. In addition the Modulation wheel can be assigned to alter some parameters of the instruments, such as attack and release.

In terms of layered sound, there are two main approaches I find particularly effective when sequencing string parts. The first technique involves the use of both sampled and synthesized patches. Sometime sampled patches can sound a bit thin and edgy, lacking some of the natural lower overtones that make an acoustic string sound so rich. A way to partially overcome this limitation is to layer a synthesized patch underneath a sampled one. The balance between the two can vary, depending on the quality of both programs, but in general I have found that a ratio of 70–80% of sampled string to 30–20% of synthesized strings works well. If you layer a few patches together, then in addition to the main track that is output to all the layered devices and MIDI channels, remember to create different volume tracks for each single MIDI channel/device. This will give you independent control over the volume of each patch and will make it easier to find the right balance between the layered patches (see Chapter 3 to review the technical aspects of this procedure).

While the combination of sampled and synthesized can be useful for improving your string parts, a second approach to layering involves the use of one or more acoustic instruments recorded on audio tracks that double one or more lines of the string sections. This is an essential tool to bring your sequenced string parts to the highest production level. The addition of live acoustic instruments to the sampled patches adds freshness, edginess, and harmonic content to the relatively flat sampled waveform. If possible, try to double the top voice of each string section (violins, violas, cellos, and basses) and the divisi sections.

The right balance between the sampled and acoustic strings depends on several factors. What I recommend is to record the live instruments as dry as possible so as to be able at the mix stage to match the sampled and live instruments with the addition of some artificial ambience. Listen to Examples 5.5, 5.6, and 5.7 on the audio CD to compare, respectively, a string part sequenced with synthesized patches only, a version with layered sampled and synthesized patches, and a final version with acoustic and sampled strings. Of course you can also combine the two layering techniques I explained in order to achieve a fuller sonority, but avoid layering too many patches together, especially with string sounds, since this can create awkward phasing effects that can spoil the final result.

One of the most challenging aspects of sequencing string parts is to recreate the natural attack of the vibrating strings set in motion by the movement of the bow. The attack of string instruments changes depending on the pressure and the intensity with which the bow is moved. A light pressure causes the strings to have a longer attack, while a more abrupt and decisive movement generates a sound with a faster attack. This property is one of the major factors that give to string instruments such a peculiar and flexible sonority. To recreate this aspect through samplers and a sequencer can be challenging and time consuming. While the aforementioned technique involving the layering of sampled and acoustic instruments can definitely improve the final result, in certain situations where the alternation of soft and loud passages is featured, a precise control over the attack of the MIDI parts is necessary. Some software samplers allow you to control the attack of the samples by using velocity-switching techniques to trigger different samples according to the velocity with which you hit the keys.

One of the most effective techniques for controlling directly the attack of the string patch you are using is to take advantage of CC 7 (volume). If you have a controller that allows you to assign a slider to a CC message (volume in this case), you can change the volume and therefore the attack of each note while you are sequencing your parts. This approach works particularly well for slow sections with long sustained notes. By moving the fader, you can accurately control the volume of each note, simulating the real behavior of the vibrating strings. While it is possible to insert and fine-tune the volume changes after you have played the parts, I recommend using the controller while you play. This will give you a much more fluent and natural feel than adding the control messages at a later stage. One of the hardest effects to recreate when sequencing the string section are legato passages, where each note needs to be seamlessly linked to the next one. This is a particular distinctive sonority of string instruments, where down-bows and up-bows can be combined in creating a continuous sound. In order to achieve the same effect with your virtual orchestra you have to carefully edit each note so that the end of the previous note overlaps slightly with the beginning of the next note (Figure 5.9a and b). By editing the notes this way and carefully controlling the *release* parameter of your patches, you will construct a very realistic legato passage.

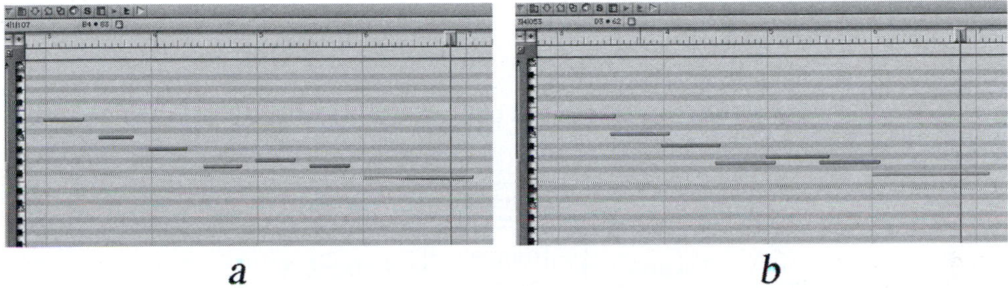

Figure 5.9 (a) Staccato and (b) legato effect for strings.

The amount of overlap each note needs to have can vary, depending on the patch you use, but as a starting point I suggest a value between 10% and 20% of the overall tick resolution of your sequencer. For example, if your sequencer is based on 480 ticks per beat (ticks are subdivisions of a beat), then you should use overlapping values of around 50–100 ticks.

5.3.3 *Panning and reverb settings*

After having voiced, arranged, sequenced, and appropriately edited your parts, the next step is to mix the string section. Two main aspects can have a pretty drastic influence on the final string sound: panning and reverb. The way you pan the different sections of the virtual orchestra can definitely contribute to enhance the realism of the MIDI parts. While the details of panning can vary from project to project and from arrangement to arrangement, a good starting point is to refer to the classical disposition of the orchestra on stage during a live performance (Figure 5.10).

While the classical disposition shown in Figure 5.10 can work, it presents a few problems related mainly to the placement of the basses, which would need to be panned hard right. This approach can in effect work effectively if the hard panning of the basses is balanced by another instrument (such as the piano) or another low-range section (such as trombones and tubas on the left). In a more practical setting you can set the basses in the middle (or slightly panned right) and spread the violins hard left, cellos hard right, and the violas either centered or slightly right to balance the violins divisi panned slightly left (Figure 5.11).

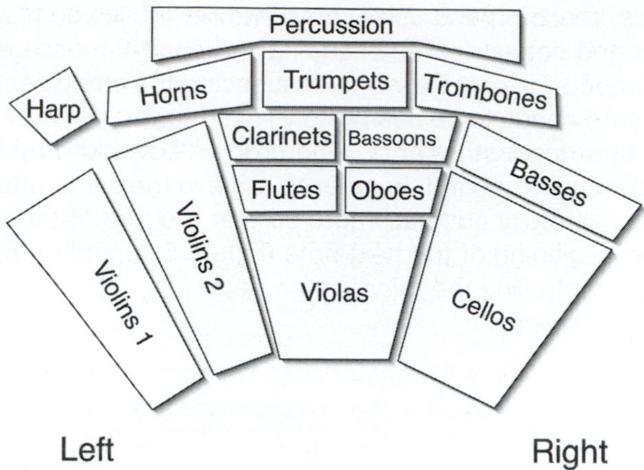

Figure 5.10 The disposition of the classical orchestra.

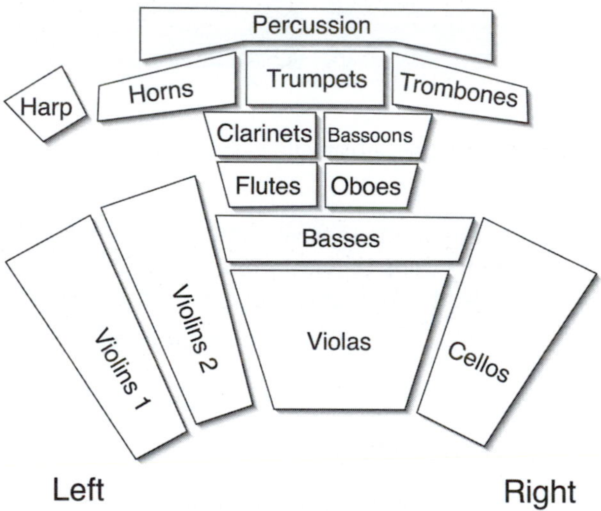

Figure 5.11 Alternative string panning setup.

If the panning allows you to control the position of the instruments in a horizontal way, then the addition of ambience effects such as reverb and delay gives you the ability to place the instruments in a bidimensional position, allowing you to have precise control over your virtual orchestra. The string section usually needs to have a little bit of reverb in order to recreate the natural environment in which a live orchestra would perform. I found that reverb lengths between 1.7 and 2.3 seconds usually work well to recreate a variety of halls. Do not overuse the reverb since this can cause a loss of definition and impact. For fast dramatic passages use shorter reverbs or reduce the volume of the return.

For long sustained passages use lower diffusion settings, while for fast and rhythmically busy passages use a higher diffusion. The diffusion parameters control the distance between bounces during the early reflection of a reverb. With low diffusion, the bounces are more separated from each other, giving a more detailed and clear reverb. A higher diffusion, on the other hand, moves the bounces closer to each other, resulting in a more even (but muddier) sonority. In general I recommend using less reverb on the low-frequency instruments, such as basses and to a certain extent cellos. Too much reverb applied to low-frequency instruments can cause your ensemble to sound muddy and without definition.

The amount of reverb you use on the string section also depends on the type of library and samples you use. Some sampled libraries are recorded with closed microphone techniques and in very dry environments in order to provide the user with a dry ensemble and leave the ambience options open during the final mix. Other libraries are intentionally recorded in more acoustically live halls in order to provide a more realistic ensemble right off the box. Which one you prefer is a matter of personal taste and a matter of musical style. New sampled string libraries are being produced constantly, so it is impossible to provide a comprehensive and updated list. Among the ones I recommend for their flexibility, sonic quality, and comprehensive sonic variety are the Garritan Orchestral Strings, the Vienna Symphonic Library, the Miroslav Vitous Symphonic Orchestra, Peter Siedlaczek's Advanced Orchestra, and the Sonic Reality Symphony Strings, just to mention a few. My favorite is the Vienna Symphonic Library, which I used for the music examples on the CD.

5.4 Wind instruments: overview

Wind instruments refers to a generic category of instruments in which the sound is generated by a column of air enclosed in a pipe. This category can be divided into several subcategories, as indicated in Figure 5.12. As you can see, the wind instruments family covers a wide area of sonorities and ranges, so it is better to analyze them according to the categories presented in Figure 5.12. Nevertheless there are some general rules when sequencing for wind instruments that apply equally to every category.

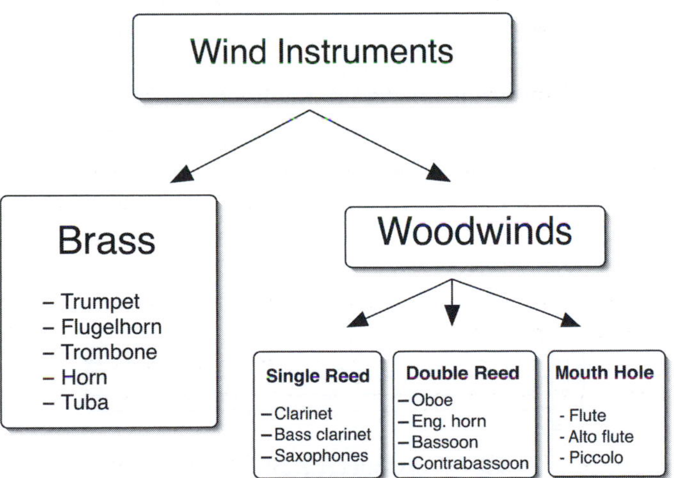

Figure 5.12 The wind instrument family.

As with the string section, one aspect you must remember is to consider your virtual orchestra as if it was a live one. Do not sequence passages that would make no sense to write for live players. This will guarantee much more credible results when dealing with sampled sounds and MIDI parts. Follow precisely the range of each instrument as indicated in the following paragraphs and the voicing that best applies to the nature of the section you are sequencing. One of the advantages the string section has over wind instruments is the freedom of phrasing. Strings can sustain very long phrasing without interruptions or rests, since the sound is generated when the strings are set in motion by the bow. For wind instruments, unfortunately, it is not that simple, since the sound is generated when the player blows into the pipe. This factor limits considerably the ability of the player to sustain long phrases.

One of the most common mistakes when sequencing wind instrument parts is to keep them going for long sections, forgetting that players need to breathe once in a while. This is mainly because in most cases you will sequence these parts on a keyboard. There are a couple of ways to efficiently avoid this mistake. The best way is to use a wind controller and to sequence each part individually, as you would play each wind instrument in the orchestra. The phrases will sound much more natural, and any phrase that would be too long for a live player to perform would stand out and could be easily fixed. These types of controllers also allow you more realistic dynamics, pitch bend effects, and phrasing. If you lack access to a wind controller, a quick solution to the "long phrases" problem can be achieved by recreating the inhale–exhale experience while sequencing the parts on the keyboard controller. Take a deep breath at the beginning of the phrase and slowly exhale while playing. If you run out of breath in the middle of a phrase it means that the phrase was probably too long and therefore wouldn't sound natural when played by the virtual orchestra.

5.4.1 The brass section: the trumpet and the flugelhorn

The brass section normally includes the trumpet, the trombone, the horn (also called French horn), and the tuba. While these instruments produce sound in a similar fashion, their tone quality, timbre, and texture vary considerably. The trumpet has a strong but poignant sonority. If used with mutes it can drastically change its sound into a more mellow texture or into a more edgy and sometimes "buzzy" sonority. The range of the trumpet is shown in Figure 5.13. Variations of the trumpet are the cornet, which has a more mellow and "shy" sound, and the flugelhorn, which has a smaller range (Figure 5.14) and a much warmer sonority.

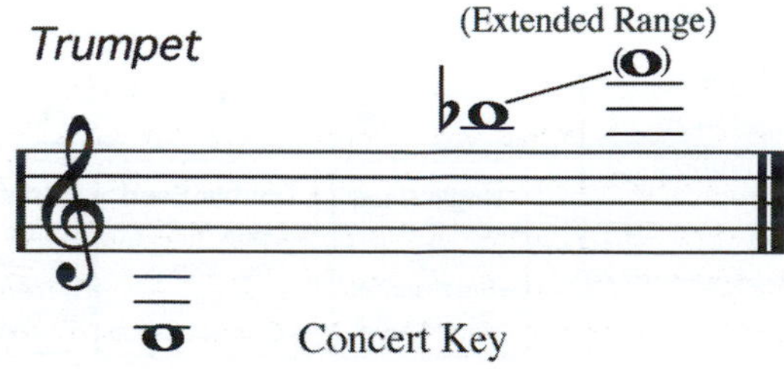

Figure 5.13 The range of the trumpet.

Figure 5.14 The range of the flugelhorn.

The trumpet section offers a fairly broad range of dynamics and sonorities. It is usually best voiced in octave, unison, or close position voicing to create a stronger sound. Usually sections can be formed using four or five trumpets, with the lead trumpet more often brighter than the others in order to cut through the other instruments better. The trumpet is a transposing instrument; that is, it is written a whole step higher than it sounds. As with all the other acoustic instruments, sampled libraries are the ones that can best reproduce the live sonority of the trumpet. I recommend using trumpet sounds from different libraries in order to recreate the live acoustic variation in sound that is peculiar to each player. Remember that, as with strings, you should assign a different MIDI track and MIDI channel for each part. This allows you greater control over the pan, volume, and intonation of each line.

To recreate a more realistic effect, you can also slightly detune each trumpet playing a different line of the voicing. Remember to detune the parts so that the overall tuning is centered on a perfectly "in tune" part. For example, with a five-trumpet section, leave trumpet 1 in tune and detune trumpet 2 by +2 cents, trumpet 3 by –2 cents, trumpet 4 by +1 cents and trumpet 5 by –1 cents. This will avoid having one part of the lines more out of tune than the others. Keep in mind that the detuning settings can vary with the quality and accuracy with which the live samples were recorded. To avoid unpleasant results, start with more moderate settings and make sure not to overdo it.

As for panning, I recommend treating each section of the brass as a unit by itself instead of "micro-panning" each instrument of each section. As a general rule try to avoid panning each trumpet individually hard left or hard right, but instead think in terms of where the section would stand in a live performance situation. The same principles of detuning and panning can be applied to the other sections of the entire woodwind family, as we will learn later in the chapter.

5.4.2 The trombone

The trombone covers the mid-low range of the brass section, usually reinforcing the trumpet section. Notice the so-called *pedal* tones, which can be used to extend the lower range of the trombone (while in theory there are seven fundamental pedal tones, ranging from B♭ down to E, only the first three are considered reliable enough to be played in a live situation). The bass trombone can actually reach the fully extended low range through the use of additional tubing and triggers. The normal ranges of the trombone and bass trombone are shown in Figure 5.15.

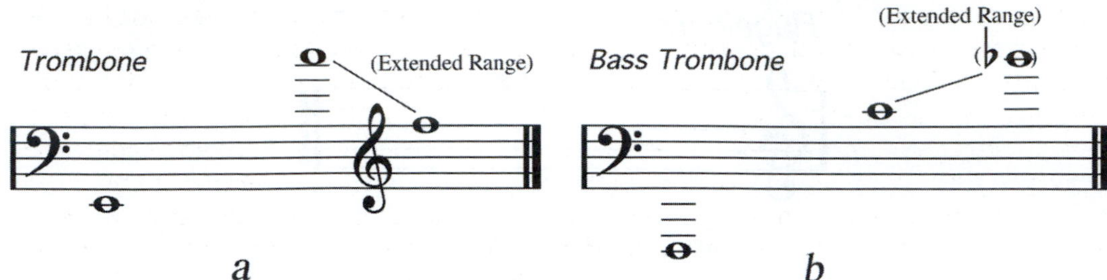

Figure 5.15 The range of (a) the trombone and (b) the bass trombone.

A normal section usually comprises four or five trombones. A common way to combine trumpets and trombone sections is to have the trombones doubling the trumpets an octave lower. This represents a starting point for basic orchestration that can be used to experiment with new sonorities and brought to more adventurous levels. Trombones are operated through a slide that can be used to create portamento effects that are typical of this instrument. Pitch Bend can generally be used to recreate this effect with mixed results, depending on the parts. The best way to reproduce such sliding effects is to use multisample libraries that allow you to create different sonorities by triggering different samples. Particularly effective are samples that reproduce the so-called *short drops* (also called "fall-off") and *long drops,* which feature a "falling" pitch at the end of a phrase. The same effect can be used for the other brass sections and in particular for the trumpet section. Trombones, as with other brass instruments, can be voiced in close position, creating a cluster of notes that is dramatic and tense, or in open voicing, which is indicated more for powerful and triumphant passages. You can apply the same rules of panning as you learned for the trumpet section, keeping in mind, though, that since the trombone section covers a lower-frequency range, it is usually better not to pan it hard left or hard right, as with any other bass-frequency-based instrument.

5.4.3 The French horn

The French horn has a beautiful sonority that can enrich the brass section by adding a nice round color. The horn can play delicate melodic passages as well as powerful section parts, depending on the dynamic used. In comparison with the trumpet and trombones, it is less powerful in terms of sonority and projection, and because of its construction it is not capable of achieving the same agility on fast passages. The horn is a transposing instrument written a perfect 5th higher than its actual sound. Its range is shown in Figure 5.16.

Figure 5.16 The range of the French horn.

The French horn can be reproduced very well by sampled-sound libraries. The quality of its sound at both low and high dynamics is very effective when sequenced with a multisample patch. Its timbre changes drastically from mezzopiano to fortissimo, going from mellow and gentle to powerful and dramatic, and for this reason it is almost inevitable to use several samples in order to cover such a wide range of sonorities. An average section can be formed by three to five horns, making it a very interesting additional color to be added to the trombones and the trumpets. Because of the nature of the instrument, the French horn is often used to underline the harmonic progression through the use of long sustained notes. In this case remember to insert in your MIDI part spaces and pauses where the musicians would breathe in order to add a realistic effect to the sequence.

5.4.4 *The tuba*

The tuba covers the extreme low register of the brass section. There are several different types of tubas, such as the B♭ tuba (also called *euphonium*), the F tuba (also called *bass tuba*), and the BB♭ tuba (also referred as *contrabass tuba*), among others. The most commonly used is the contrabass tuba. Its range is shown in Figure 5.17.

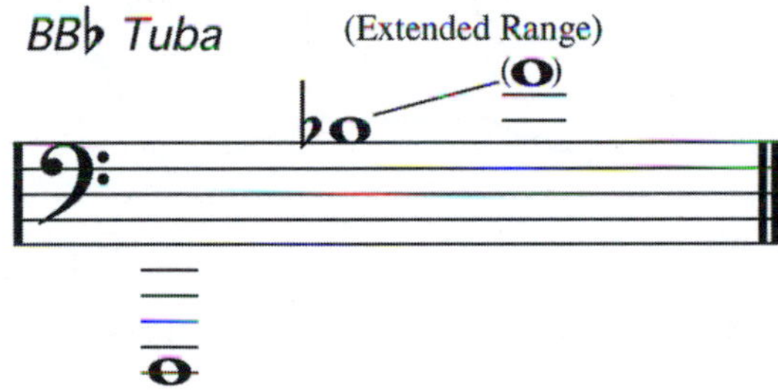

Figure 5.17 The range of the BB♭ tuba.

Sampled libraries can effectively reproduce the sonority of the tuba. Usually it blends extremely well with a four-horn section, providing the perfect complement to their mid-low range sonority. When sequencing tuba parts, pay particular attention to the phrasing you use, due to its somewhat limited ability to play rhythmically complicated passages at fast tempos.

5.4.5 *Sequencing brass: libraries, pan, and reverb*

I have few recommendations on how to sequence for brass. Always strive for clarity and balance. As I mentioned earlier, think in terms of real instruments and not in terms of electronic samples. It is crucial that you follow the range of the acoustic instruments and not write passages that would be unrealistic on a live instrument. In terms of panning and positioning, if scoring for a larger ensemble such as a studio or symphonic orchestra, use the diagrams in Figures 5.10 and 5.11 as a starting point.

If scoring for brass only, you can use alternative settings such as the one shown in Figure 5.18. This should be only a starting point to be altered and modified depending on the type of project and the type of instruments that are sequenced at the same time. If you are sequencing a brass-only ensemble, for example, then you have more freedom in terms of panning, since the real essence and main focus of the sequence are the brass sections. If, on the other hand, you are sequencing the brass sections as background pads for a more complicated ensemble, then I recommend keeping a more conservative approach to panning. As with the string section, try as much as possible to double (or substitute) the top line of each section with a live acoustic audio track. This will help to add dynamics and a much more realistic color tone to the MIDI parts. Among the sound libraries that I particularly enjoy using and recommend to sequence brass parts are the Dan Dean Brass Ensembles, the Miroslav Vitous Symphonic Orchestra, Peter Siedlaczek's Advanced Orchestra, and the Vienna Symphonic Library.

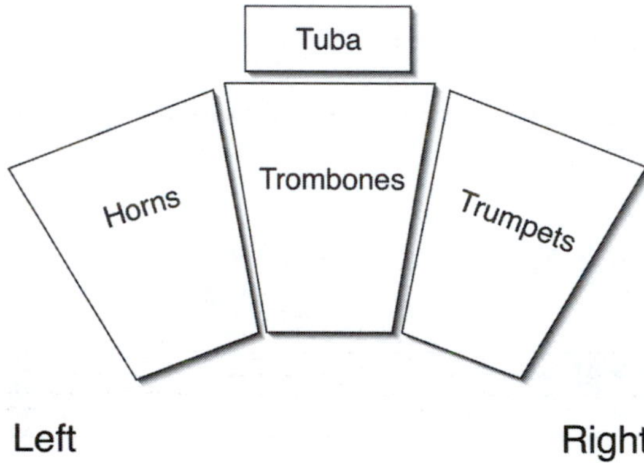

Figure 5.18 Suggested panning disposition of a brass ensemble.

5.5 The woodwinds: overview

The woodwinds includes the saxophones, the flutes, oboe, English horn, clarinets, bass clarinets, bassoon, and contrabassoon. The color palette provided by the woodwinds is wide and extremely versatile. The saxophones, for example, can be used equally effective in sequencing fast, swinging passages and velvety, slow, pad-like parts. The flutes, clarinets, and oboes can be used in addition to the saxophone section to add a sparkling and edgy timbre. The combinations are literally countless, but in the modern studio orchestra a few settings have emerged and have been standardized as part of the current live and recording setup. Keeping these formulas in mind will help you in sequencing more realistic and credible woodwinds parts.

5.5.1 The saxophones

The saxophone family includes soprano, alto, tenor, baritone, and bass. For modern orchestration you will most likely find the alto, tenor, and baritone to be the main voices. The soprano is often used as a solo instrument or as an addition to a regular section. A common saxophone section includes the use of two altos and two tenors, even though there are many variations of this setting, such as one alto and three tenors or a five-saxophone section that features the addition of the baritone. It is important to understand which ensemble you are sequencing for in order to reserve enough MIDI tracks and channels for each line. The range of the saxophones is shown in Figure 5.19.

The saxophones are transposing instruments. Table 5.3 lists their transposition keys.

The saxophones are very versatile in voicing. They sound equally well and balanced in unison, octaves, or closed or open voicing. When sequencing for saxophones, try to use lines that feature realistic and smooth voice leadings, just as a real player would have to perform them. This will guarantee a convincing sonority for the saxophone section. Unfortunately,

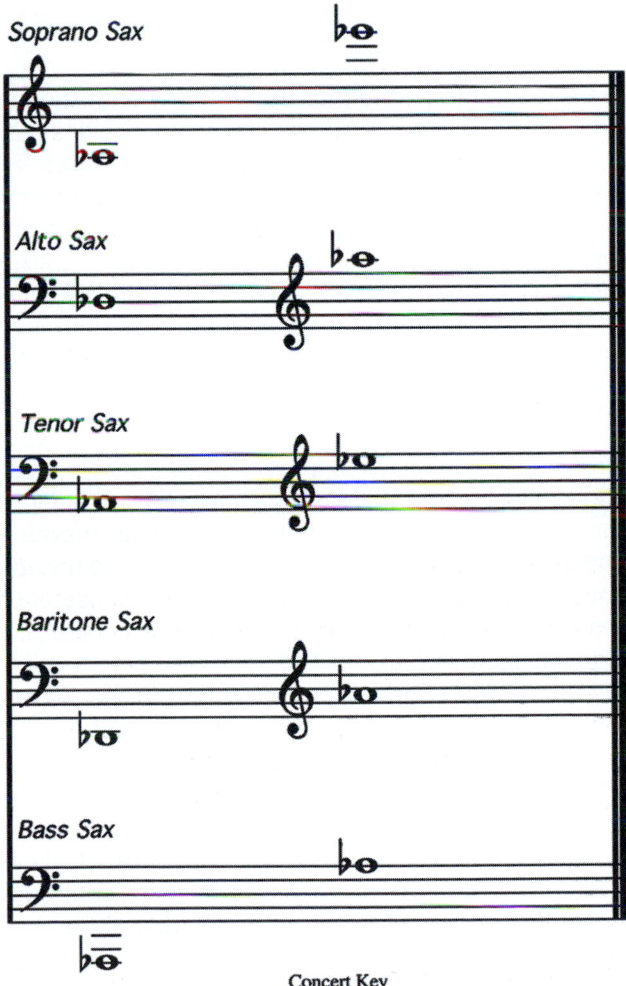

Figure 5.19 The range of the saxophone family.

Table 5.3 Transposition for the saxophone family

Instrument	Transposition
Soprano	Written a whole step higher than it sounds
Alto	Written a major sixth higher than it sounds
Tenor	Written a major second plus an octave (major ninth) higher than it sounds
Baritone	Written a major sixth plus an octave higher than it sounds
Bass	Written a major second plus two octaves higher than it sounds

the saxophone in general is one of the hardest wind instruments to reproduce with either a synthesizer or a sampler. While synthesized-saxophone patches sound very stiff and fake, sampler patches do a much better job in reproducing its original tones and nuances. The main problem, though, is that the saxophones feature sonorities and harmonic content that are very hard to reproduce in their full range through a sampler. Again, as seen earlier, a multisample patch can help us in recreating the complex sonorities of a saxophone.

Among the main libraries available on the market, I recommend using the Quantum Leap Brass library (even though it is labeled as "Brass" library, it includes realistic samples and patches of soprano, alto, tenor, and baritone solo saxophones and sections) and Saxophones 1 by Vienna Symphonic Library. When sequencing saxophones, more than with any other wind instrument section, it is advisable to include one or more audio tracks of live recorded saxophones along with the MIDI parts. The phrasing (especially for jazz and swing projects) that a live saxophone can achieve cannot be reproduced through MIDI without the introduction of artificial and mechanical effects. If you use a two-alto, two-tenor setup, you should either layer or substitute the lead alto and the lead tenor. If you sequence for two altos, two tenors, and one baritone, then you can opt for layering the lead alto and the baritone.

The remaining instruments of the woodwind family (flutes, clarinets, oboes, and bassoons) can very effectively be sequenced using sampled libraries. The simpler nature of the waveform of these instruments (the clarinet and flute in particular) makes their reproduction through synthesis and sampling techniques fairly easy. Their tone quality is pretty stable in terms of sonorities and dynamics. Particularly effective are the use of oboe and bassoon. These two instruments can easily be placed in a sequence without the need to replace them with live performances. Among the libraries that can reproduce the aforementioned woodwinds with particular fidelity I would mention the Vienna Symphonic Library, the Miroslav Vitous Symphonic Orchestra, and Peter Siedlaczek's Advanced Orchestra.

5.5.2 The flutes

The flute is the most agile of the woodwinds and is therefore a fairly easy instrument to sequence. The flute in C is probably the most common instrument belonging to the flute family, although it is not rare to sequence for piccolo flute and more seldom for the alto flute. Their ranges are shown in Figure 5.20.

The flute can be used in sections (open voicing usually works better than clusters), as solo instrument, or as reinforcement to the string or brass section. It blends particularly well with the other woodwinds, especially if voiced at the unison or in octaves. The flute can

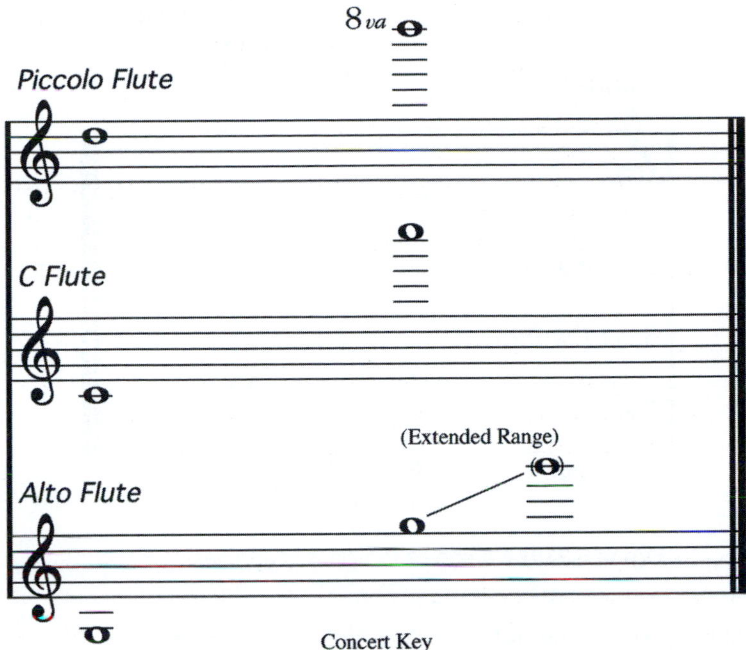

Figure 5.20 The range of the flute family.

be employed in an extremely wide range of styles and situations, from pop to jazz, from classical to ethnic.

5.5.3 The clarinets

The clarinet is a very agile instrument (second only to the flute) and can be used in both classical and jazz ensembles. It has a dark and melancholy sonority that can also be used in sparkling and up-tempo passages. Since its waveform resembles a square wave, even basic synthesizers can reproduce a convincing clarinet sound. The tone quality depends mainly on the range used. The lower range has a dark quality, the middle register has a more cold sonority, while the higher register features a bright and exciting sound. The clarinet is a transposed instrument in B♭ (it is written a whole step higher than it sounds). The bass clarinet covers the lower rage of the clarinet, providing a deep and pointed sound. It is a transposing instrument, written a major ninth above where it sounds. The range of the clarinet and bass clarinet is shown in Figure 5.21.

As with the flutes, the clarinets are very versatile instruments in section or solo. In section it is better to use an open voicing setting, while as solo it can be sequenced as double of the lead part (unison or octave) of pretty much any combination of brass and woodwinds. In order to achieve a more realistic sound in a MIDI setting I would recommend using two patches of the same sonority and detuning one of them by a range between 1 and 3 cents. By alternating the use of the two patches you can recreate a realistic detuning effect like one that would occur in a live situation. This technique can be also applied to the other woodwinds, especially the oboe, which has a rather difficult intonation. Be cautious,

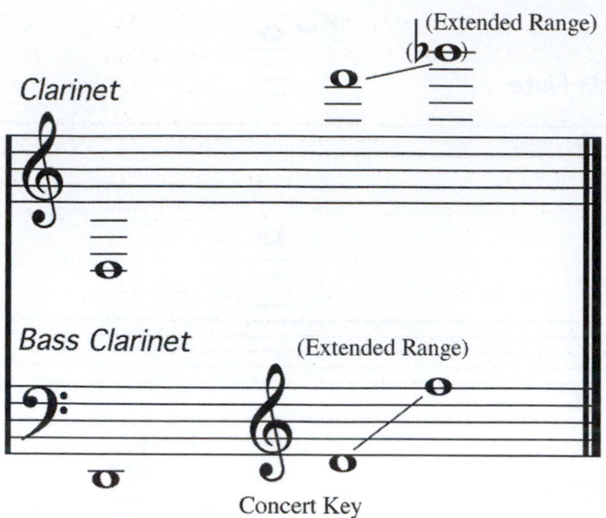

Figure 5.21 The range of the clarinet and bass clarinet.

however, in applying any of the detuning techniques explained so far. Misuse could make your virtual orchestra sound like a bunch of drunk musicians!

5.5.4 The oboe and the English horn

The oboe and the English horn are, among the woodwinds, the instruments that can be particularly well reproduced by a sampler. Their sound is melodic and expressive, especially in the middle register. The oboe is a nontransposing instrument (it sounds where it is written), whereas the English horn is written a perfect fifth higher than it sounds. Their range is shown in Figure 5.22.

Figure 5.22 The range of the oboe and English horn.

One of the hardest elements to reproduce when sequencing wind instruments in general, and the oboe and the English horn in particular, is the natural and acoustic release of the instrument. While this aspect of the sound in an acoustic performance is controlled by the change in air that is blown into the pipe, in a MIDI setting one way to control the natural release of such instruments is to alter or reprogram the patch being used by lengthening the release parameter. While this will offer you the perfect solution for long and sustained passages, it will create problems in fast passages, where a short and responsive patch is needed.

A better solution is to take advantage of the flexibility of the Control Change messages and in particular of CC 7, which controls the volume. In passages where a staccato sonority alternates with longer sustained phrases, you can use CC 7 to gently fade the sustained notes as an acoustic instrument would do. You can achieve this effect either by controlling the volume through a slider programmed to send CC 7 while you are sequencing the part or by inserting the volume changes from the graphic editor after sequencing the part. In fact I suggest using a combination of the two techniques by recording rough volume changes while playing the part and then using the graphic editor to smooth out the details of the fades. In Examples 5.8 and 5.9 on the Audio CD you can listen to the difference between a woodwind part sequenced without volume changes and one that was sequenced by using the volume change technique just described.

5.5.5 *The bassoon and the contrabassoon*

The bassoon and contrabassoon are part of the same family of double reeds as the oboe and the English horn. The bassoon is a nontransposing instrument written in bass clef for the middle to lower range and in tenor clef for middle to high register. In the lower register

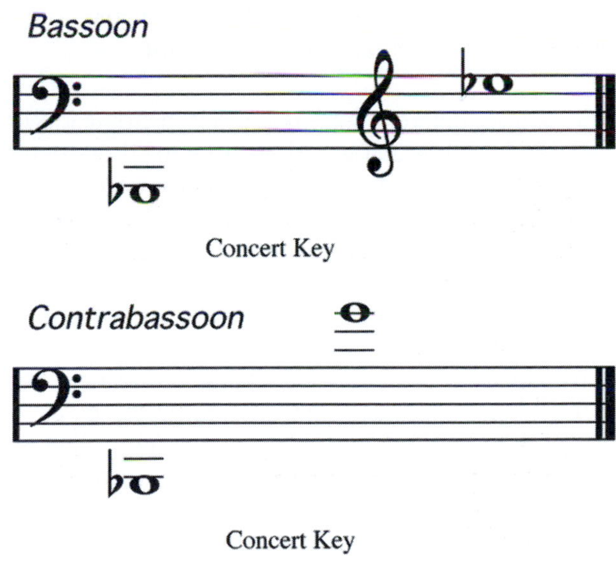

Figure 5.23 The range of the bassoon and contrabassoon.

it has a robust sound that gets more vibrant and edgy in the middle and high range. It can be effectively combined with the other woodwinds to give more support to the lower register of the section. It can also be replaced by the bass clarinet, which has a more present overall sonority. The contrabassoon is pitched an octave lower than the bassoon. It is used mainly to reinforce the bassoon in lower passages or to give more strength to the double bass or to the woodwinds. Its low range is strong and powerful, while its high range is preferably avoided or substituted with the bassoon. The range of the bassoon and contrabassoon is shown in Figure 5.23.

5.6 The synthesizer: overview

Even though the acoustic instruments just analyzed play an important role in the modern MIDI orchestra setup, most of the contemporary productions feature synthesized sounds that have little or no connection with acoustic instruments. However, one of the most interesting parts of having synthesized instruments available in your palette as composer and arranger is their variety and endless resources in terms of color and sonorities. In fact, as we saw for the acoustic instruments, synthesizers also can be grouped in families or categories, depending on the type of sound generator and technique used to produce the waveforms. The boundaries between these categories, called synthesis techniques, in reality are much less clear than those in the acoustic realm of instruments.

Nowadays the main synthesizers include hybrid versions of several techniques in order to produce machines that are more versatile and well rounded than ever. Nevertheless each synthesis technique has a distinctive sonority and is mainly used to produce a certain category of sounds and patches. As we learned earlier in this chapter, different categories of wind instruments can be arranged and combined to produce effective ensembles. In the same way, synthesizers and synthesis technique can effectively be mixed and combined to achieve a versatile and powerful contemporary orchestra.

We already learned how to layer different sounds in order to create innovative sonorities and how to combine acoustic and synthesized sounds in order to program realistic acoustic/MIDI parts. In this section of the chapter we are going to learn more about some of the most popular synthesis techniques and their sonic features in order to be able to make the right choice when selecting a synthesizer or sound module for our studio or our projects. The type of synthesis used by a particular machine can have a big impact on the color, timbre, and final sonority of the patches produced. Among the many types of synthesis, I will briefly discuss the ones that are most commonly found in contemporary synthesizers: subtractive, additive, frequency modulation, wavetable, sampling, physical modeling, and granular. The goal of this section is not to provide a complete guide to synthesis and its many features, a subject that would take a manual of its own. Instead it is to render an overall picture of the aforementioned types of synthesis available on the market for the modern composer and producer to enable them to make the right decisions when selecting sounds and devices for their projects. For more specific and detailed information on synthesis I recommend reading *Sound Synthesis and Sampling* by Martin Russ and *Computer Sound Designing: Synthesis Techniques and Programming* by Eduardo Miranda, both published by Focal Press, UK.

5.6.1 Hardware and software synthesizers

Before getting to the specifics of different types of synthesis it is important to understand the difference between two ways of conceiving the role of the synthesizer. Up to the late 1990s the word *synthesizer* meant, to most composers and producers, some sort of hardware component (sound module, keyboard synthesizer, drum machine, etc.) able to both reproduce acoustic sonorities and generate completely new waveforms, such as pads and leads. As hardware synthesizers became more and more sophisticated and demanding of hardware power and complicated to program, a new breed of synthesizer started to become popular: the *software synthesizer*. This new tool, based on the combination of a computer and a software application, takes advantage of two main factors: (1) In recent years, computers became more and more powerful, and (2) hardware synthesizers became mainly a combination of internal processor (CPU) and software written for that particular type of processor.

The main difference between hardware and software synthesizers consists in how the sound is generated. While the former utilize a dedicated CPU to produce the sounds, the latter take advantage of the CPU of a computer and specially written software to create the waveforms. The advantages of software synthesizers are many. After the initial investment in the computer, the price of a software synthesizer is much lower than its hardware counterpart. In addition, the continuous evolution of algorithms, raw waveforms, and patches can easily be integrated with a new version of the software, while the upgrade of hardware synthesizers would be more problematic.

Another important aspect that makes software synthesizers the synthesis option of the future is their seamless integration with sequencers and other music software. Through the use of plug-ins and standalone versions it is nowadays possible to create, sequence, record, mix, and master an entire project completely inside a computer, without the audio signal having to leave the machine. Of course one major drawback of this type of synthesis is that it requires very powerful computers to run several plug-ins at the same time on top of the MIDI and audio sequencer. Software synthesizers are becoming the standard for every contemporary production in both the studio and the live situation. The availability of portable computers that are almost as powerful as their desktop counterpart contributed greatly to the spread of software synthesizers.

Evidence of the increasing interest in the development of the software synthesizer is that even major manufacturers such as Korg, EMU, and Roland are developing new synthesis engines based on software-only algorithms. Even more striking is that more and more companies are bringing back vintage hardware synthesizers as software versions, making available the warm sonorities of the old analog devices with the power and the possibilities of the new software-based technology. The types of synthesis I will describe in the following pages are not platform dependent, meaning that the actual way the waveform is produced doesn't basically change from hardware to software. The idea behind each synthesis technique, at its most basic level, is not affected by the type of synthesizer (hardware or software). In the same way the sonorities typical of each synthesis technique are not substantially changed by the fact that the waveforms are produced by a software or hardware synthesizer. In general, though, the software approach opens up possibilities that are usually not conceivable at affordable prices on hardware platforms.

A software synthesizer is usually available in two formats: as a plug-in or as standalone. The former requires a *host application* to run. Without this application, the synthesizer is unable to launch. All four sequencers analyzed in this book function as host applications for the main software synthesizers available on the market. Plug-in software synthesizers come in different formats, depending on the application used to host them. Table 5.4 lists the plug-ins format supported by DP, LP, CSX, and PT. The same plug-in formats in fact are utilized for regular effects such as reverb, delay, and chorus. As noted in the "Comments" column of Table 5.4, through the use of a so-called "wrapper," certain formats unavailable natively for a certain application can be used. A "wrapper" is a small application (a sort of software adapter) that is able to translate one plug-in format into another. In recent years a few host applications, not sequencer based, have emerged. These programs (such as Steinberg V-Stack) are used only as plug-in hosts for software synthesizers. The advantage of such hosts is to provide availability and use of multiple plug-ins without the need of running an entire MIDI/audio sequencer. They are particularly well suited for live performances, where stability and streamlined settings are extremely important.

Table 5.4 Plug-in format supported by DP, LP, CSX, PT LE, and PT TDM

Application	Plug-in format	Comments
DP	MAS AU	VST format plug-ins can be used through a VST-to-MAS "wrapper". DP version 4.5 can also host RTAS
LP	AU Proprietary Logic Format	VST format plug-ins can be used through a VST-to-AU "wrapper"
CSX	VST DirectX (Windows Only)	
PT LE	RTAS	VST format plug-ins can be used through a VST-to-RTAS "wrapper"
PT TDM	TDM RTAS	VST format plug-ins can be used through a VST-to-RTAS "wrapper"

An alternative to the plug-in format is the standalone application. A standalone software synthesizer is a separate independent application that doesn't need a host to run. This option has the advantage of being slightly less demanding in terms of CPU power, since it doesn't require an entire sequencer to run. One of its drawbacks, though, is a lower integration in terms or routing and interaction with other components of your virtual studio. Most software synthesizers are available as both plug-ins and standalone versions. The choice between one or another format really depends on how you will use the software synthesizer. For live settings I recommend the standalone version, since it is usually more stable. For a MIDI/audio studio situation, where the sequencer becomes the central hub of both MIDI and audio signals, I recommend a plug-in system, which is more flexible in terms of signal routing.

5.6.2 Synthesis techniques

The majority of the synthesizers (both hardware and software) available on the market use one or more synthesis techniques that have been developed since the 1960s. Each approach to synthesis has a peculiar way of creating (or, better, synthesizing) the waveforms. This factor contributes to linking a specific synthesis technique to a unique sonority. This is particularly important to keep in mind when selecting a synthesizer and integrating it in your studio. The right choice of machine or module to buy or to use in a particular project depends mainly on the type of sonority and patches you need. For example, you would hardly choose an analog synthesizer to program a realistic string ensemble, just as you probably wouldn't choose a sampler to program a multistage complex synthesized pad. Let's take a look at the most common and popular types of synthesis to gain a better understanding of their features and typical sonorities.

5.6.3 Analog subtractive synthesis

A synthesizer is a device capable of generating electrical sound waves through one or more *oscillators* (VCO — voltage controlled oscillator), which are electronic sound sources (analog or digital) used to produce sound waves, which can be simple or complex depending on the level of sophistication of the oscillators used. Analog subtractive synthesis constitutes one of the oldest types of synthesis, made commercially available and marketed in the 1960s. In the case of analog subtractive synthesis, one or more oscillators generate a basic and repetitive waveform: sine waves, triangular waves, square waves, saw tooth waves, pulse waves, and noise. This section constitutes the so-called "sound source" or "generator" section of a synthesizer. The other two main areas of a synthesizer based on the subtractive approach are the "control" section and the "modifiers" section. The former comprehends the keyboards, Pitch Bend, Modulation wheel, foot pedals, etc.; the latter includes the filter section, which is used to alter the basic and repetitive waveform generated by the oscillators.

The name *subtractive* comes from the process involved in creating more complex waveforms from a basic and repetitive sound wave. After being generated, the simple waveform is sent to the modifiers section, where a series of high-pass and low-pass filters (whose number and complexity vary with the sophistication of the synthesizer) "subtract" (or, better, remove) harmonics to the waveform, producing more interesting and complex wave shapes. The altered sound is then sent to the amplifier section, called a VCA (*voltage-controlled amplifier*), which is still part of the modifiers, where the amplitude of the waveform is amplified. The VCA can be altered through the use of an *envelope generator* (EG), which is a multistage controller that allows the synthesizer to control the amplitude of a waveform overtime.

The most basic EG has four stages: attack, decay, sustain, and release. In more modern synthesizers the EG can be divided into more stages to allow higher flexibility. In addition to controlling the VCA, the EG can be assigned to control how the filters change overtime, enabling the synthesizer to produce even more complex waveforms. To be able to introduce some variations into the repetitive cycle of the subtractive synthesizer, one or more auxiliary oscillators are introduced. These oscillators, called *low-frequency oscillators* (LFOs), have a rate much lower that the one used to generated the waveforms. They can be assigned to control several parameters of the synthesizers, such as the filters section or the pitch of the main oscillators. Usually the LFO can be altered by changing its rate (the speed at which the LFO changes overtime) and its attack time. The signal flow and the interaction among the sections and components of a subtractive synthesizer can be seen in Figure 5.24.

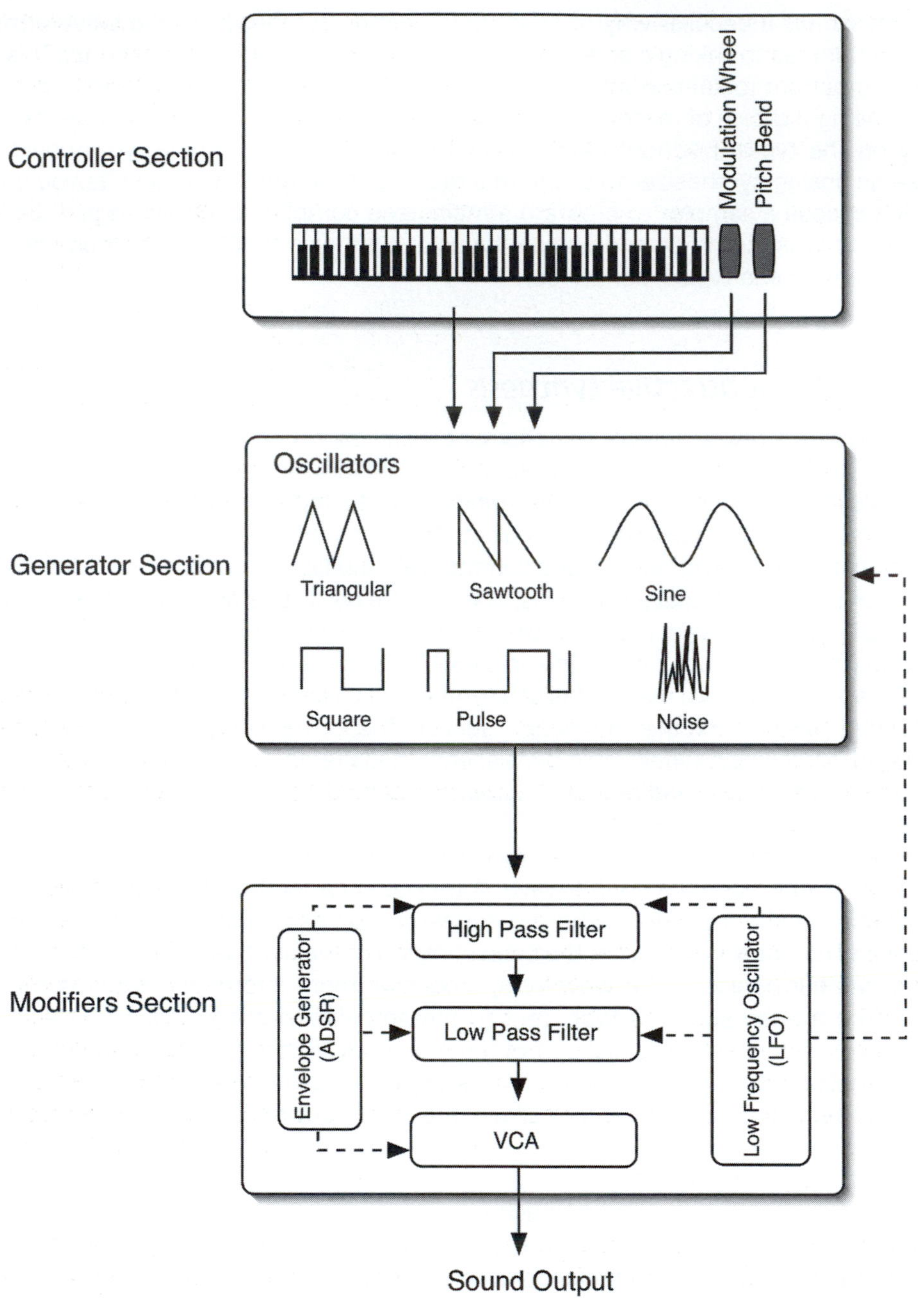

Figure 5.24 Components and sections of a subtractive synthesizer.

Some of the most important synthesizers that used subtractive synthesis as their main sound generator techniques are the famous Minimoog, the Prophet 5 by Sequential Circuits, and the Juno series by Roland. However, the most important aspect to consider when discussing this type of synthesizer is the sound quality and the overall sonority they are capable of producing. Subtractive synthesis is particularly suited for low and deep analog basses and edgy and punchy analog leads. These are among the sounds and patches that made subtractive synthesis famous and that are still largely used in contemporary dance productions. Another area in which subtractive synthesis is capable of producing original and interesting sonorities is synth pads that are rich and thick.

Subtractive synthesis, because of its limited basic waveforms, is usually not suited to recreating acoustic instruments such as strings and woodwinds, even though it can be somewhat effective in producing synthesized brass. Even though in the 1970s and '80s this type of synthesizer was used to sequence some orchestral parts, they can't compete with more advanced sample-based synthesis options. For some audio examples of analog synthesis generated by a subtractive synthesizer listen to Examples 5.10 through 5.13 on the audio CD. I always recommend having one or two subtractive synthesizers as part of your setup in order to be able to use some vintage "analog sounds." It is important to have as many sounds and patches available in your studio as possible to have a rich palette to choose from, and the vintage sonority offered by subtractive synthesis is a must-have.

5.6.4 Additive synthesis

Additive synthesis is based on the basic concept that even the most complex acoustic waveforms can be reproduced by summation of multiple sine waves. Additive synthesis, by using multiple oscillators generating sine waves, tries to reproduce complex and sophisticated sonorities by adding each waveform according to frequencies and amplitudes determined by the programmer.

A graphic example of this approach can be seen in Figure 5.25. In this example a simple sine wave (the fundamental) of amplitude 1 and frequency 1 is added to a sine wave (3rd harmonic) of amplitude 1/3 and frequency 3 (the numbers used are simply illustrative and have no reference to real frequencies and amplitudes). The result is a complex waveform compiled from the addition of sine wave components. By adding other harmonics at different amplitudes, the original sine wave will be changed even further, producing a completely new sonority. The example illustrates the power of this type of synthesis. One of the main drawbacks of the additive approach is that many oscillators are needed to produce complex waveforms, and therefore synthesizers based on the additive approach require a very powerful sound engine.

One of the first commercially available additive synthesizers was the legendary Synclavier, produced by New England Digital in the mid-1970s. Another fairly successful example of commercially available additive synthesizers was the Kawai K5, which used a 126-harmonic system to produce complex waveforms. In the case of additive synthesis there are mainly three sections, just as in the case of subtractive synthesizers: controller, generator, and modifier. The same structure, in fact, can be applied to almost any type of synthesis. The main difference among several synthesis techniques resides in the way complex waveforms are generated. The controller section is very similar to any synthesis approach. The modifier part can change and get more and more sophisticated as more modern filtering techniques becoming available.

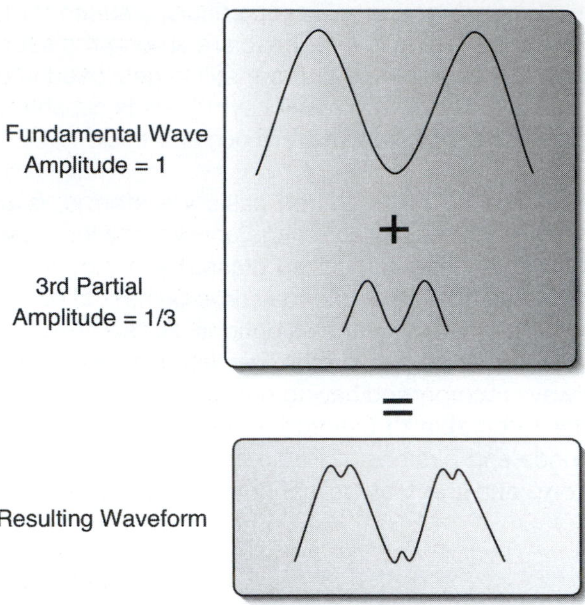

Figure 5.25 Example of basic additive synthesis.

Additive synthesizers are usually able to produce convincing thin sounds, mainly related to vibrating strings, such as guitars and harps. They are not particularly effective in reproducing percussive sonorities with fast transients, such as drum sounds. A new series of software synthesizers based on an evolution of additive synthesizers has recently emerged. These applications use the concept of *resampling* to resynthesize waveforms that are fed to the software as audio files. In other words a resampler analyzes a complex waveform provided as an audio file and then tries to synthesize it by reconstructing its complex wave through the summation of sine wave components. The results are particularly effective not only in creating fairly accurate acoustic sounds but mainly in generating synthesized pads and leads that are based on acoustic "genetic" audio material.

5.6.5 Frequency modulation synthesis

As we have learned up to this point, subtractive synthesis offers limited possibilities in terms of creating complex sonorities starting from elementary waveforms, and additive synthesis is capable of generating more advanced patches but with stringent requirements in terms of processing power. Another answer to the need for synthesizing more complex and sophisticated waveforms came from yet another innovative approach to synthesis: *frequency modulation* (FM). This technique involves a multiple-oscillator system such as in the case of additive synthesis. But instead of using the oscillators in a "passive way," where the resulting waveform of each oscillator is passively added to the others, in FM each oscillator influences and changes the output of the others.

In its most simple setup, FM synthesis uses two oscillators, one called the *modulator* and the other the *carrier* (advanced FM synthesizers utilize three or more oscillators to generate more complex waveforms). The modulator changes and alters the frequency of the carrier by constantly modulating the basic frequency at which the carrier operates. In a multi-oscillator setup there are usually several matrixes available that control the flow of the signal path among modulators and carriers according to predetermined algorithms. When you program an FM synthesizer, you can usually select several options as to which matrix to use to produce a certain complex waveform.

One of the most commercially successful FM synthesizers ever made was the Yamaha DX7, which was released in the early 1980s and saw many successful revisions and improvements over the years. FM synthesis can usually produce "glassy" and bell-like tones with good, fast transients and a stringlike quality. FM synthesizers are considered best at generating patches such as electric pianos, harps, synth electric basses, bells, and guitars. One of the drawbacks of FM synthesis is that it is fairly hard to program.

As with the other synthesis techniques analyzed so far, I recommend having at least one FM synthesizer available in your studio. There are a few software synthesizers based on FM synthesis, such as the classic FM7 by Native Instruments, which is based on the engine of the original DX7 but with new controls and matrixes available to create more sophisticated sonorities faster and more easily. Listen to Examples 5.14 through 5.17 on the audio CD to get an idea about the quality of the timbres generated by FM synthesis.

5.6.6 Wavetable synthesis

While all the synthesis techniques we have analyzed so far use basic waveforms with complex filters and algorithms to produce more sophisticated and complex sonorities, wavetable synthesis starts with complex waves that are sampled from both acoustic and synthesized instruments. The sampled waves are stored in tables in the ROM (read only memory) of the synthesizer to be recalled and used to build complex patches. This approach became more and more popular among synthesizer manufacturers in the mid- and late 1980s, when the price of memory started decreasing and the size of the chip started increasing. Wavetable synthesis has several advantages, the main one being that it can reproduce acoustic instruments with surprising fidelity since it actually utilizes sampled versions of the acoustic waveforms. A generic wavetable synthesizer stores the attack and the sustain parts of the waves. When the sample is triggered from the keyboard, the synthesizer will play the first part of the waveform once (the attack) and then will keep looping the sustained portion until the key is released. Depending on the amount of memory available, the length of the sustained part can vary considerably. Earlier wavetable synthesizers used to hold only few seconds of samples, and therefore the loop would be fairly short, affecting the overall acoustic accuracy of the waves.

Modern wavetable synthesizers can hold several minutes of samples, thereby reducing the loop effect and rendering a much more accurate timbre. Complex sounds can also be achieved by layering several waves in order to generate richer pads and sustained patches. Each layer is assigned to a *partial* that has separate control in terms of envelope, filters, effects, etc. The number of partials available changes from machine to machine, but generally in commercially available synthesizers you can find structures that use four or eight partials. The initial set of waves available with the machine can be increased through the use of expansion slots or cards that can be installed inside the device or inserted in the front panel. The expansion cards vary greatly in size and content.

One of the main problems that early wavetable synthesizers had to face was the limitation of ROM initially installed onboard. This had a clear impact on the overall quality of the sounds, since the loop section had to be very small. Later models could afford a much larger memory and therefore be more accurate. The samples generated are usually processed through a series of filters and envelope generators, similar to the ones we saw with subtractive synthesis, to further shape the waveform. Wavetable synthesizers are particularly appreciated for their overall flexibility, comprehensive lists of patches, all-around use, and good to excellent sound quality. These types of synthesizers are the most valuable in a studio since they can reproduce with good results pretty much any type of sonority. I highly recommend having at least two machines based on this type of synthesizer as MIDI "workhorses." Some of the most successful synthesizers that use this approach are the Roland JV series (1080, 2080) and their evolution, the XV series (3080 and 5080, and the "Fantom" series).

5.6.7 Sampling

A sampler is a particular type of synthesizer. It utilizes a similar approach to that seen in the wavetable technique, but instead of being limited to a table of small samples stored by the manufacturer in the machine's ROM, it stores the sample in RAM (random access memory), which can be erased and "refilled" with new samples at the user's will. When you turn on a sampler, it contains no samples in RAM; you have to either load a bank of samples from an HD or CD-ROM or sample your own waveforms. Before turning the unit off, you have to remember to save to HD the samples and the changes made to the banks in order to quickly be able, at a later time, to reload the settings for future sessions. Samplers are the best options to reproduce acoustic instruments, as I explained in the first part of this chapter. The amount of RAM available on a device reflects the number and length of samples you can load and use at the same time. The samples recorded and stored in the memory are mapped to the keyboard in order to have the full extension and range of the instrument we sampled. The higher the number of samples that form a particular patch, the more accurate the patch will be, since for every key that doesn't have an original sample assigned the machine will have to create one by interpolating the digital information of the two closest samples available. As in the case of wavetable synthesizers, the sampled waves can be altered through a filter section similar to the one found in subtractive synthesizers. Some of the modern and most advanced hardware samplers can go up to 512 MB of RAM, allowing you to use extremely accurate and high-quality multilayer patches.

Software samplers are becoming more and more popular, mainly for their ability to take advantage of large memory sizes and (for certain types, such as Kontakt and GigaStudio) their unique feature of being able to stream the samples directly from the HD, basically putting an end to the limitations created by the RAM-based architecture of hardware samplers. This type of approach to sound creation constitutes the core of the modern MIDI and audio studio. The majority of acoustic instruments that are sequenced nowadays are recorded using sample-based devices. Hardware samplers are slowly fading out, but their software counterparts are becoming more and more popular, delivering more advanced possibilities in terms of sophistication of the sounds and integration with the other devices involved in the production process. Among the most used sampler applications are the MachFive by Mark of the Unicorn, Kontakt and Kompact by Native Instruments, ESX24 by Apple/Emagic, and GigaStudio by Tascam.

5.6.8 *Physical modeling synthesis*

Physical modeling (PM) synthesis is relatively newer than the other types of synthesis analyzed so far. With physical modeling synthesis, instead of trying to recreate and produce a waveform starting from an analysis of its original acoustic counterpart and then trying to reshape a synthesized version of it through the use of filters and modifiers, the synthesizing process involves the analytical study of how a waveform is produced and of all the elements and physical parameters that come into play when a certain sound is produced. The sound-producing source is the key element here and not the resulting waveform.

Physical modeling synthesis is based on a series of complex mathematical equations and algorithms that describe the different stages of the sound-producing instrument. These formulas are derived from physics models designed through the analysis of acoustic instruments. Behind physical modeling lies the principle of the interaction of a vibrating object (strings, reeds, lips, etc.), the medium (air), an amplifier (e.g., the bell of a trumpet, the body of a piano, the cone of a speaker in an amplifier) in producing a particular sonority. Mathematical models that describe how the several elements of any sound-producing instrument interact are stored in the synthesizer that generates the waveform, calculating in real time the conditions, relations, and connections between each part involved in creating a particular sound. The number of algorithms present in a PM synthesizer determines its versatility and sound-generating power. An example of such models is the famous Karplus-Strong algorithm, which describes in mathematical terms a plucked string. A diagram representing the various stages in which PM would dissect the process involved in the production of a violin waveform is given in Figure 5.26.

One of the drawbacks of PM is that it requires an incredible amount of CPU power to process in real time the calculations necessary to generate the waveform. Because of this, PM has only fairly recently become one of the most advanced and interesting types of synthesis implemented in both hardware and software synthesizers. The strengths of PM are many, and they all impact the final users (meaning the programmers, composers, and producers) in one way or another. The programmer is presented with parameters that are usually fairly easy to understand and deal with, since they mostly reflect real acoustic parameters. No longer do you have to "guess" which parameter (or parameters) will have an impact on the way a sound will change when sending a higher-velocity value or how another sound will change when using a certain vibrato technique. With PM, sounds are programmed using real parameters that apply to real acoustic instruments and that are therefore much easier to relate to a predetermined sonic result.

PM synthesizers require much less RAM than wavetable devices and samplers. The former can calculate the waveform for the entire range of the instrument without requiring a multisample setup, while the latter need several samples not only to cover the full range but also to provide all the variations necessary to reproduce several tonal colors. In order to slim down the amount of calculation a PM synthesizer needs to do in real time, the original algorithms are often simplified and translated in sequence of filters already programmed. Other "tricks," such as cycle loops, are often used to simplify even further the computational process.

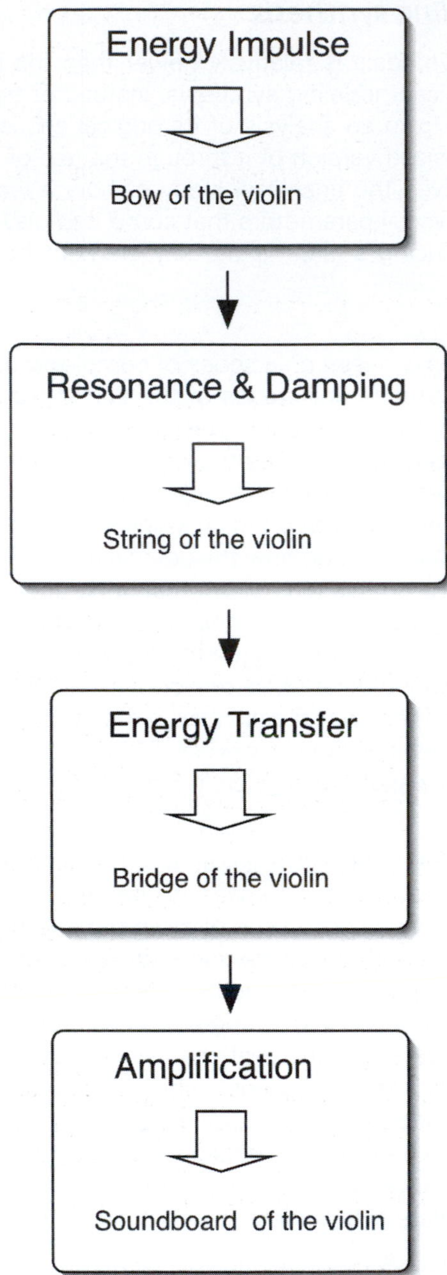

Figure 5.26 Example of physical modeling for a violin sound.

One of the most active manufacturers in developing accessible and marketable solutions for musicians based on PM synthesis has been Yamaha, which in the mid-1990s introduced one of the first commercially available and fairly affordable PM synthesizers, the VL1, which had a two-note polyphony limit.

As with other synthesis techniques, PM is progressively taking advantage of the software synthesizer revolution. The extreme flexibility and adaptability of software synthesizers make them perfect companions for a type of synthesis such as PM. As we saw earlier, the number of algorithms preprogrammed in a PM synthesizer determines the capabilities of the machine. While most hardware devices are based on a pretty "closed" architecture that makes software upgrades rather complicated and limiting the possibilities of expendability in terms of new models, a software synthesizer allows for a much higher degree of flexibility, due to its intrinsic open architecture.

One aspect about PM that is probably more intriguing and fascinating is in fact not related to the reproduction of real acoustic instruments but instead to the unlimited possibilities in generating "morphed" instruments that are constructed by mixing and matching models and stages from several sources. Imagine having an "energy-generation" module of a violin fed into the "resonance and damping" section of a drum and then sent to an "amplifier" stage of a trumpet. The resulting sonority will be something totally new yet extremely musical and inspiring!

5.6.9 Granular synthesis

This type of synthesis is also fairly new as compared with some of the more established and widely used techniques analyzed so far. Even though the concept on which granular synthesis (GS) is based can be traced back to the end of the 19th century, actual practical use in synthesis was developed only around the late 1940s and early '50s. The main idea behind GS is that a sound can be synthesized starting from extremely small acoustic events called *grains*, which are usually between 10 and 100 ms in length. Granular synthesizers start from a waveform (it can be any type of waveform, such as FM, wavetable, or sampled) that is "sliced" in order to create a so-called "cloud," which is made up of hundreds or even thousands of grains. A completely new and complex waveform is then assembled by reordering the grains according to several algorithms that can be randomly based or functional.

One advantage of GS is that the sounds created from such small units can constantly change and evolve, depending on the grain-selection algorithm. Since the grains forming the sound cloud are so small and so numerous, their repetition is perceived not as a loop (as in the case of wavetable synthesis or sampling) but instead as "quantum" material that constitutes the basic substance of which sounds are made.

Grains and the resulting waveforms can be processed using more standard modifiers, such as envelope generators and filters. Usually GS is particularly suited for the creation of sophisticated and fresh-sounding synthesized pads and leads (check out Examples 5.18 and 5.19 on the audio CD). One of the drawbacks of GS is the CPU power required to calculate the grains and the clouds in real time. This factor limits the implementation of GS as part of software synthesizers.

This overview of some of the most popular and tested synthesis techniques was intended to give you a sense of how different approaches to synthesis can affect, in practical terms, the work of the modern composer and producer. Each technique has strengths and weaknesses that make it the ideal choice for a certain style, texture, and sonority. To understand their fundamentals is key to building versatile and efficient studios as well as to producing coherent and balanced projects. When buying or considering using a particular synthesizer (software or hardware), think about which texture and sonority you are looking for, keeping in mind the strengths of the synthesis techniques just analyzed, and make a decision based on facts and goals. While the synthesis techniques we learned can be specifically and individually found in devices on the market, it is more likely that you will have to choose between products that will incorporate several approaches to synthesis at the same time. In fact most of the devices (and applications) available today feature three or more techniques integrated with one another, expanding their flexibility and power.

5.7 Summary and conclusion

In the previous chapters I covered the sequencing techniques related to the MIDI and audio aspects of music production. While these techniques are extremely important in the overall environment of the modern producer and composer, the actual orchestration techniques related to the interaction of acoustic, MIDI, and synthesized instruments represent an even more valuable area of contemporary music production. The study of the different components that come into play when sequencing is crucial to achieving more advanced and professional results. We analyzed the main sections of a contemporary virtual orchestra: rhythm section, string section, wind instruments, woodwinds, and synthesizers.

The rhythm section, usually drums and percussion, guitar, bass, and piano, constitutes the core of many contemporary productions. The piano is among the instruments that are usually easier to sequence and reproduce in a MIDI setting, because most controllers used to input the MIDI data are keyboard based. Sampled libraries are usually the best option to choose from in terms of sounds and patches, mainly because of their multisample capabilities, which allow you to render with extreme fidelity all the colors and nuances of such a dynamic instrument. If possible, one of the best options when sequencing piano parts is to take advantage of acoustic pianos that are integrated with MIDI interfaces, such as the Yamaha Disklavier. This way you will be able to combine the realistic acoustic sound of the piano with the editing capabilities of the MIDI sequencer.

The guitar and bass can be effectively reproduced and sequenced through the use of guitar- and bass-to-MIDI controllers. The realism achieved with such devices is much superior to anything else programmed using a keyboard controller. This approach will allow you to use convincing and realistic voicing that would be hard to reproduce on any other type of controllers. This concept applies particularly to parts that involve chord-strumming passages. While arpeggios and single-line parts can be effectively sequenced from a keyboard controller, strummed passages are very hard to reproduce. If you lack access to a guitar-to-MIDI converter, you can edit the notes belonging to a certain chord in order to have each note of the chord slightly delayed by a few ticks, as it would be for an acoustic guitar. In order to get more realistic sonorities for both bass and guitar sequenced parts, I recommend rerecording with a microphone the MIDI tracks as audio after sending the output of your synthesizer (or sampler) through a guitar or bass amp. This will add a realistic ambience and tone to the sometimes-sterile sonority of the synthesized patch. As with most acoustic instruments, samplers represent the best option to effectively reproduce the sophisticated sonorities of the guitar. The electric bass sound is a little bit more forgiving and can be well reproduced by the most modern and advanced synthesizers.

The realism of sequenced drums and percussion parts greatly depends on the type of music and style you are programming. Usually, contemporary rhythms and "loop"-oriented grooves are easier to program, while more acoustic and feel-based parts (such as jazz and orchestral ensembles) represent a challenge when it comes to MIDI sequencing. Sound and groove (or rhythmic feel) are the two main aspects you have to focus on when it comes to drum sequencing. To achieve a higher degree of realism I recommend using MIDI drums or MIDI pads to record the parts. In the case of a drum kit, avoid having passages where more than four drum pieces are playing at the same time, since this probably wouldn't be possible in a live situation (unless you have multiple drummers). In order to improve the feel of a drum MIDI part, try to include one or two percussive instruments recorded live, such as shakers or tambourines. To improve the acoustic ambience and interaction with other tracks of drums and percussion, try to rerecord through a couple of good condenser microphones, on separate audio tracks, the MIDI material played through the main speakers of the studio (this technique is basically similar to the one explained for guitar and bass parts). Loops constitute a valuable alternative to MIDI parts but can sound too repetitive if used without variation. Use loops recorded in multitrack settings in order to provide more variety.

The string section is among the hardest and most controversial sets of instruments to reproduce with a MIDI system. The modern string section includes violin, viola, cello, and bass. Because of the sonic differences among these four instruments it is usually better to avoid using a generic "strings" patch. Instead use specific patches for each section and also for divisi passages. Sequencing strings in a MIDI setting is no different than writing the part for their acoustic counterpart. Therefore use the same voicing and intervals as for a real set of players. Avoid close voicing, especially in the higher register of the violins, unless you are looking for a particular harsh sonority (thirds, fourths, and fifths usually work better and can provide a more robust sound). Sequence each part and line individually and on separate MIDI tracks and MIDI channels. This will give you more flexibility when editing the data.

Sampled libraries are the first choice when it comes to string sounds. If your score makes use of alternate bowing techniques such as sul tasto, sul ponticello, col legno, tremolo, pizzicato, and glissando, use different MIDI channels for each sonority. To improve the realism of string patches, you can layer sampled sounds with synthesized ones in a ratio of roughly 70% to 30%, respectively. Another technique involves the layering of sampled instruments with real acoustic instruments. This approach is also valid for brass and wind instruments in general. If you choose to use this technique in your sequences, double the top voice of each section with an acoustic instrument recorded on audio tracks.

In order to recreate the natural attack produced by the bow on the strings of violin, viola, cello, and bass, use volume changes while playing the parts. If your controller allows it, assign one of its sliders to CC 7 and while playing create short crescendos for each note. This technique works particularly well for long and sustained passages. For legato passages, extend the end of each note to the beginning of the following so that they overlap by a few ticks (I suggest a value between 10% and 20% of the overall tick resolution of your sequencer). When mixing the string section, you should follow the natural configuration of a live orchestra, as shown in Figures 5.10 and 5.11. Adding a medium reverb to the strings can improve the overall texture and mix among the instruments. Reverb length between 1.7 and 2.3 seconds works usually well for a wide range of styles, remember to use a low diffusion setting for sustained passages and a higher setting for fast staccato parts.

The wind instruments category includes the brass section and the woodwind section. The latter is further divided into single reed, double reed, and mouth-hole instruments. Even though the wind instrument category covers a wide range of sonorities and sonic textures there are some common sequencing techniques that can be applied to all these instruments. As we saw in the case of string instruments, you have to sequence as you would write for acoustic instruments, that is, respecting range, voicing, and performance techniques that could be applied to live performances. Keep in mind that wind instruments have limited phrase length capabilities because the sound is produced by the air blown into the pipes. Therefore always check and avoid phrases that wouldn't be realistic and performable in a live setting. When possible, use a wind MIDI controller to sequence wind instrument parts. Brass instruments include trumpet, flugelhorn, trombone, bass trombone, French horn, and tuba. Each of these instruments has a wide range of sonorities and textures that can even be increased by the use of mutes. For classical brass ensembles, you can use effectively sampled libraries to reproduce their entire range of dynamics and sonic nuances. For more modern and rhythmic brass parts, synthesized brass patches can provide a useful alternative. In terms of voicing, try to avoid clusters and intervals smaller than or equal to a major second in the low register of the brass section (fifths and octaves work better). In the higher register, clusters are effective and can provide a nice, "punchy" sonority.

The woodwinds category includes the saxophones (soprano, alto, tenor, baritone, and bass), the flutes (C, piccolo, and alto flute), oboe, English horn, clarinets, bass clarinets, bassoon, and contrabassoon. The woodwinds are extremely flexible in terms of sonorities and combinations. The saxophones, for example, can be used equally effectively to sequence fast, swinging passages as well as velvety, slow, pad-like parts. The flutes, clarinets, and oboes can be used in addition to the saxophone section to add a sparkling and edgy timbre.

The saxophones sound equally well balanced in unison, octaves, or closed or open voicing. When sequencing for saxophones, try to use lines that feature realistic and smooth voice leadings, just as a real player would have to perform them. Saxophones are usually extremely hard to reproduce in a MIDI setting. Sampled libraries are definitely the way to go to get convincing and realistic results. In the case of the saxophone section, I highly recommend layering acoustic and sampled tracks to increase the realism and phrasing accuracy of the sequenced parts.

The remaining instruments of the woodwinds can be very effectively sequenced using sampled libraries. The simpler nature of the waveform of these instruments makes their reproduction through synthesis and sampling techniques more acceptable and easier to accomplish. Several combinations of different woodwinds can ignite a spark in your sequences. The flute, for example, blends particularly well with the other woodwinds, especially if voiced at the unison or in octaves. The clarinet is a very agile instrument (second only to the flute) and can be used in both classical and jazz ensembles, mainly because of its multifaceted sonority (dark in the low register, bright and exciting in the high register). To recreate the natural detuning effect that often characterizes woodwind solo instruments and sections, you can apply a fine detuning of around 1–3 cents. The use of volume changes allows you to effectively recreate the natural release of wind instruments. This technique is particularly telling when slow, sustained passages alternate with fast, staccato parts.

The synthesizer gets the same consideration, respect, and use in the modern orchestra/ensemble as any other instrument. In fact most contemporary productions feature synthesized sounds that have little or no connection with acoustic instruments. The use of synthesizers can be seen by the contemporary composer and producer as an additional color palette to play with when arranging and orchestrating a new project. The tone, color and texture possibilities of a synthesizer can vary considerably, depending on the type of synthesis adopted to generate the waveforms and patches. Among the many types of synthesis I discussed, the ones most commonly found in contemporary synthesizers are subtractive, additive, frequency modulation, wavetable, sampling, physical modeling, and granular. Table 5.5 gives the main features and the pros and cons of these synthesis techniques.

Table 5.5 Types of synthesis, their main features, and their pros and cons

Type of synthesis	**Description**	**Pro**	**Con**	**Best used for**
Subtractive	A three-section synthesis: controller, generator (using basic repetitive waveforms), modifier (filters, envelopes, amplifier)	Doesn't require a lot of CPU power	Somehow limited in terms of creating complex and changing waveforms	Synth basses, synth leads, synth pads
Additive	Based on the summation of sine waves at different frequencies and amplitudes to create more complex sonorities	In theory, capable of reconstructing any acoustic waveform with precision	Requires a lot of CPU power to run the necessary calculation in real time Hard to program	Acoustic and synthesized instruments

(Continued)

Table 5.5 (*Continued*)

Type of synthesis	Description	Pro	Con	Best used for
FM	Based on the interaction of two or more oscillators (modulator and carrier). The modulator interacts and modifies the carrier to create complex waveforms	Allows you to create complex waveforms	Complex to program	Bells, synth basses, electric pianos, plucked instruments
Wavetable	Based on a series of complex waveforms stored in ROM and modified through filters similar to those found in subtractive synthesis	Well suited for a wide range of instruments Expandable via expansion cards	Because of the limitation of ROM size, the complex waveforms are limited in length and are looped	Very versatile Can effectively produce acoustic and synthesized sonorities
Sampling	Based on samples of complex waveforms loaded and stored in RAM	Users can sample their own waveforms Patches can be updated via sampler library CDs and DVDs	Not ideal for complex and constantly changing synth pads Hardware sampler limited by RAM size	Best option for acoustic instruments, such as strings, winds, guitars, piano, percussion
Physical modeling	Based on mathematical descriptions (algorithms) of the different sound-production stages of instruments	Extremely powerful and innovative	Requires a very powerful CPU	Particularly suited to wind instruments, voice-based patches, guitars, and string instruments
Granular	Based on a simple or complex waveform subdivided into small "grains" forming a "cloud" of audio material. The grains are combined in patterns following mathematical or random algorithms to generate new, complex waveforms	Constantly varying and innovative sonorities	Requires a very powerful CPU Somewhat unpredictable results	Particularly suited to synth pads and leads

5.8 Exercises

Exercise 5.1

Using as reference the categories of instruments analyzed in this chapter and the ones provided here, make a list of the best patches available in your studio for each category. Make sure to note the device and patch number, and write some comments about each chosen patch.

Category	Device	Patch/bank no.	Comments
Piano			
Electric piano			
Electric guitar			
Acoustic guitar			
Electric bass			
Acoustic bass			
Drums			
Generic strings			
Violins			
Violas			
Cellos			
Basses			
Generic brass			
Trumpets			

Category	Device	Patch/bank no.	Comments
Flugelhorns			
Trombones			
French horns			
Tubas			
Clarinets			
Bass clarinets			
Soprano sax			
Alto sax			
Tenor sax			
Baritone sax			
Oboe			
English horn			
Bassoon			
Contrabassoon			
C flute			
Piccolo			
Alto flute			
Synth pads			
Synth leads			

6 The Final Mix

6.1 Introduction

Practical sequencing techniques and orchestration principles constitute the core of modern music production. The way you sequence and arrange the various parts of your project is crucial to achieving professional-sounding results. The previous chapters provided information and hints on how to improve the craft of sequencing, MIDI editing, and creative MIDI orchestration. To keep the highest standard in every respect and at every stage of the music production process is extremely important. As a composer and producer you probably have always focused your efforts mainly on the creative aspects of music production, sometimes forgetting the technicalities involved in this process. By now I hope you have learned how to balance the two by always keeping in mind how creativity and technology can go hand in hand. The latter is a tool that serves the former. Do not forget, though, that technology can also be an important source of ideas and inspiration to spark the creative process.

The fact that you have learned the techniques presented so far is meritorious and at the same challenging. Knowing more about what your studio can do for you can be useful and intimidating at the same time. What is often forgotten during the production of a project is that each stage, each moment is crucial and that the way we approach it has an impact on the final result. This is particularly true for the mix and the mastering (or premastering) stages. Often the composer and producer are so focused on the creative process that when it comes to mixing and delivering the final product they lose interest or they are not as focused anymore as they should be (this is true particularly for composers). The days of the composer as the person in charge of only the writing and arranging stages of music production are gone. The composer today most often is also the producer, programmer, performer, and audio engineer. This is why I want to dedicate the last part of this book to the final two stages of the music production process: mixing and premastering.

These two aspects are too often disregarded as "technicalities" by the composer/producer. A good mix can make a project shine and can improve its overall quality. A bad mix can damage irreparably a beautiful piece of music in the same way that bad sequencing can completely undermine a well-written composition. This is why I want to devote the last chapter of this book to presenting some fundamental principles pertaining to the mixing and premastering stages of music production. While this chapter doesn't pretend to cover extensively such a vast subject, it is true that most composers are unfamiliar with some of the most important principles of mixing. Therefore I consider it important for them to be exposed to some of the fundamental procedures and approaches involved at the mixing stage of music production. With just a few concepts and hints, your productions will sound better and will improve considerably in balance and clarity.

6.2 The mixing stage: overview

All the effort you put into composing, sequencing, editing, and orchestrating your project deserves the best final results. The mixing stage is where everything comes together. All your tracks, eq., effects, etc. are channeled down to a stereo track, or to a multichannel mix in the case of a surround project. Let's look at how to approach the final mix and how to solve a few of the most common issues that often puzzle the modern composer/producer.

The first step in approaching the final mix for a generic hybrid project composed of MIDI and audio tracks is to choose how to handle the MIDI tracks. If your session contains a combination of both MIDI and audio tracks, then first decide whether the MIDI tracks need some extra "tweaking" from an audio (eq., reverb, effects in general) point of view. The sound associated with the MIDI tracks is generated externally by the MIDI devices from the computer (unless the tracks are assigned to a software synthesizer); therefore to add effects to them you have to use the external processing power of your mixing board and outboard gear or the built-in effects of the MIDI devices.

This latter technique is definitely faster and can save you time, but it is also limited by the number of channels and features of your mixing board and by the number of audio outputs present on your MIDI devices. If you opt for this solution, then the procedure is fairly simple. The audio and software synthesizer tracks of your sequencer will be mixed by taking advantage of the internal plug-in effects installed within the applications, while the MIDI tracks assigned to external MIDI devices will be processed and mixed using the equalizers and effects connected to your mixing board or, in the case of a digital board, through its internal effects.

After having applied all the effects and made the needed changes, you will *mix down* the main output of the mixing board to a stereo audio track of your sequencer. As I mentioned before, while this technique is probably the fastest, it has some limitations and also presents a few problems. For example, it is usually pretty hard to match the reverb, eq., and dynamic effects of the plug-ins hosted by the sequencer and those of the outboard gear. By using two different sets of effects (plug-ins and external gear), most of the time you will in fact accentuate the already-noticeable difference between audio and MIDI tracks. In addition, if your MIDI devices don't feature enough separate outputs, it is hard to effectively apply eq. and effects to single MIDI tracks.

A better solution consists of "converting" all the MIDI tracks in audio and then mixing the entire project inside the audio sequencer. This approach will guarantee a more coherent and balanced mix. Some of the drawbacks of this technique, though, are related to the fact that the MIDI-to-audio conversion is very time consuming since it has to be done in real time and for every single track independently. If your project is 5 minutes long and you have to convert 20 MIDI tracks, it could take up to 100 minutes to dump all the MIDI tracks as individual audio tracks. Keep in mind that this figure can vary, depending on how many inputs and busses are available, respectively, on your audio interface and your mixing board. The higher the number of inputs on the interface and the higher the number of busses available on the mixing board, the faster the process will be.

Hard drive space and computer performance is another issue. By adding 20 more audio tracks (the ones converted) to the already existing ones, you will stress the CPU more, limiting its capacity in terms of number of plug-ins. Nevertheless this technique is preferable, especially if you have a powerful enough computer to handle a high number of audio tracks and plug-ins. By recording each MIDI track individually as audio material, you will have greater control over eq. and effects such as reverb and dynamics. In addition, for each track you will be able to take advantage of the automation techniques we learned in Chapters 2 and 3. To record each MIDI track as a separate audio track, you have to *solo* the MIDI track you want to record, record-enable an empty stereo track in the sequencer, route the channels of your mixing board receiving the output from the selected MIDI device to a bus, and connect the output of the bus to the input of the audio interface attached to your computer. If you have multiple busses available on your mixing board, you can actually simultaneously transfer several tracks (one for each bus). If your mixing board lacks a bus system, you can use its main output as a temporary bus and connect the main output of the board to the input of the audio interface (make sure not to send the output of the audio interface back to the board, to avoid a feedback loop). While this description can sound a bit intimidating, take a look at Figures 6.1 and 6.2 to compare the two options.

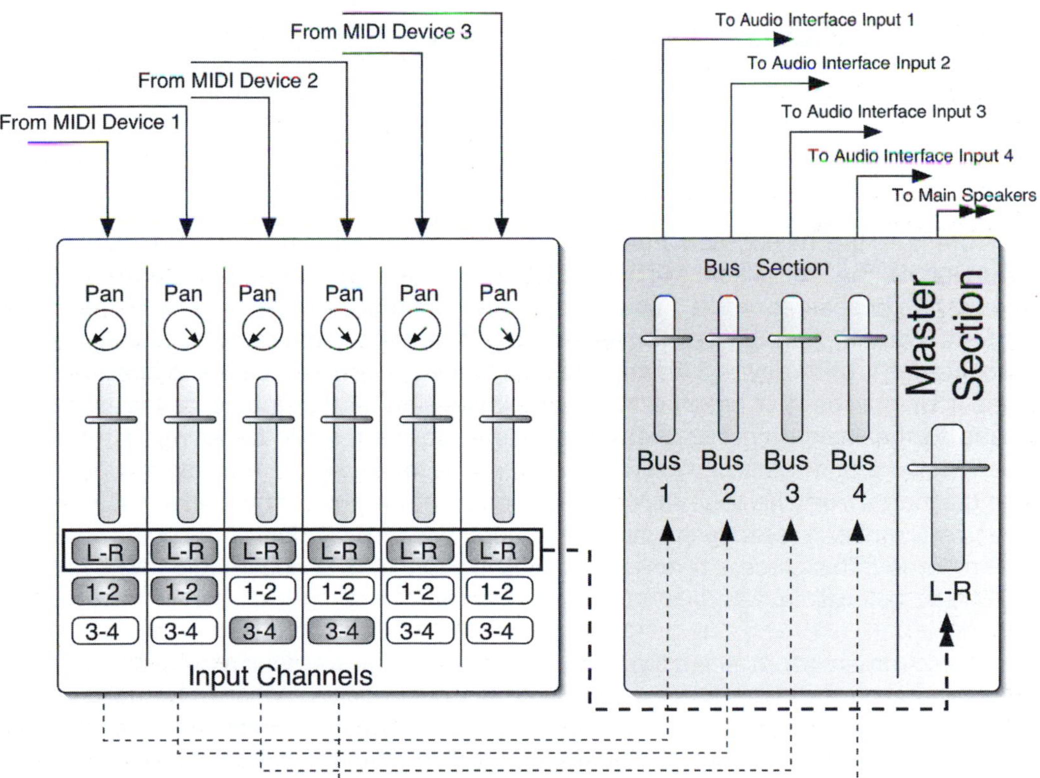

Figure 6.1 Signal routing to record MIDI tracks as audio with a multibus mixing board.

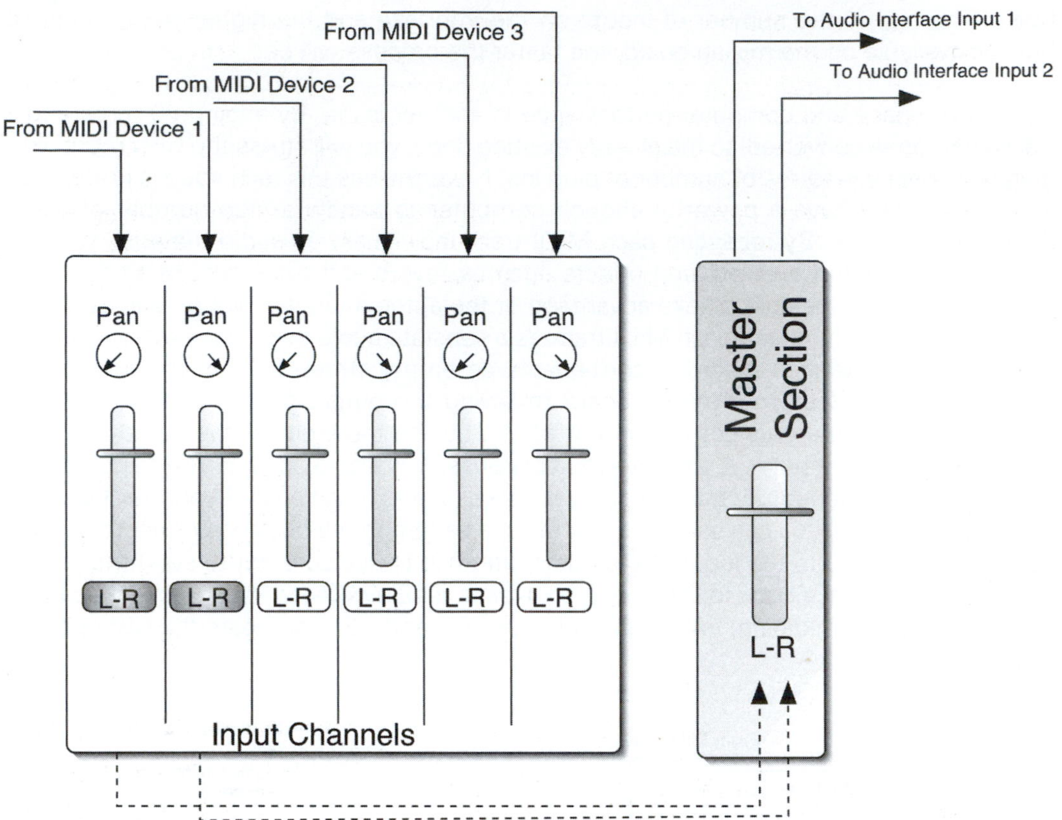

Figure 6.2 Signal routing to record MIDI tracks as audio with a mixing board without busses.

In Figure 6.1, the mixing board features a main set of outputs (L-R) and four busses (1 through 4). The inputs from MIDI device 1 are bussed to busses 1 and 2 and then sent (through the busses) to inputs 1 and 2 of the audio interface. In the same way, the inputs from MIDI device 2 are bussed to busses 3 and 4 and then sent to inputs 3 and 4 of the audio interface. MIDI device 3 will have to be bounced in a second pass and bussed to any of the four busses that become available. Notice how all the inputs of the board are bussed to the main output so that we can listen to them while bouncing. If the board doesn't have a bus system, you will temporarily have to use the main stereo output to send the inputs from the MIDI devices to the inputs of the audio interface. In this case you have to remember to send to the main L-R output only the channels of the board you want to be recorded. This process takes longer since, if you want individual audio tracks for each MIDI track, you will have to record them one by one.

In any case remember to record the signal on the audio tracks as loud as possible but without distortion. In the digital domain distortion can be a disaster. While in the analog world a little bit of distortion or saturation can create a nice warm and compressed effect, in the digital domain as soon as the signal goes over the 0-dB level you will get a nasty digital distortion noise that you won't be able to remove from the audio track. Usually try to keep your recording meters between −8 and 0 dB. When it comes to recording volume settings, before transferring the MIDI tracks as audio there are two approaches you can follow. If you have a fairly simple automation setup, where basically you only used CC 7 and CC 10 to set up initial volume and pan settings for the MIDI devices, I recommend suspending the automation for the MIDI tracks and resetting all the MIDI volumes to 127 (maximum) to be able to get the most output from each MIDI device. You will have to recreate the automation for volume and pan later, at the final mixing stage. If, on the other hand, you have created a complex automation for your sequence in order to recreate the natural attack and release of live instruments, as explained earlier in this book, I recommend leaving the automation "On" and dealing with overall volume issues later, at the mixing stage.

6.2.1 Track organization and submixes

Once you have transferred all the MIDI tracks to audio but before the start of mixing, take a few minutes to reorder the audio tracks, which will definitely speed up the process of mixing later. I recommend ordering the tracks by section: woodwinds, brass, strings, synthesizers, piano, guitar, bass, and drums/percussion. Keeping a standardized order for the tracks is particularly useful over time, since you will start getting used to it and it will get easier and easier to orient yourself in complicated mixing sessions. The order I suggest works well because it follows the standard order with which the sections are arranged in a score layout. With this template everybody can easily look at your sequence and be able to find any instrument in a matter of seconds.

After transferring the MIDI tracks as audio tracks, make sure to keep the MIDI data; do *not* delete them. They will come handy if you ever want to go back to the original MIDI tracks to make changes. Make sure you mute the MIDI tracks after each transfer, and move them to the bottom of the track list. Also keep in mind that, depending on the sequencer, you might be limited in the number of audio tracks or voices you can use. PT LE has a limit of 32 audio tracks, while DP and CSX theoretically have no limitation in the audio track count, their practical limit depends on the amount of RAM and on the CPU of the computer on which they are running. LP has a maximum of 255 stereo or mono tracks and an unlimited number of output channels in playback. If the track count is limited, you might have to create submixes of certain instruments to be able to handle a high audio track count. Even though I highly recommend keeping as many separate audio tracks as possible, Table 6.1 lists possible instruments that can be grouped without creating too many limitations in mixing capabilities.

From Table 6.1 you can probably discern the four main factors that allow us to group certain instruments together without having to sacrifice too much flexibility at the mixing stage: frequency range, reverb (and more generally "overall effects"), equalization, and pan.

The frequency range covered by each instrument and section plays an important role in grouping tracks. Usually instruments that cover a similar frequency range share similar settings of pan and reverb and therefore can be submixed on the same stereo track. The frequency range thus has a direct impact on the reverb settings and panning of the parts. Usually instruments that cover the lower frequencies of the audible spectrum need less reverb (or none at all) because high reverb would cause the mix to be muddy and unclear.

Table 6.1 Suggested submixes should you run out of tracks during a mixing session

Instruments/groups	Comments
High woodwinds	Can be grouped together because of their similar sonorities and acoustic response to reverberation; includes clarinet, flutes, oboe, and English horn
Low woodwinds	In general require less reverberation to maintain a clear and intelligible mix; includes bass clarinet, bassoon, and contrabassoon
Saxophones	Can be set as a separate category mainly because of their distinctive phrasing and sonority, even if they cover a wide frequency range
Trumpets	Constitute the higher range of the brass section, so they usually require different settings of reverb and equalization
Trombones, french horns	Cover the middle and low range of the brass section; ideally you want to keep the two separated for better control over the equalization of each sonority
Violins, violas	Even though there are substantial differences in overall sonorities, they cover the high and middle-high range of the string family and thus can be treated together in a mix where track count is an issue
String basses, cellos	Cover the low and middle range of the string family and usually require substantially different settings of equalization, panning, and reverb than the violins and violas
Synthesized leads	Can be grouped together unless their sonic features and purposes are completely different
Synthesized pads	Usually can be effectively mixed together since they can share most of the panning and reverb settings
Piano, electric piano, keyboard synthesizers	Acoustic piano would preferably be on a separate stereo track, if possible
Acoustic guitars	Make sure to separate acoustic guitars from the electric guitar since they have different needs of reverb and equalization
Electric guitars	See acoustic guitars
Electric/acoustic bass	
Bass drum	Needs to be separated from the rest of the drum kit mainly because it needs to be fairly dry (no reverb) and because of its peculiar equalization
Snare drum, toms	Ideally the snare would be separated from the toms, but you can sometimes have them on the same track because they both usually require a similar amount of reverb
Hi-hat, cymbals, shaker, tambourine	Can be assigned to the same groups because of their frequency range and reverb settings

For the same reason, high frequencies usually can handle and need more reverb; therefore instruments that cover this area of the audible spectrum can be submixed on a common stereo track.

It is a bit more complicated to group several tracks based on their equalization needs, though. Usually each instrument has peculiar frequency characteristics that are specific to its sonority. If you have to make a choice, though, try to keep instruments inside the same section separated according to the frequency range to which they belong (low, middle, or high).

6.2.2 The "rough" mix

Once you have the audio tracks in order and (if necessary) submixed according to the criteria just listed, it is time to start getting an overall balance among the tracks. This is an important step. Before starting to add effects and working on equalization, create an overall good balance between tracks. Even though there is no specific approach or technique for creating a draft mix, I recommend starting from the ground up, meaning from the foundation of the orchestration. Begin with the bass drums and then move to bass, then to the rest of the drum kit, followed by the piano and keys, guitars, strings, brass, woodwinds, and then leads (which could include synthesizers, acoustic instruments, or vocals). Every time you add an instrument, go back and adjust the volumes of the other tracks. If you are doing things right, you will probably have to apply small changes to the tracks that are playing back already.

To achieve a perfect balance among instruments and tracks, keep in mind a few important aspects. Each style and composition has a certain feel and overall sonority that needs to be achieved. Since you are the composer and producer of the project, you have the advantage of knowing extremely well the material being mixed. Focus on the elements important to that particular project.

Mixing these days is very much a companion to orchestration. By raising the volume of a certain section or instrument, you basically rewrite the dynamics you had in mind while writing and sequencing that part. I like to think of mixing as "conducting" the orchestra by indicating crescendos and decrescendos, sforzando and pianissimo, etc.

Once you've set in place a decent rough mix, I recommend saving it as a snapshot or storing each level as automation data; this will speed things up later if you have to program automation data for some of the tracks. When trying to get a good balance, look for two or three elements of the mix that represent the most characteristic and important features of that particular project, and work the other instruments and parts around it. This is a good approach to avoid a massive global wall of sound where nothing is clearly distinguishable and to reach overall clarity. Usually, lead instruments and especially vocal tracks are a good starting point for featured tracks. Always have two or three parts that you consider more predominant and exposed than the other background parts.

6.3 Panning

The way you position your instruments in the horizontal left and right axes has an impact not only on the stereo imaging of the virtual orchestra but also on the balance between sections and instruments. Even though panning can change drastically from project to project, there are a few tips I find particularly useful. You will find that by changing the panning you will have to adjust the volume of some tracks. Usually instruments that are panned hard left or hard right tend to be more exposed and separated than instruments left more in a central position.

6.3.1 Balance

There are two main criteria to follow when panning instruments on the stereo image: *balance* and *frequency placement*. Even though balance may seem an obvious consideration, it is often forgotten; if it is overlooked during the mix process it will have to be fixed later at the mastering stage.

Balance is achieved by panning sections and instruments so you reach steady equality between the left and right channels. You definitely want to avoid having one side continually more active or predominant than the other. Check the balance between the two channels by listening carefully to the mix and by controlling the meters of the stereo master track of your sequencer.

A good starting point for maintaining balance is to have a clear idea of the instrumentation featured in the project and the importance of each instrument to the arrangement. Avoid extreme pan settings for lead vocal and instruments that are featured continually in the piece unless they can be counterbalanced by similar parts, such as a vocal countervoice or another lead instrument. For short and temporarily featured solos or passages, it is ok to have instruments hard panned, especially if you can alternate sides between several featured solos (e.g., a short electric guitar solo panned left followed by a short saxophone solo panned right). Before deciding on the panning, I recommend sketching out graphically the instruments featured in the project and their placement in order to come up with a sort of blueprint of the mix. On the other hand, avoid placing too many instruments in the center without taking advantage of the clarity and spatial openness offered by the stereo field. Pads, keyboards, and strings offer a good starting point to open the stereo image by either recording the patches in stereo and panning the stereo track hard left and right or carefully panning each instrument using more extreme settings. For strings, brass, and woodwind sections, refer to Figures 5.10, 5.11, and 5.18 as a starting point for your mixes. As I mentioned earlier, each project is very different, so I suggest taking the settings shown as only a general starting point to be adapted to your needs.

6.3.2 Frequency placement

The second principle to keep in mind when working on panning and instrument placement is based on the frequencies featured in each part. The so-called *frequency placement* approach is based on the fact that low frequencies are harder to place in space than high frequencies. This is because the brain perceives sounds in space according to the difference in phase of the waveforms received by the two ears. Since low frequencies have longer periods, a small difference in phase is harder to perceive and, therefore, low frequencies are more difficult for the brain to place precisely in space. Thus usually it is more natural for the ears to listen to audio material that features low-frequency instruments placed in the middle of the stereo image rather than panned hard left or right. You should also avoid extreme panning settings for low-frequency instruments, such as bass, bass drum, and string basses.

Remember, rules can be broken if there is a good reason. For example, if your project features a virtual ensemble mimicking a live ensemble such as a jazz quartet or a classical orchestra or a string quartet, then it is better to use the placement that the live ensemble would follow in a live performance setting. A typical example would be a jazz quartet comprising piano, bass, drums, and saxophone. Here you can place the instruments with more freedom by panning the piano slightly left, the drums in the center, and the bass slightly right. The saxophone could be placed either in the middle or slightly to the right to counterbalance the piano. The same can be said for a string quartet, where the cello (which covers the lower end of the spectrum in such an ensemble) can be panned right, just as it would be in a live performance setting.

As you can see, panning, like all the other elements, from sequencing to orchestration, can have a clear impact on how "real" your virtual ensemble sounds. Try to keep realistic settings in panning as well, and always try to reproduce a real live performance situation as much as possible, no matter which style or ensemble you are sequencing for. Keep in mind that you can always fine creative and original solutions when it comes to panning and that rules can be broken for creative reasons.

Whereas low frequencies usually sound more natural if panned in the center, high frequencies can be panned at any degree, ranging from center to hard left or right. An example is the hi-hat and cymbals of the drum set. These instruments can help open the stereo image if panned with more extreme settings. Crash and ride cymbals can be panned, respectively, hard left and hard right (or vice versa), just as in an acoustic drum kit. The same can be said for shakers, tambourines, and high-pitched percussion in general.

The final goal of successful panning is to open up the stereo image of your mixes without creating unbalanced positioning of the instruments. By taking advantage of the entire stereo image and accurately planned and precise panning settings, you can greatly improve the intelligibility and clarity of the production.

6.4 Reverberation and ambience effects

Another important aspect of mixing is the type and amount of reverb you apply to single instruments and sections to place them correctly in a natural ambience. Natural reverberation is produced by a buildup and complex blend of multiple reflections of the waveforms generated by a vibrating object (e.g., the string of a violin or the head of a snare drum). The reflections are generated when the waveforms bounce against the walls of an enclosure, such as a room, a stadium, or a theater. Depending on the size of the enclosure and the type of material that covers the walls, reverberation can change in length, clarity, and sonic characteristics. In a digital audio workstation, reverberation is usually added at the mix stage in order to place instruments accurately and realistically in a predetermined acoustic environment. The style of the project and the type of instrument are crucial in determining and properly choosing the best reverb type and in setting its parameters.

While pan allows you to control the positioning of the instruments along the horizontal axis, reverb, along with volume, allows you to control the positioning of the instruments bidimensionally (Figure 6.3). As you can see in the figure, by controlling the balance between reverb, volume, and pan you can precisely position any instrument in any place on the bidimensional "canvas" of your mix. In general, by adding more reverb (wetter signal) and by lowering the volume of the dry signal you can position an instrument in the background of a mix. If you lower the reverb level and slightly raise the dry volume, you can bring the same instrument closer to the listener and therefore inside the middleground area. A louder and dryer signal will sound very much in the front of the mix (foreground). Listen to Examples 6.1 through 6.4 to hear how volume, reverb, and pan can be used to place an instrument in space. In addition you have all the panning settings I described earlier to move instruments and sections on the horizontal axis. The way you program the parameters of reverb is crucial to getting the best out of it. Table 6.2 lists the most important parameters present on a reverb unit or plug-in and a brief explanation of the impact they have on the overall sound. To appreciate how these parameters affect reverberation, listen to Examples 6.5 through 6.12 on the audio CD.

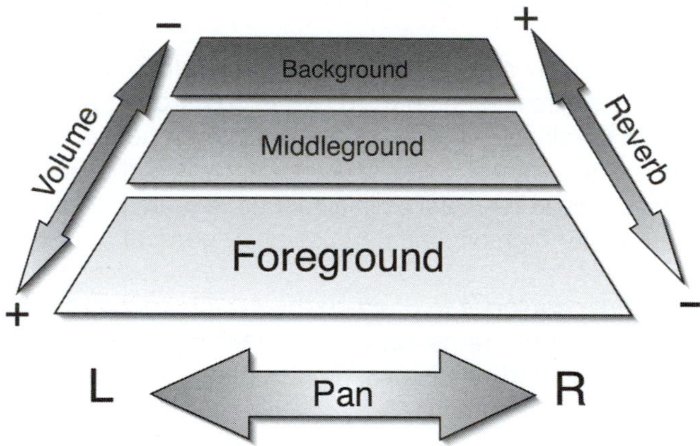

Figure 6.3 Bidimensional placement of instrumentation through the use of pan, reverb, and volume.

The parameters listed in Table 6.2 are generic indications of some of the main settings found on the majority of reverb units (hardware and software). Some units are simpler, providing more basic parameters (for example, you might not find a control for the diffusion in low-end reverbs); other units might give you more control parameters in filtering and room size capabilities. Remember that reverbs, in terms of CPU power, require a lot of calculation and therefore are among the plug-ins that necessitate more resources. When possible, use reverbs through an auxiliary send and not as insert (see Chapter 4 for a review on how to use effects such as insert and send). This will save CPU power because you will need fewer reverbs and can share them among several tracks.

When considering which reverb to use and how to choose the right settings, let your ears be your guide. Each scenario is different and can call for very different configurations. Try not to use the standard presets that come with your plug-in. They can be a useful starting point, but in my many years of composing, sequencing, and production I have never found a preset that was 100% useful without customization.

Table 6.2 Generic reverb parameters

Parameter	Description	Comments
Type	Describes the type and overall sonority of a reverb	Typical types include hall, chamber, plate, gate, and reverse
Length (decay)	Represents the decay or length of the actual reverb tail	Usually expressed in seconds or milliseconds More sophisticated reverbs allow for control of the decay parameter for two or more frequency ranges
Room Size	Controls the size of the room in which the reverb is created. Bigger rooms are usually associated with longer reverberation times	Sometimes, depending on the type of plug-in, you can also find a control that allows you to control room shape
Early Reflections (ER)	Controls the first bounces that follow the dry signal before the actual reverb	Can have several subparameters, such as ER size, gain, and Pre-Delay
Pre-Delay	Controls the time between the dry signal and the early reflections	
Delay	Controls the time between the dry signal and the actual reverb tail	
Diffusion	Controls the distance between the reverb's reflections	A low diffusion features more scattered and distant reflections; a high diffusion features closer and more frequent reflections Low diffusion is usually indicated more for sustained pads and passages (e.g., string sections), high diffusion more for percussive instrument and fast-transients audio material
High Shelf Filter	Cuts some of the high frequencies of the reverberated signal	Not found in all reverbs Useful for avoiding excess high frequencies in the reverberated signal to recreate a more natural sound
Mix	Controls the balance between dry and wet signals	If the reverb plug-in is used as insert, set the Mix parameter to 50% or lower. If you use the reverb as a send, set the Mix parameter to around 100%

When setting the parameters of a reverb always think in terms of bidimensional placement of the audio material you are working on. Longer reverbs have the effect of "pushing" the instruments farther back from the listener; shorter reverbs move the instruments closer to the listener. As a general rule (remember, rules are to be broken if necessary!), try to keep the reverb length between 1.5 and 2.5 seconds, considered a good starting point for generic reverberation for vocals (between 2.0 and 2.5), drums (1.5–2.0), piano, strings and pads (1.7–2.2).

As we saw with panning, reverberation too is affected by frequency-related issues. Instruments that cover mainly the low end of the frequency range, such as bass and bass drums, usually require much less reverberation (or none), while on instruments that cover the middle- and high-frequency range, such as guitars, HH, and even snare drum, you can apply more reverb. This is mainly because reverb in general adds muddiness to the mix through the tail created by its reflections. Adding too much reverb to instruments that cover the low-frequency range reduces the definition of the overall mix, especially for the electric bass and the bass drums, which can quickly lose sharpness and clarity. For sections that cover a wider range of the audible spectrum, such as the string section and the brass section, I recommend using slightly different reverb settings, depending on the specific instruments forming the section. With strings, use slightly less reverb for the basses, a bit more for the cellos and violas, and a wetter mix for the violins.

The best way to accurately control the amount of reverb assigned to each track is to use the aux send available on each track, as I explained in Chapter 4. The same is true for the brass section, where the tuba can use very little reverb, while the trombones can handle medium reverberation settings. Trumpets and French horns are the brass instruments that can use a good amount of reverb to blend nicely with the other instruments.

6.4.1 Specific reverb settings for DP, CSX, LP, and PT

All four sequencers come bundled with one or more plug-ins to simulate reverberation effects. Some of the included plug-ins offer basic controls, such as Reverb A in CSX and D-Verb in PT, whereas eVerb of DP and PlatinumVerb of LP both feature a comprehensive list of parameters. Of particular interest is the addition of Space Designer, which is bundled with LP. This new reverb plug-in is based on an innovative technique and algorithm that, instead of "synthesizing" a reverb effect by calculating the reflections through standard algorithms, utilizes a more realistic approach called the *convolution process*. This plug-in is able to generate the reflections of the reverberated signal by merging the input dry signal of a track with a sampled reverberation signal that was prerecorded in a real acoustic space, such as a church, a recording studio, a theater, or a generic room. The quality of reverbs built on the convolution process is much higher than those generated via conventional algorithms. In addition, users technically could "sample" their own acoustic environments and build their own reverb libraries. I will discuss Space Designer in detail later in this chapter.

Most of the parameters explained in Table 6.2 are present in each of the reverbs I just mentioned, and therefore you should be fairly familiar with them already. Other, more specific parameters change according to the plug-in used. In Tables 6.3 through 6.6 you will find brief descriptions of the parameters of each reverb, including LP's Space Designer so that you can completely customize the settings and adapt each parameter to your projects and needs.

Two extra parameters can be revealed and adjusted by clicking on the "Full Message" button in the top left corner of PlatinumVerb (Figure 6.7, page 258): Diffusion and Early Reflections Scale. The former controls the distance between the reflections of the reverb

Table 6.3 DP's eVerb parameters (Figure 6.4, page 254)

Parameter	Description	Comments
Mix	Same as the generic description (Table 6.2)	
Initial Reflection: Size, Level and Pre-Delay	Size, Level (in decibels), and Pre-Delay of the early reflection environment	Pre-Delay controls the time delay between dry signal and early reflections
Reverb: Delay, Level	Delay and Level of the actual reverb tail	Delay controls the time delay between early reflections and the beginning of the reverb tail
Reverb Time: Low End, High End, and Crossover	Low End and High End allow you to set two separate reverb lengths for two separate ranges of the frequency spectrum. Crossover controls the split point between the two frequency areas	
Diffusion	Same as the generic description (Table 6.2)	
Shelf Filter: High Cut, High Damp	Allows you to cut some of the high frequencies of the reverb tail. Cut (0.5–18 kHz) selects the frequency; Damp selects the value (dB) (from 0 to −40) of the damping	
Color	Allows you to send some of the early reflections into the reverb tail adding tonal coloration to the overall sound	A very subtle parameter that, when activated, morphs the parameters of the early reflections into the reverb tail
Hi-Q Link	Connects the High Frequency Reverb time with the Shelf Filter parameters so that when you increase the former, the latter will cut more of the higher frequencies	Functions to create a reverberation as natural as possible. If turned Off, the reverb will sound brighter; if turned On, the reverb will sound darker and in most cases more realistic

tail as explained in Table 6.2, the latter controls the scaling factor for the Early Reflections Predelay.

6.4.2 Convolution reverb: Logic Pro's Space Designer

As I mentioned earlier, the convolution reverb that comes with LP (Space Designer) utilizes sampled responses of specific environments (such as theaters, churches, and studios as well as outdoor environments) to produce (or, better, reproduce) artificial reverberation

Figure 6.4 eVerb in DP.

(Figure 6.8, page 259). This approach has the indisputable advantage of creating an extremely realistic reverberation. Some of the drawbacks, though, include the high CPU power needed to calculate the reflections in real time and also, to some extent, the limited changes you can apply to the original settings without obtaining unwanted sonic artifacts.

While some of the parameters used in Space Designer should look familiar by now (Low Shelving EQ, Stereo Spread, Crossover), others need further explanation. For example, the familiar Mix parameter in Space Designer is controlled using the two sliders on the right. One controls the dry signal ("direct") and the other one controls the signal of the reverb ("reverb"). The input section on the left of the control panel is used to set how Space Designer routes the input from the channel on which the effect is inserted. Pretty

Figure 6.5 Reverb A in CSX.

Table 6.4 CSX's Reverb A parameters (Figure 6.5)

Parameter	Description	Comments
Mix	Same as the generic description (Table 6.2)	
Pre-Delay	Same as the generic description (Table 6.2)	
Room Size	Same as the generic description (Table 6.2)	
High Cut	Similar to High Shelf Filter (Table 6.2)	
Low Cut	A Low Shelf Filter that allows you to cut some of the low frequencies of the reverb tail	
Reverb Time	Same as reverb Length (Table 6.2)	

self-explanatory are *stereo* and *mono*, receiving the input in stereo or mono from the channel on which the effect is inserted. *Xstereo* receives a stereo input but inverts the left and right channels. For a more detailed description of the other Space Designer's parameters, consult Table 6.7 (page 259).

If you are in IR Sample mode you have control over the envelope parameters of Volume and Filter, pretty much like in the modifier section of a synthesizer. You can switch between the two different edit modes from the top of the graphic display of the sampled reverberation. To alter the parameters, either you use the graphic display and grab/drag the insert points of the envelopes, or you change their values in the parameter list section at

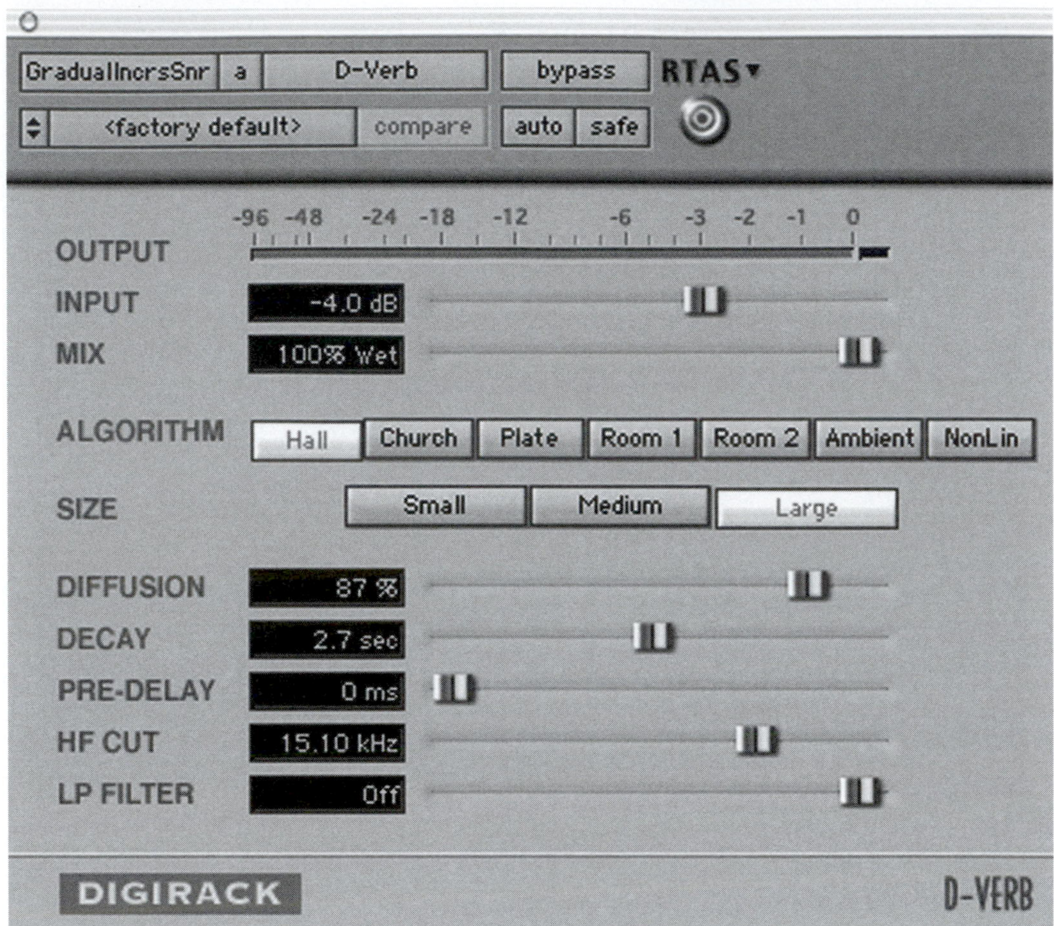

Figure 6.6 D-Verb in PT.

Table 6.5 PT's D-Verb parameters (Figure 6.6)

Parameter	Description	Comments
Input	Controls the input volume of the dry signal	
Mix	Same as the generic description (Table 6.2)	
Algorithm	Same as reverb Type (Table 6.2)	
Size	Determines the size of the room	
Diffusion	Same as the generic description (Table 6.2)	
Decay	Same as reverb Length (Table 6.2)	
Pre-Delay	Same as the generic description (Table 6.2)	
HF Cut	Similar to High Shelf Filter (Table 6.2)	
LP Filter	A low-pass filter that allows you to drastically cut frequencies higher than the cutoff frequency set	

Table 6.6 LP's PlatinumVerb parameters (Figure 6.7, page 258)

Parameter	Description	Comments
Pre-Delay	Same as the generic description (Table 6.2)	
Room Shape	Determines the shape of the room in which the reverberation is created; value goes from 3 to 7, representing the number of corners of the room	
Stereo Base	Controls the positioning of the virtual microphones used to capture the reverb generated	
Room Size	Same as the generic description (Table 6.2)	The value represents the length of the walls
Initial Delay	Same as Delay (Table 6.2)	
Spread	Controls the stereo image of the reverb	With a value of 0% you get a mono reverb (same reverberation on both channels); with a value of 100% you get a full stereo effect
Crossover	Allows you to set the split point at which two separate settings for the reverb can be made, transforming the effect in a real dual-band engine	
Low Ratio	Controls the deviation from the main Reverb Time section for the lower frequencies below the crossover point	With a value of 100% the settings reflect exactly those of the main Reverb Time section; with values below 100% the reverberation time for the lower-frequency section gets shorter; for values above 100% it gets longer
Low level	Indicates the overall level for the reverb for frequencies below the crossover point	
High Cut	An LP filter that cuts the high frequencies above the frequency set	
Density	Controls the density of the reflections forming the reverb tail	A denser reverb usually sounds better but requires more processing power
Reverb Time	Same as the generic description (Table 6.2)	
Mix	Same as the generic description (Table 6.2)	
Balance ER/Reverb	Controls the balance between the early reflections and the reverb tail	

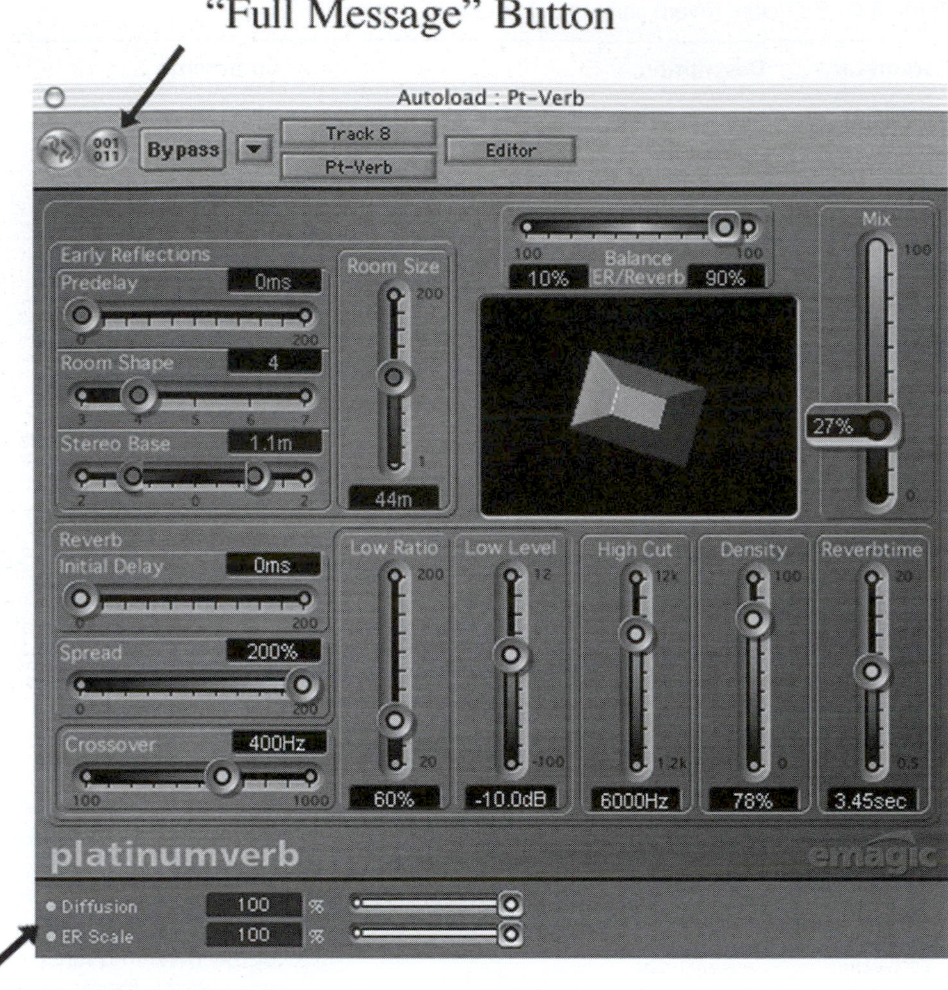

Figure 6.7 PlatinumVerb in LP.

the bottom of the window. The same procedure can be used in Synthesized IR mode, with additional control over the Density envelope, which allows you to precisely control the change in density of the reflections over time. The Pre-Delay of the reverb is controlled from the window right underneath the graphic display of the sampled waveform. The "Reverb Volume Compensation" button allows you to automatically match the volume of different Impulse Response files. The "Latency Compensation" option allows you to automatically delay the dry signal of the necessary amount of samples it takes the reverb to calculate the reflections. If you want to create your own Impulse Response files, you can use the "Deconvolution" button in the top right corner of the reverb window. For a detailed description on how to prepare the files used by Space Designer to create successful Impulse Response files, I recommend reading the Space Designer manual's comprehensive section on this topic.

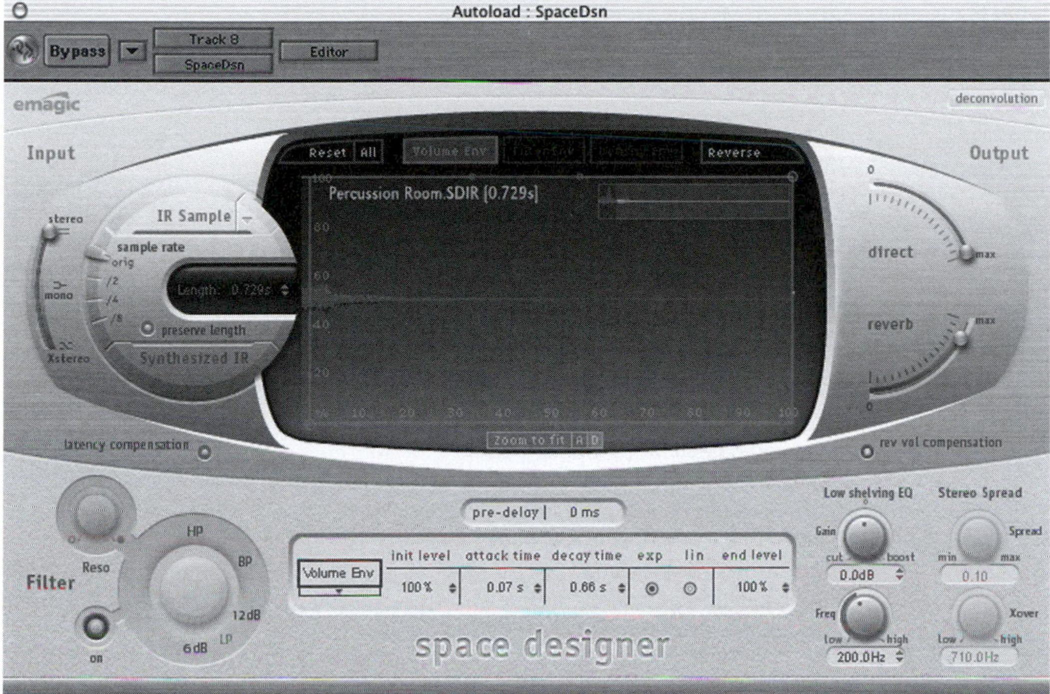

Figure 6.8 Space Designer in LP.

Table 6.7 LP's Space Designer parameters

Parameter	Description	Comments
IR Sample	Allows you to load new Impulse Response files or to switch back to a Sampled response from a Synthesized one	Click on the drop-down menu and select either "Load" or "Load and Initialize the parameters" to open a new response
Sample Rate	Determines the sample rate at which the Impulse Response works	The starting sampling rate is the one at which LP operates. If you choose /2, /4, or /8, the effect operates at the respective fraction of the original frequency. If you lower the frequency, the CPU is freed up, requiring less computational power. Unless "Preserve Length" is selected, the length of the reverb will get longer or shorter depending on the ratio chosen (e.g., /2 will set the reverb length to double its original time)

(*Continued*)

Table 6.7 (*Continued*)

Parameter	Description	Comments
Preserve Length	If checked, changing the sample rate ratio will affect not the length of the reverb but only its quality	
Synthesized IR	Generates a synthesized Impulse Response (instead of a sampled IR) based on the current settings	
Filters	Allow you to control the color of the reverberation HP: high-pass filter to reduce low frequencies BP: band-pass filter to reduce middle frequencies 6-dB LP: a gentle low-pass filter 12-dB LP: a more drastic low-pass filter for warmer reverberation Resonance: boosts some of the frequencies around the cutoff frequency	

6.5 Equalization

An equalizer, in its broad sense, allows you to boost or cut the volume of specified frequencies. During the mix, equalization can be effectively used in different ways either to correct problems that were created during the recording session or problems that arise because of incompatibility among instruments or simply in a creative way to produce original effects.

No matter what you are going to use an equalizer for, there are a few ideas you should know when embarking on an equalization session. Equalizers generally need to be used as inserts on the channel and not as auxiliary sends. You have to be familiar with the most common types of equalizers in a digital audio workstation setting. Among the several types of equalizer available nowadays there are five main categories that have proven to be the most useful in a mixing situation: peak, high shelf, low shelf, high pass, and low pass. Table 6.8 describes their parameters and some of their most common uses. In Figure 6.9 you can see the symbols with which they are usually indicated.

Remember that equalizing is mainly a "problem-fixing" procedure. This means there's no point in starting to play around with the settings if you're not clear about what you want to achieve with it and how the final result should sound. A good approach to equalizing is to listen carefully to the soloed track and to come up with a list of things you might want to improve or correct. The next step is to bring the gain up and sweep across the frequency range until you find the exact cutoff point you want to cut or boost. Finally set the gain as desired.

Table 6.8 Characteristics and parameters of the most common types of equalizers

Type of equalizer	Description and parameters	Typical uses
Peak	Allows you to cut or boost frequencies around the center frequency Center frequency: determines the frequency to cut or boost Gain: positive gain boosts, negative gain cuts Q point: determines the "shape of the bell," or how wide the area around the cutoff point is going to be—the lower the value, the larger the bell, and vice versa, the higher the value, the smaller the bell. The Q parameter can usually (but not always) vary from a value of 0.7 (equal to a two-octave frequency range) to a 2.8 (1/2 octave)	A Peak eq. is extremely versatile. It can be used to pinpoint and cut/boost a very precise frequency or it can be used in a broader way to correct wider acoustic problems. It is usually utilized in the middle of the frequency range
High Shelf	Cuts or boosts the frequency at the cutoff and all the frequencies higher than the set cutoff point. Has only two parameters: the cutoff frequency and the gain	Usually used in the middle-high and high end of the spectrum. Can effectively serve to brighten up a track via a positive gain of 3 or 4 dB and a cutoff frequency of 10 kHz or higher (be careful because this setting can increase the overall noisiness of the track). Can also be used to reduce the noise of a track by reducing by 3 or 4 dB frequencies around 15 kHz or higher
Low Shelf	Cuts or boosts the frequency at the cutoff and all the frequencies lower than the set cutoff point. Has only two parameters: the cutoff frequency and the gain	Usually used in the low-middle and low range of the audible spectrum to reduce some of the rumble noise caused by microphone stands and other low-end sources
High Pass	Cuts all frequencies below the cutoff point. Has only one parameter, the cutoff frequency	A very drastic filter, it is often used to cut very low rumble noise below 60 Hz
Low Pass	Cuts all frequencies above the cutoff point. Has only one parameter, the cutoff frequency	A very drastic filter, it is often used to cut very high hiss noise above 18 kHz. Use with caution so as to avoid cutting too much high end of the track.

Keep in mind when equalizing that you will have to make small adjustments every time you are adding tracks to the mix, since the way an instrument sounds is affected by the frequencies and respective ranges of the other instruments. The most important concept here is to be able to emphasize the characteristic frequencies of the track you are working on and to

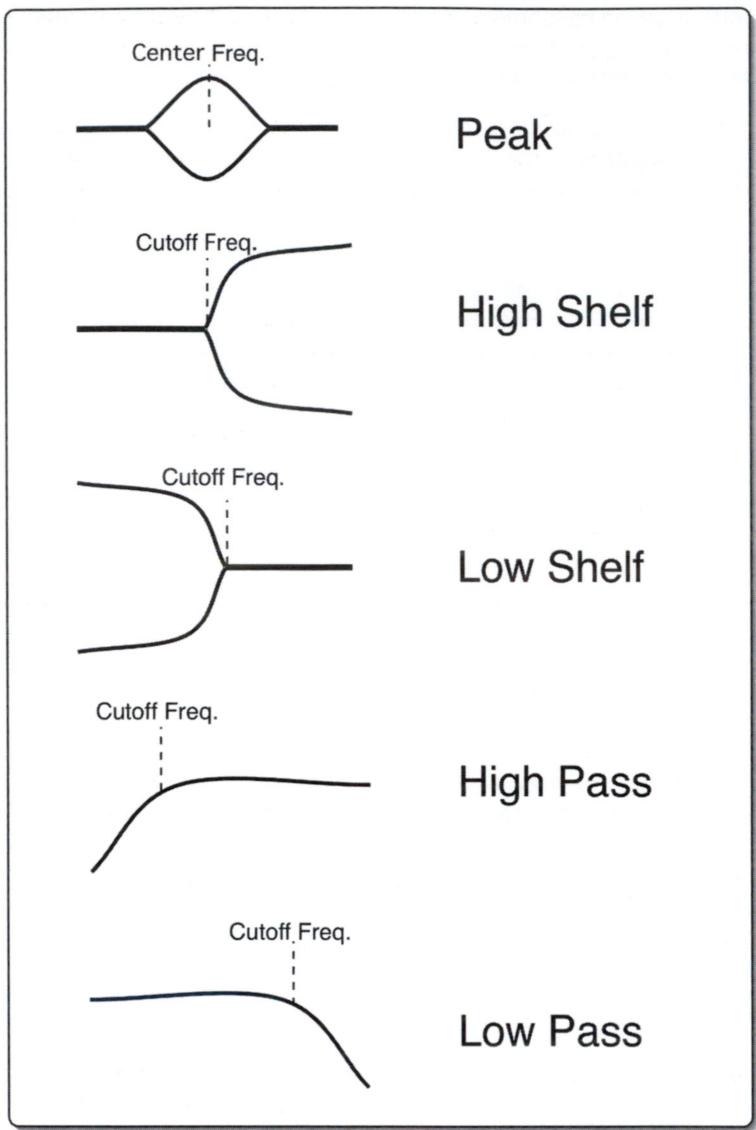

Figure 6.9 Conventional symbols for most common types of equalizers.

eliminate frequencies that do not enhance its sonic features in any particular way. In fact you should be able to "carve" a small niche within the audible range for each instrument and section so that it is clearly intelligible and not masked by other instruments. If the mix sounds muddy and cluttered, start trying to focus on which instruments contribute to the lack of clarity. Then try to use the equalizer to add clarity by gently shifting the center of each instrument involved so that there is no overlap. As a general rule it is always better to cut than to boost, mainly because the human ear is more used to a reduction than to an augmentation in the intensity of frequencies. While, as I mentioned earlier with reverb settings, it is hard to generalize, there are a few common settings that make a useful starting point during an equalization session. I sum them up in Table 6.9.

When equalizing, you must pay attention to some of the most common mistakes that even the seasoned engineer sometimes makes. First of all always try to keep your equalization gain parameter to a reasonable level. As a general rule, avoid cutting or boosting by

Table 6.9 Generic frequency ranges and their features used in an equalization session

Frequencies	Application	Comments
20–60 Hz	Cut to reduce rumble and noise related to electrical interference	It is a good idea to always reduce this area by 4–6 dB to lower the low-frequency noise
60–80 Hz	Boost to add fullness to low-frequency instruments, such as bass and bass drums	
100–200 Hz	Boost to add fullness to guitars, French horns, trombones, piano, snares Cut to reduce "boomy" effects on mid-range instruments	This frequency range effectively controls the powerful low end of a mix
200–300 Hz	Cut to reduce low and unwanted resonances on cymbals Boost to add fullness to vocal tracks	Be careful not to boost too much of this frequency range so as to avoid adding muddiness to the mix
400–600 Hz	Cut to reduce an unnatural "boxy" sound on drums Boost to add presence and clarity to bass	This frequency range can also be effective to boost the low range of the guitar
1.4–1.5 kHz	Boost for intelligibility of bass and piano	
2.8–3 kHz	Boost to add clarity to bass Boost to add attack and punch to guitars	This range can also be used effectively to add clarity on vocal parts
5–6 kHz	Boost for vocal presence Boost for attack on piano, guitars, and drums	A general mid-range frequency area to add presence and attack
7.5–9 kHz	Cut to avoid sibilance on vocal and voice tracks Boost to add attack on percussion Boost to add clarity, breath, and sharpness to synthesizers, piano, and guitars	A middle-high-range area that controls the clarity and the attack of the middle-high-range instruments
10–11 kHz	Boost to increase sharpness on cymbals	High-range section that affects clarity and sharpness

(*Continued*)

Table 6.9 (*Continued*)

Frequencies	Application	Comments
	Boost to add sharpness on piano and guitars Cut to darken piano, guitars, drums, and percussion	
14–15 kHz	Cut to reduce sharpness on cymbals, piano, and guitars Boost to add brightness on vocals Boost to add real ambience to synthesized and sampled patches	
18 kHz	Cut to reduce hiss noise Boost to add clarity to overall mix	A delicate high-range section that should require drastic positive or negative gain settings only in extreme situations

more than 6 dB unless absolutely necessary. If for some reason you see that some of your eq. settings go over this limit, try to question why and see if there is a better solution to the problem. The same can be said for situations where you end up boosting (or cutting) several frequencies at the same time whose only effect is a raising (or lowering) of the overall volume of the track with no real change to its sonic content. In this case try to bypass the equalizer and experiment with volume changes instead. All four sequencers provide comprehensive equalization tools that can be inserted on any audio track. In the next section I provide a brief description of some of the equalization tools that each sequencer features, to give you a better idea of the possibilities they offer.

6.5.1 Equalizers in DP, CSX, PT, and LP

DP is bundled with a comprehensive multiband equalizer (Figure 6.10) that can work as a two-, four-, and eight-band eq. Each band can be set to work as either peak, shelf, or band-pass filter, providing great flexibility.

The parameters of each band can be changed either graphically by clicking and dragging the respective band number in the graphic window or by altering the values in the lower-right part of the equalizer window. Each frequency can be bypassed independently. In CSX the equalization section (a four-band eq.) is automatically inserted on each audio channel of the virtual mixing board. The only action you have to take to use the equalizer is to activate the On button on each band (Figure 6.11).

If you open the mixing window in CSX you can choose to look at the eq. section in three different ways: a traditional view, with the controls shown as concentric rotary dials; a linear view, with the parameters shown as horizontal bar lines; and graphically, with the controls shown as curves. The type of view can be changed by selecting it from the sidebar

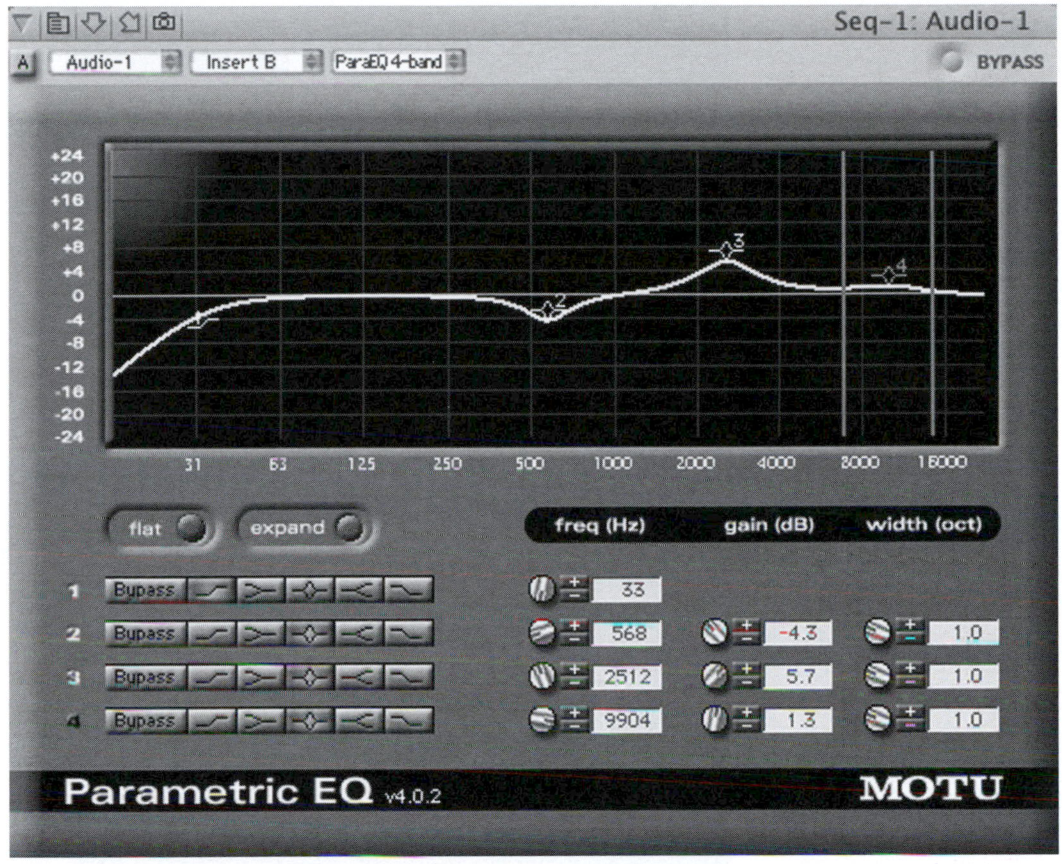

Figure 6.10 A four-band equalizer in DP.

menu of the mix window (Figure 6.11). The equalization plug-in bundled with PT is very straightforward. You can choose between a one-band and a four-band equalizer (Figure 6.12). The latter is the more flexible, providing high and low shelf plus two peak equalizers.

LP is among the four sequencers that offer the most comprehensive equalization bundle options. The list of plug-ins dedicated to equalization includes the most flexible and complete "Channel EQ," "DJ EQ," "Fat EQ," "SilverEQ," "Parametric EQ," and a series of single-band "high and low" equalizers. The "Channel EQ" is the most advanced. It provides an extensive number of settings and eight bands, which include High Pass, Low Pass, High Shelf, Low Shelf, and four peak equalizers (Figure 6.13). The parameters for each band can be set by either moving the mouse over the band you want to edit in the graphic window and simply clicking and dragging to make changes or by using the numeric fields located in the lower part of the window. The parameters in the field can be changed by moving the mouse over the desired field and clicking and dragging (an upward motion

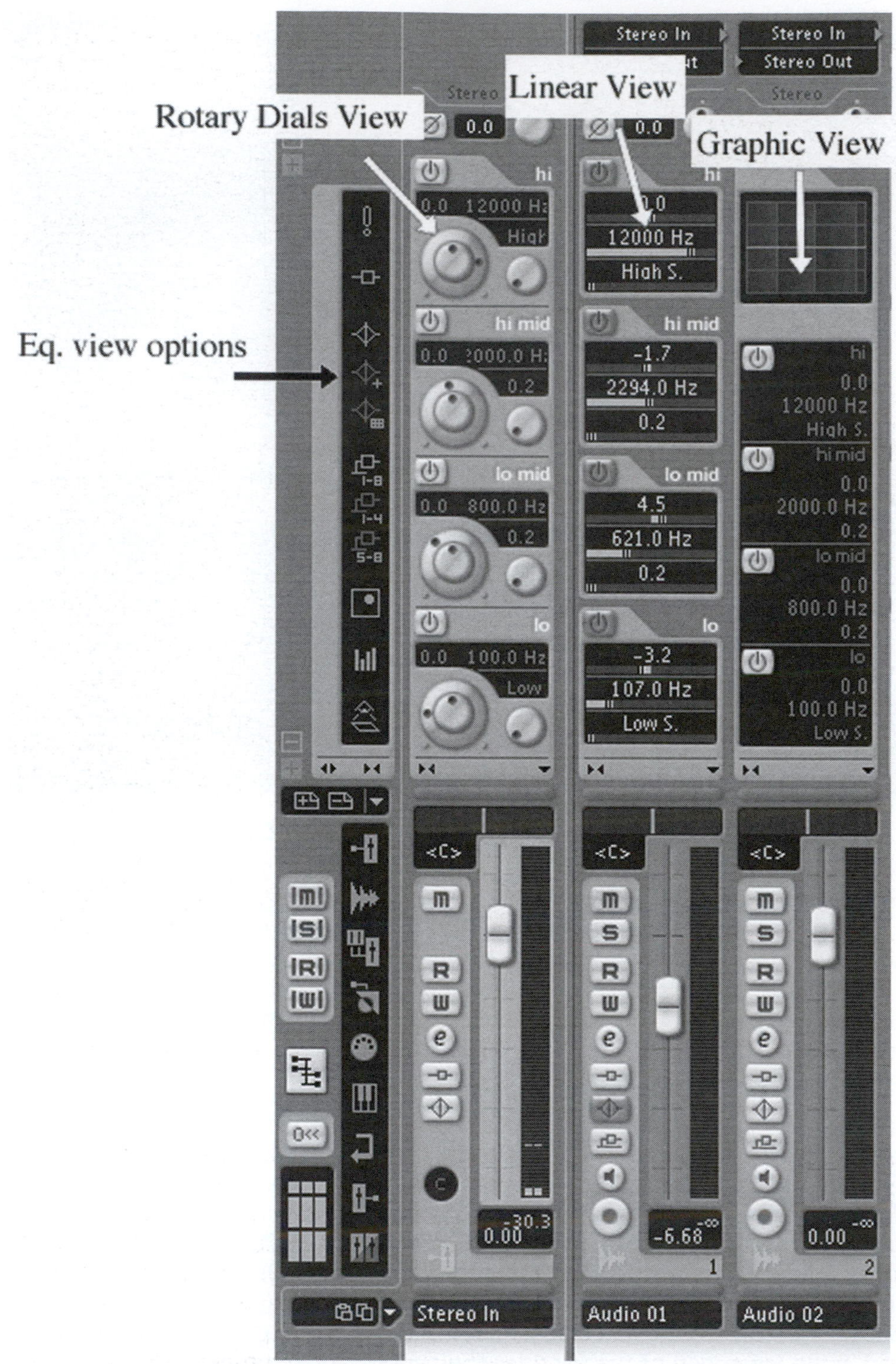

Figure 6.11 A four-band equalizer in CSX showing the three different view options.

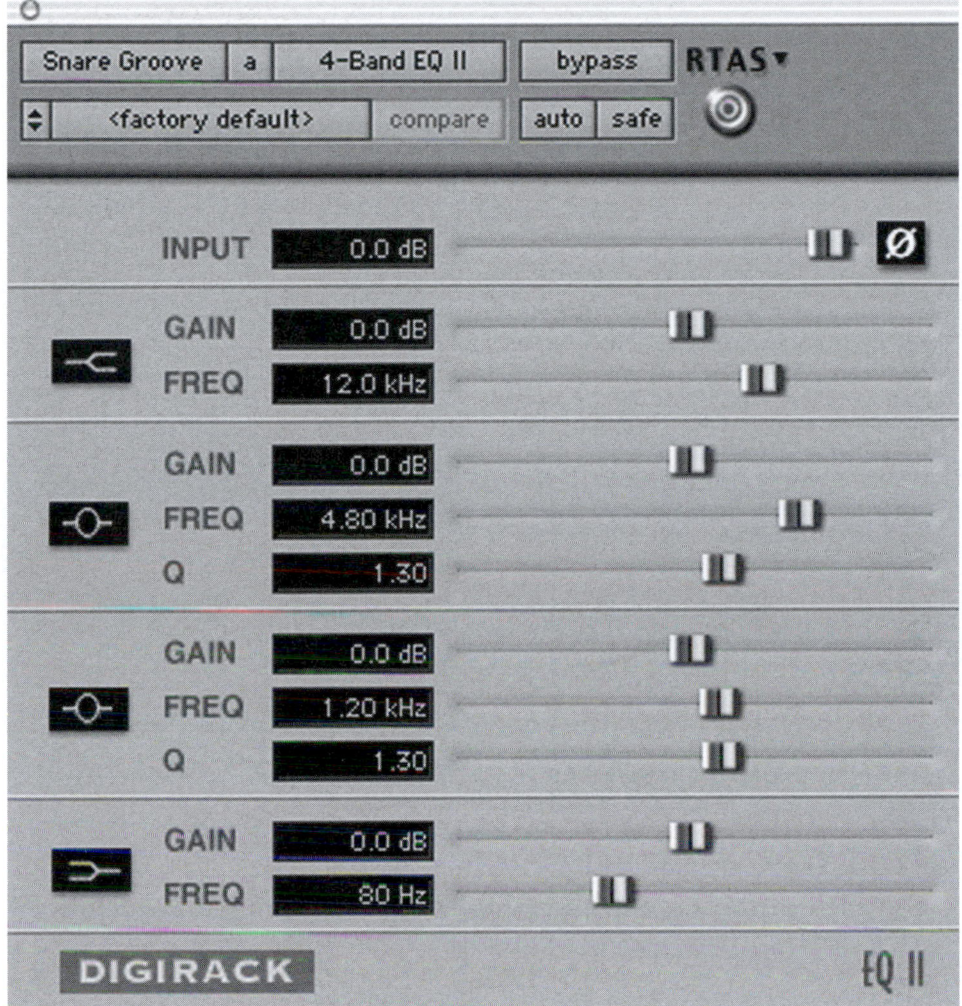

Figure 6.12 Four-band equalizer in PT.

increases the value; a downward motion decreases it). To reset a band to its original default setting, *option-click* on its field.

What sets this equalizer apart from the others we have analyzed so far is the presence of a built-in real-time spectrum analyzer that shows the frequency content of the track on which the effect is inserted. You can activate the analyzer by clicking the button to the left, located underneath the "Master Gain" slider. To change the resolution of the analyzer, click on the field marked "low," "medium," or "high," depending on the default settings, and change it to the desired value. The resolution controls how detailed the analyzer is. A "low" setting uses very little CPU power but gives you a more approximate reading; a "high" setting gives a more detailed representation but uses more CPU power. You can monitor the analyzer either pre- or postequalizer by selecting the respective setting in the field underneath the "Analyzer" button.

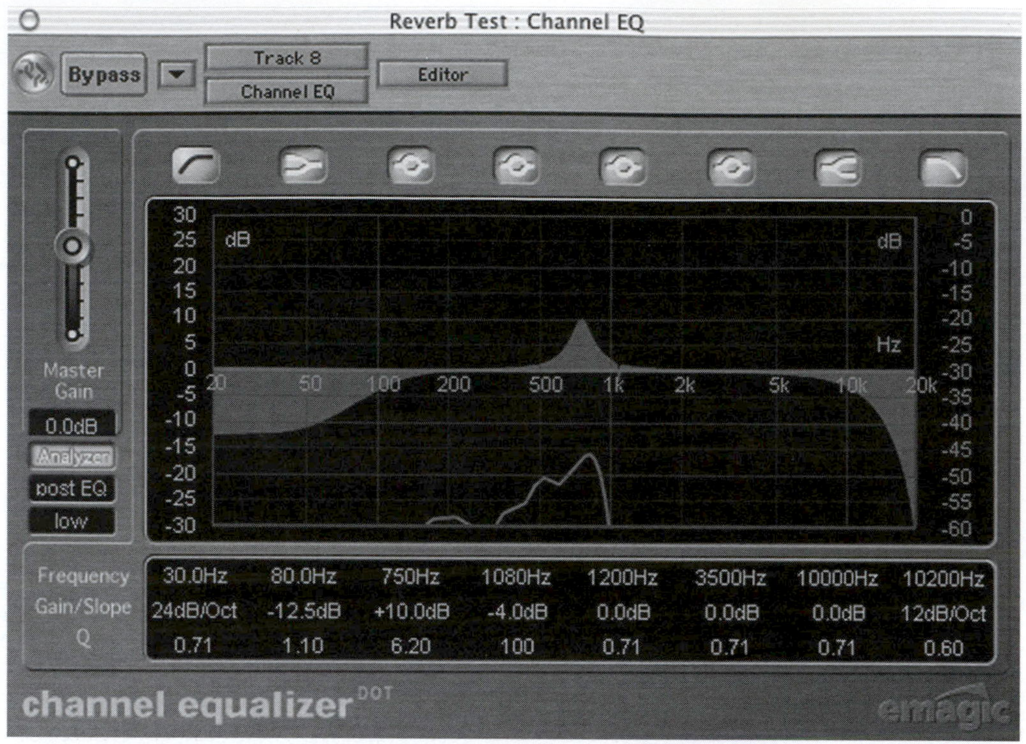

Figure 6.13 The "Channel EQ" in LP.

6.6 Dynamic effects: compressor, limiter, expander, and gate

Through reverb and panning you can control the positioning of an audio signal and through equalization you can control its shape in terms of frequency response. The last type of effects I want to discuss in this chapter are the so-called *dynamic* effects. Their name comes from the fact that they are designed to alter, in one way or another, the dynamic range of an audio signal. They are important because they allow you to control over time the ratios between high and low peaks in the dynamics of an audio track. These types of effects are more often used on live acoustic instruments than on synthesizers; therefore their application in a sequencer/MIDI environment is somehow limited. Nevertheless I would like to introduce you to their practical uses and parameters.

There are four main dynamic effects: compressor, limiter, expander, and gate. Table 6.10 lists their main features and applications. Among the dynamic effects shown in the table, the compressor is the one most commonly used on acoustic live tracks. Usually its effectiveness is somewhat reduced when inserted on audio tracks that contained material recorded from a synthesizer, since their audio material is for the most part well balanced in terms of amplitude variations. I recommend using a compressor sparsely on synthesized tracks and only if really needed. In fact synthesized tracks in general can benefit from a higher dynamic range and not a reduced one. Reducing their dynamic range even farther can sometimes flatten their sonority too much, compromising their realism. On the other hand, a compressor can help an acoustic track to "seat" better on a complex mix with the other MIDI tracks.

Table 6.10 "Dynamic" effects

Effect	Description	Comments
Compressor	Allows you to reduce the dynamic range (difference in amplitude between high and low peaks) of an audio signal	Useful in situations where the audio signal has a very high dynamic range
Limiter	Represents a drastic version of a compressor, where the "Ratio" parameter is set extremely high	Used mainly during tracking to avoid distortion and during mastering to maximize the overall volume of the mix
Expander	The exact opposite of a compressor; allows you to increase the dynamic range of an audio signal	Sometimes useful to "re-generate" a track that was overcompressed
Gate	An extreme application of an expander, where the ratio is set extremely high	Very useful to reduce unwanted noise during "silent" passages of an audio track

As mentioned in Table 6.10, a limiter is very similar to a compressor. The only difference is that a limiter yields a more drastic reduction of the highest peak in dynamics, practically setting a "ceiling" over which the signal cannot go. It can be effectively used to increase the overall volume of a signal without getting distortion. The parameters of compressor and limiter are listed in Table 6.11.

Table 6.11 The parameters of compressor and limiter

Parameter	Description	Comments
Threshold	Sets the level (in decibels) above which the effect starts reducing the gain of the signal	If the signal is below the threshold, the effect doesn't affect the signal. As soon as the level goes over the threshold, the effect starts reducing the gain of the signal
Ratio	Sets how much the gain of the signal is reduced after the level goes over the threshold	It is usually set as an *x:y* value. With a setting of 1:1, the level is not altered; at 30:1 the level is highly compressed; at 100:1 the effect is considered a soft limiter. In certain limiters the Ratio parameter is omitted, implying a ratio set to infinity. In this case the effect can be referred as a hard limiter
Attack	Sets how quickly the effect reacts after the signal goes over the threshold	

(*Continued*)

Table 6.11 (*Continued*)

Parameter	Description	Comments
Release	Sets how quickly the effect reacts after the signal returns below the threshold	
Knee	Controls the curvature during the transitional moment from below and to above the threshold point. A "Soft-knee" allows for a more gentle transition; a "Hard-knee" generates a more drastic transition	This parameter is not found on every plug-in
Gain	Allows you to control the overall gain of the signal after compression/limiting.	Since the clear effect that compression and limiting have on the signal is a reduction of amplitude, the Gain parameter allows you to boost back the compressed signal by the amount specified. Some dynamic plug-ins feature an "Auto Gain" option that automatically sets the level of amplification so that the output after compression matches the input before compression

In Figure 6.14 you see a diagram showing the effect that a compressor has on a generic waveform. Notice how the gain parameter is set to boost the overall level of the signal, in order to bring back its highest peak to its original value.

Expanders and noise gates have effects opposite to those of compressors and limiters. An expander increases the dynamic range of a signal by lowering the level of audio material that is below a set threshold. The amount of reduction is set by the ratio. If an extreme ratio setting is chosen (such as negative infinity), then the effect is a noise gate. This means that when the signal goes below the threshold, its amplitude is turned to negative infinity, meaning the signal is basically muted. While expanders are usually not very common in a MIDI studio, noise gates can be effectively used to reduce the noise in between "silent" passages. Look at Figure 6.15 to see how a noise gate would alter a generic waveform.

6.6.1 Dynamic effects in DP, PT, CSX, and LP

All four sequencers come bundled with dynamic plug-ins. In DP and PT the dynamic plug-ins include compressor, limiter, expander, and gate. Their parameters feature the same options that I described in Table 6.11. The dynamic plug-in the CSX offers compressor, gate, and limiter inside a single instance (Figure 6.16).

In the compressor bundled with CSX, all the parameters should look familiar, the only exception being the RMS/Peak button. This option (also present in the compressor plug-in

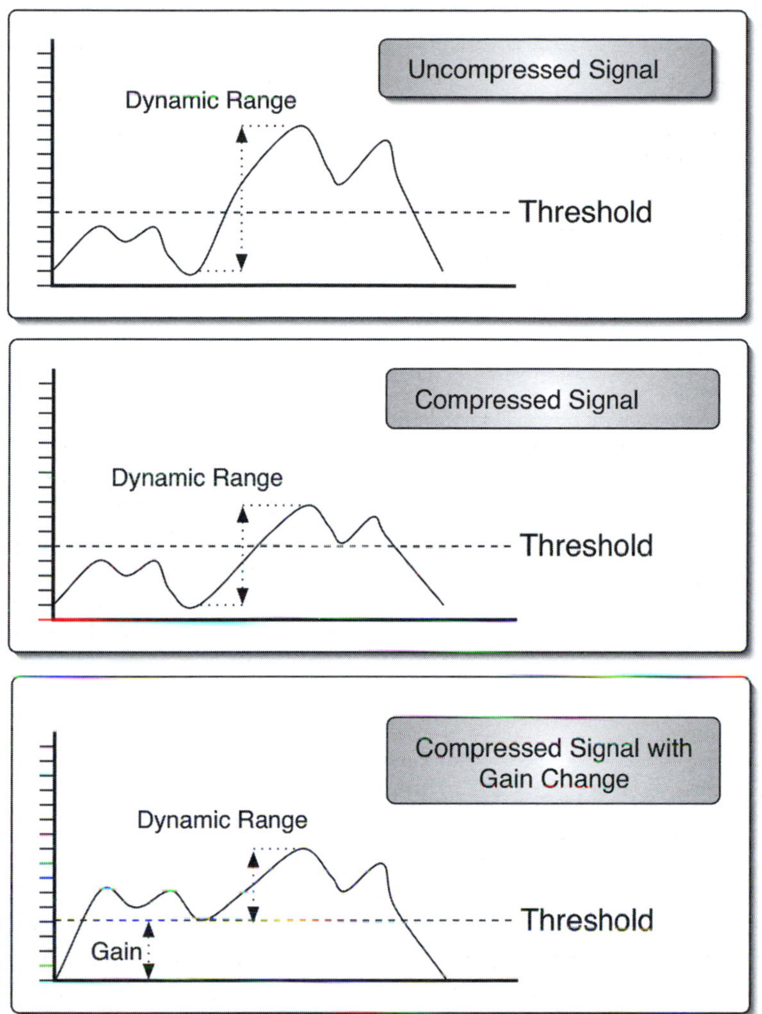

Figure 6.14 Example of a compressor.

of LP) allows you to choose the dynamic-peak detection algorithm utilized by the plug-in. RMS uses the average power in amplitude of the signal as a basis for determining if the threshold is reached. With the "Peak" option selected, the plug-in will determine the level based on a peak-by-peak analysis. In general use RMS for sources, such as vocals and strings, that do not feature a high number of fast transients. On the other hand use "Peak" for percussion and drums parts with fast transients.

The most interesting part of the dynamic plug-in in CSX is the "Autogate" section. While most of the parameters are similar to those found in a regular noise gate, what sets the Autogate apart from the other dynamic effects is the presence of the Trigger Frequency Range options. This parameter allows you to specify the exact frequency range used to detect if the signal is above or below the set threshold. The "Listen" option lets you find the exact frequency range you want to use to trigger the gate; by moving the boundaries,

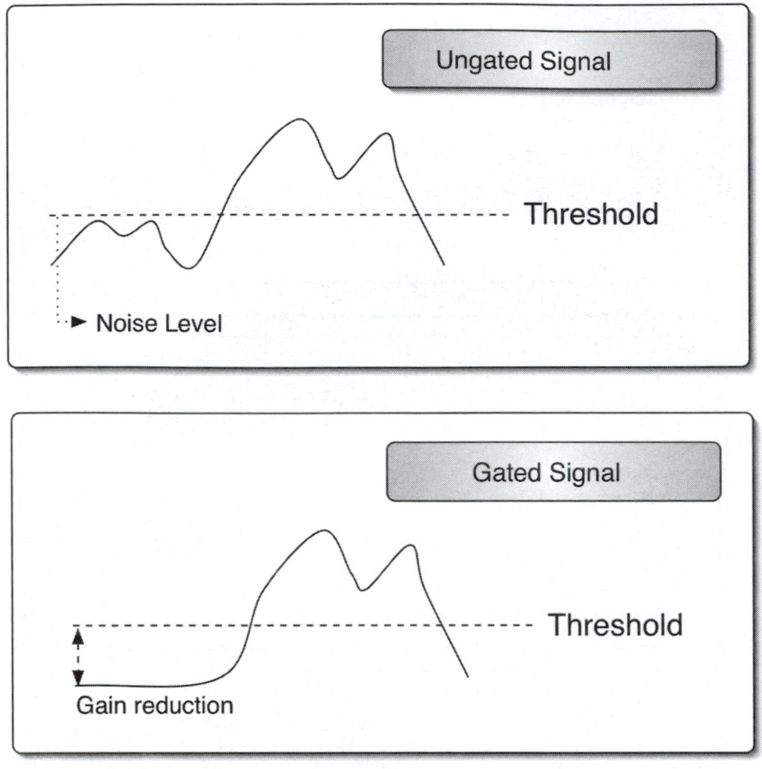

Figure 6.15 Example of a noise gate.

you will filter the signal after the gate. After having found the exact range, turn the Autogate to the "On" position. To bypass it, set it to "Off." The "Calibrate" function is used to automatically set the threshold level of the gate in order to reduce low-level or hiss noise. To set it, find a section of the audio that features only the noise (the longer the section, the more accurate the results). Click the Calibrate button and after a few seconds the threshold will be automatically set. Notice also that the Attack parameter has a special option called "Predict." This option, sometimes also called "Lookahead," allows the computer to "look" in advance for incoming peaks in dynamics that would go over the threshold and therefore trigger the gate. The "Routing" option allows you to set how the three dynamic effects are ordered in the "virtual rack."

LP offers an extensive and flexible suite of dynamic plug-ins that range from compressors to limiters to de-esser to expanders and gates. The most comprehensive and peculiar among all the different dynamic plug-ins available in LP is the "Multipressor" (Figure 6.17). This dynamic effect allows you control compressor, expander, and noise reduction settings for up to four different bands (you can also choose to use it for only two or three bands, saving some CPU power). In practical terms the "Multipressor" is a multiband compressor/expander that allows you independent control over the compression/expansion parameters of each band individually. This type of dynamic effect is particularly useful in mastering situations where detailed and precise control over the entire spectrum is a crucial feature.

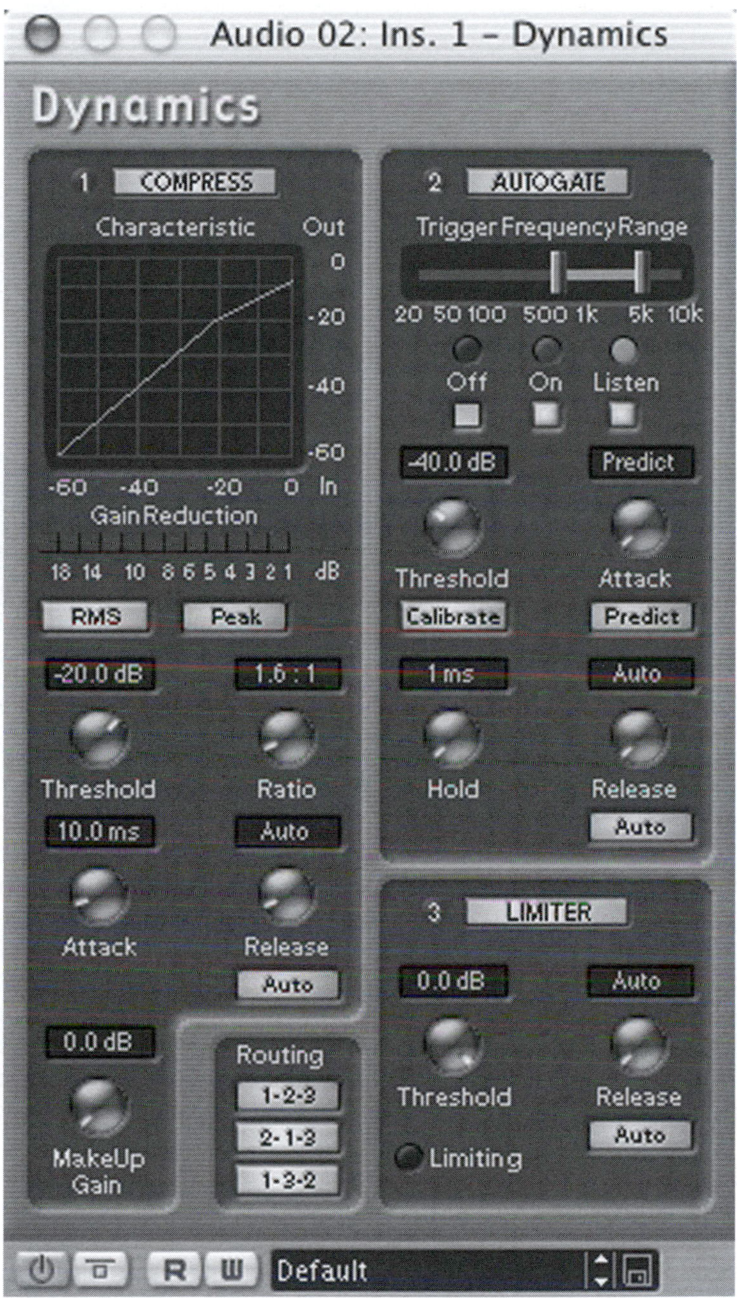

Figure 6.16 The dynamic plug-in in CSX.

In the "Multipressor" the parameters for each band are the same. The range of each band can be adjusted by clicking and dragging horizontally the separation lines between the bands located in the upper left area of the "Multipressor" window. Each band has a separate volume that can be adjusted by clicking and dragging up and down on the vertical lines

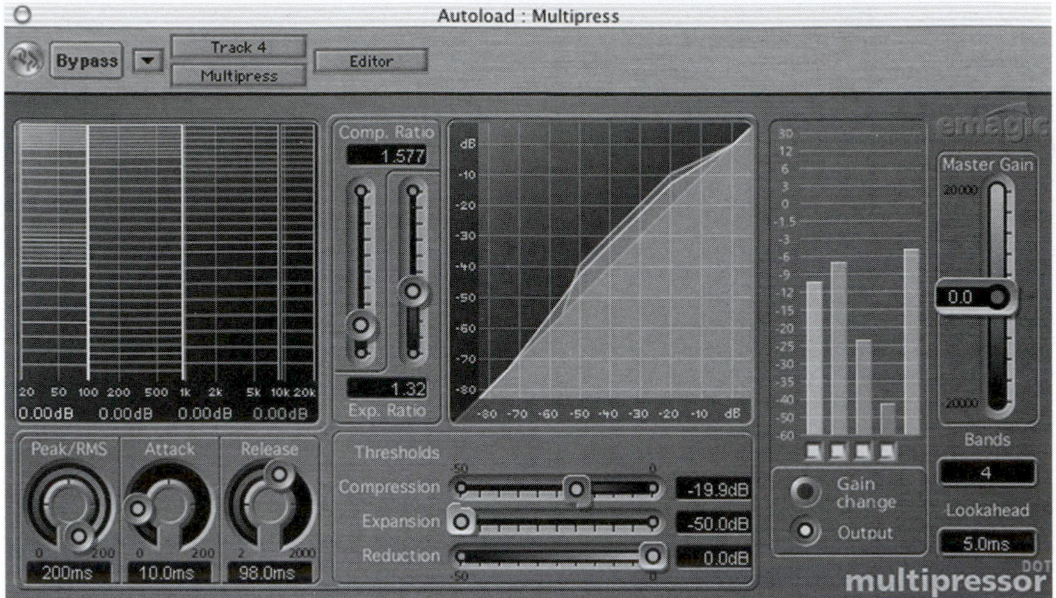

Figure 6.17 The "Multipressor" in LP.

of a specific band. To select and edit the specific parameters for a desired band, simply click anywhere inside the area for that particular band in the upper left area of the window.

The parameters of each band are typical of those found in a compressor/limiter, including Attack, Time, RMS/Peak, Ratio, and Threshold. The field marked Reduction allows you to specify the floor level used for noise reduction. If you move the slider all the way to the left, maximum reduction will be applied; if you move it all the way to the right, the reduction filter is bypassed. The "Lookahead" parameter allows you to specify how far in advance the "Multipressor" looks for incoming peaks. Usually the longer the value of the Lookahead parameter, the better it is, since it allows the computer more time to react to incoming peaks. The "Multipressor" gives you two options for monitoring the signal of the audio processed: Gain Change and Output. The former shows the compression applied to the signal for each band, while the latter indicates the overall output after compression/expansion.

As I mentioned with the other effects, it is almost impossible to come up with presets that would work in every situation. This is particular true for dynamic effects. Each piece of audio material is different and therefore calls for specifically targeted parameters and settings. Compressors and expanders seem the hardest to set since their effect is very often subtle and difficult to identify. In general when working with such effects, I always recommend starting with very mild settings and then working your way toward more drastic and aggressive changes. In the case of a compressor I suggest starting with a low ratio and a fairly high threshold. This allows the compressor to "kick in" sporadically and to reduce the volume very gently. Then start lowering the threshold and raising the ratio until you start hearing undesired audio artifacts, such as "pumping" effects. At that point, back up a little by raising the threshold and lowering the ratio again just a notch. This should give you a good starting point in basic settings. I also recommend not overusing dynamic effects.

They can be very useful if applied with parsimony, but they can be disastrous if unnecessarily placed everywhere.

6.7 Bounce to disk

When your mix is ready it is time to mix down the multitrack session into a stereo file (or multiple mono files in the case of a surround mix). This process is often referred as *bounce to disk*. This means that the sequencer does a digital mix-down of all the audio tracks present in a sequence and creates a mix on a stereo file (for regular stereo projects) or multiple mono files (for surround projects). The entire operation is done in the digital domain and in some cases at a speed that is faster than real time. The way the four sequencers handle this operation is very similar. The main thing to keep in mind is that basically the mix-down that will be created reflects exactly the mix you are hearing from your sequencer, so make sure to check carefully that the mix is exactly how you want it before creating the bounce. Let's take a look at how each sequencer handles the bounce-to-disk operation.

In DP make sure that all the tracks you want to be bounced are playback-enabled. Select the boundaries of the bounce by clicking and dragging over the tracks and the bars in the track list window. Include all the tracks and bars you want in the bounce. Select the "Bounce to disk" option from the Audio menu. In the window that appears you can specify several parameters that will affect the bounce. Starting from the top, the first drop-down menu allows you to select if your bounce will be a stereo file (*split stereo*), a regular mono file (*mono no attenuation*), or mono with attenuation (*mono with 3.5 dB attenuation*). This last option prevents your mono mix from clipping and distorting when the two channels (left and right) are combined.

If you took advantage of the multichannel capabilities of DP, you can also select a multichannel option for the mix-down. Select for resolution 16 bit if you eventually want to burn an audio CD or 24 if you want to work further on the stereo file (for premastering and mastering applications). With the "Import" option you can select if you want the file to be imported directly into the sequence (*Add to sequence*) or added only to the Soundbites window (*Add to Soundbites window*) or not imported at all. The "Source" menu specifies the virtual output of DP that will be used to collect the audio material to be bounced, usually the main output to which each audio channel is assigned. You can choose the destination of the files on your HD by clicking on the "Choose" button at the bottom of the window.

Once all options are set, give a name to the bounce and simply click the "OK" button. The time at which DP creates the bounce depends on the number of plug-ins used and on how powerful your computer is. The default file format in which DP bounces is Sound Designer 2. In DP version 4.5 you can also bounce in the following formats: AIFC, AIFF, Next/Sun, Wave, and MP3.

A similar procedure is followed by CSX. To bounce to disk in this sequencer, make sure the Left and Right Locators are set to the boundaries you wish to bounce (you can set them either by punching in the start and end points in the transport window or by clicking and dragging their icons in the ruler at the top of the Arrange window). Make sure all the tracks you want to include in the mix are not muted, and select the "Audio Mix-down" option from the Export submenu found in the File menu. The window that appears allows you to specify several settings that will affect the bounced file. In Table 6.12 I sum up all the different parameters available in CSX for the "Audio Mix-down" operation.

In LP the bounce operation is triggered by pressing the "Bounce" button located on every audio output track in the mix window (Figure 6.18). Clicking on the "Bounce" button will open the "Bounce Dialog" window. This window offers you parameters similar to those we analyzed with CSX. Table 6.13 has a summary of the parameters used by LP during the bounce operation.

Table 6.12 Audio Mix-down options in CSX

Parameter	Description	Comments
Format	You can choose to have the bounce automatically converted to AIFF, MP3 (MPEG1 Level 3), Ogg Vorbis File, SD 2, Wave 64, Broadcast Wave File, or Wave	For a more detailed description of the most common audio file formats, refer to Table 6.15
Channels	Mono creates a mono mix Stereo Split creates a stereo mix formed by two separate files, one for each channel Stereo Interleaved creates a stereo mix where both channels are stored in a single file N. Chan Split is used if you want to export a surround mix as independent split files or create mixes from busses N. Chan Interleaved is the same as the preceding option, but it generates interleaved stereo files	
Resolution	Sets the bit resolution of the mix-down from 8-bit up to 32-bit floating point	If you select a compressed file format such as MP3, the Resolution parameter is substituted by the Quality & Attribute option
Sample Rate	Sets the sample rate of the mix-down from 8 kHz up to 96 kHz	
Outputs	Selects the output bus that will collect the audio bounced to disk	
Real-time Export	If checked, the bounce will be done in real time	Choose this option if you have used plug-ins that require intensive calculations that need to be executed in real time and not at a faster speed
Update Display	If checked, CSX will update the meters on the mix window during bounce	Enable it if you want to make sure that clipping has not occurred during bounce
Import to	Pool: the bounce will be automatically imported in the Audio Pool Track: the bounce will be automatically imported in an audio track	

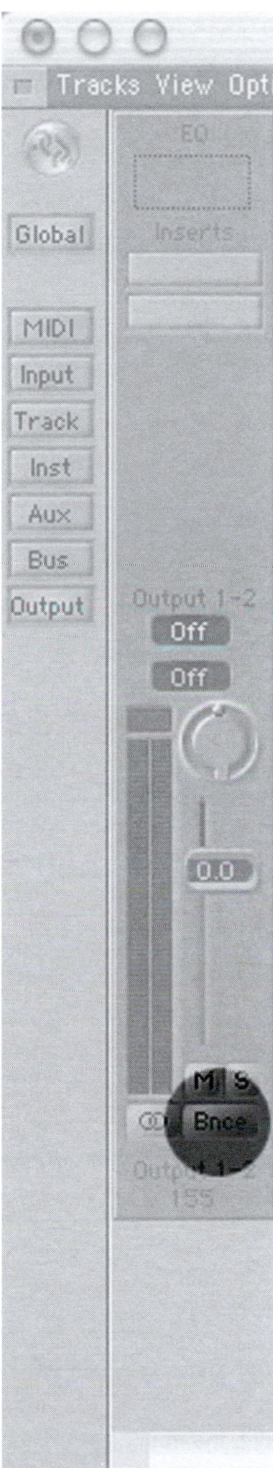

Figure 6.18 The "Bounce" button location in LP.

Table 6.13 The options available in the "Bounce" window in LP

Parameter	Description	Comments
Start/End Position	Selects the range of bars that will be included in the bounce	
File Format	You can choose among four formats: AIFF, WAV, SD 2, and MP3	
Resolution	Sets the bit resolution of the bounce (8, 16, or 24)	If you selected MP3 as file format, the bounce options will change, allowing you to change all the parameters related to MP3 encoding, including Bit Rate, VBR (Variable Bit Rate), Quality, High-Pass filter at 10 Hz and Stereo Mode (Joint or Normal)
Stereo File type	Split creates a stereo mix formed by two separate files, one for each channel Interleaved creates a stereo mix where both channels are stored in a single file	
Surround Bounce	Allows you to bounce all the separate channels of a multichannel mix as individual files	The available formats are: LCR, Center Only, Quadro, ProLogic (with and without center channel), 5.1 (with and without center channel), 7.1 (with and without center channel), and EX (with and without center channel) The settings for all the multichannel file formats available can be set up under the "Audio Preferences" submenu found in the Audio menu
Dithering	Allows you to apply dithering to the bounced file in order to reduce quantization noise when going from a higher bit resolution to a lower one (e.g., files recorded at 24 bit that are bounced at 16 bit)	Dithering should be used only when going from a higher bit resolution to a lower one during the bounce process. There are three different types of dither in LP, based on POW-r (Psychoacoustically Optimized Wordlength Reduction) technology: – POW-r 1: Dithering – POW-r 2: with wide frequency range Noiseshaping – POW-r 3: with 2- to 4-kHz Noiseshaping
Bounce Mode	Real-time: the bounce is executed in real time Offline: it speeds up the bounce	The Offline option is not available for audio outputs that use DSP cards The Offline option automatically mutes the MIDI and the Input Objects during bounce

After setting the parameters for the mix-down, click on the "Bounce" button. If you want to add the bounced file to the current audio file list of the Audio window, click on "Bounce & Add." Dithering is often used when downgrading the bit resolution during the bounce. Using a lower bit resolution introduces audio artifacts that can greatly deteriorate the sound. This is because a low bit rate is responsible for the introduction of higher quantization noise during the conversion, especially for signal at extremely low amplitudes. When the amplitude of a digital signal is very low, the ratio between amplitude and quantization noise is very low and therefore even a small quantization artifact can considerably damage the quality of the sound. Dither can reduce this problem by adding to the signal a very low white noise that is able to both mask and reduce the distortion created by the quantization process. Since the introduction of the noise can be annoying, another filter, called *Noiseshaping*, can be introduced. This filter moves the noise generated by the dithering process toward the extreme ranges of the audible spectrum, reducing the overall perception of the white noise.

In PT, to mix down your project, select the option "Bounce to Disk" from the File menu. The range, in terms of time/bars included in the bounce, depends on the region selected in the Edit window. If nothing is selected, then the entire sequence will be included in the bounce. The options included in the "Bounce to Disk" window are summed up in Table 6.14.

Table 6.14 The "Bounce to Disk" options in PT

Parameter	Description	Comments
Bounce Source	Selects the output that will be used to collect the audio material	It can be set as an audio output or a bus
File Type	Selects the type of file format that will be used for the bounce	The options are: AIFF, SD2, BWF, MPEG-1 Layer 3 (MP3), QuickTime, and Sound Resource
Format	Selects the type of channel format (Stereo/Mono) of the bounce: Mono (Summed): a mono bounce Multiple Mono: a multichannel mix-down with each file stored separately (noninterleaved) Stereo Interleaved: a stereo mix where both channels are stored in a single file	
Resolution	Sets the bit resolution of the bounce (8, 16, or 24)	
Sample Rate	Sets the sample rate of the mix-down from 44.1 kHz up to 192 kHz	You can also choose Pull up/down rates, which are slight variations of the main sample frequencies

(*Continued*)

Table 6.14 (*Continued*)

Parameter	Description	Comments
Conversion Quality	You can choose between: Low, Good, Better, Best, and Tweak Head. The last one gives you the best results but is the slowest	This option is available only if you choose to bounce at a sampling frequency and bit resolution that are different from the one at which the session was recorded
Convert During Bounce	If selected, PT converts the files while they are bounced	This option allows for a faster bounce, but it affects the accuracy of the timing of plug-in automation
Convert After Bounce	If selected, PT converts the files after they are bounced	This option allows for a slower bounce, but it provides the highest accuracy in the timing of plug-in automation
Import After Bounce	Allows you to automatically reimport the bounced files in the current session for further editing and processing	

6.7.1 Audio file formats

As we saw in the previous section, each sequencer can bounce in different file formats, which can be used for different applications and in different situations. In Table 6.15 I list these file formats, their main features, and their uses.

Table 6.15 Major audio file formats supported by DP, LP, PT, and CSX during the bounce operation

Audio file format	Features	Uses and comments
AIFF	*Audio interchange file format* is an uncompressed format used mainly on Macintosh machines but compatible with any major operating system. It is an interleaved format	Mainly to prepare audio files for audio CD burning on Macintosh platforms
WAV	Mostly used on Windows machines, it is compatible also with Macintosh machines. It can be either compressed or uncompressed. The uncompressed version is also referred to as PCM/ uncompressed. The compression formats depend on the CODEC installed with your operating system. It is an interleaved format	In its uncompressed format, mainly to prepare audio files for audio CD burning on PC platforms

(*Continued*)

Table 6.15 (*Continued*)

Audio file format	Features	Uses and comments
Sound Designer II (SD2)	Developed by Digidesign, it is among the most popular audio file formats used by audio professionals. It can be either interleaved or noninterleaved.It can be time-stamped, meaning that an audio file is stamped with the original SMPTE position at which it was recorded. In addition to its original time-stamp location, a so-called "User" location can be added for easier handling during editing	The perfect choice for exchanging audio files among applications since the time-stamping option allows you to set each imported file to its original location
Broadcast Wave Format (BWF)	Has the same features as Wave file but lacks the compression option. In addition it can be time-stamped	
Wave 64	A format developed by Sonic Foundry that is identical to the regular Wave format but capable of handling much larger files (over 2 GB)	Because of its capability of storing large files, can be effectively used to record live concerts or very long sessions
MPEG-1 Layer 3 (MP3)	A compressed format based on lossy-compression algorithms. The data reduction of the algorithm depends on the bit rate chosen. As a general indication, a bit rate of 128 kbit/s reduces the original size of the uncompressed file down to 1/10	Should be used only for quick reference mixes or for mixes that need to be posted on the Internet or sent by e-mail. A bit rate of 128 kbit/s gives a fairly good quality and a fairly small size file. For more accurate compression (and slightly larger files) choose either 160 kbit/s or 192 kbit/s, for smaller files (but lower quality) choose 96 kbit/s or 80 kbit/s. Depending on the software installed on your computer there are other parameters you can control when bouncing/encoding in MP3 format. Here is a list of the most common options: • Variable Bit Rate (VBR): sets the encoder to constantly change bit rate according to the complexity of the audio material being processed • Stereo Mode: if set to "Normal" the encoded file will hold information for each channel in two separate tracks; if set to "Joint Stereo" the file will have one track holding information common to both channels and the other holding information unique to each channel

(*Continued*)

Table 6.15 (*Continued*)

Audio file format	Features	Uses and comments
		• Filter: usually allows you to cut all the frequencies below 10 Hz, reducing the size of the file without compromising its quality
QuickTime	The proprietary format created by Apple for their multimedia applications, it can handle both audio and digital movies. In PT, for example, you can use it to "Bounce to Movie" in order to create a new movie file with the edited audio material from the current session	
Ogg Vorbis	A patent-free compressed audio format similar to MP3 that provides good audio quality with relatively small file sizes	
Sound Resource	Used by the Apple Operating System and by other Macintosh applications. If you plan to use the bounce file for import to Macintosh applications that do not support SD2, you might try this format	Not very common these days in the audio industry
Windows Media	Developed by Microsoft, it is the equivalent of QuickTime for Windows machines. It is included as a default encoding option in PT and CSX for PC but not in the Macintosh version	A compressed format based on a lossy-compression system similar to MP3
Windows Media Pro	A more recent evolution of the Windows Media audio format that allows you to compress the audio files and bounces with a lossless algorithm. In addition it supports multichannel formats	At the moment, available only as default in CSX for Windows
RealAudio	A highly compressed format developed by RealAudio and used mainly for streaming audio and video over the Internet	Available only as default in CSX for Windows and in PT for Macintosh for System 9

6.8 Premastering: introduction

Once your mix is ready, all your tracks are balanced, well placed in space, and equalized, and the final mix has been bounced to disk, your project is almost ready to be delivered. I say "almost" because one of the steps involved in the production process that leads to the final product is premastering. This step is sometimes overlooked by the modern composer because his or her focus is mainly on the composition and orchestration stages.

Before discussing any further what a premastering session involves, I would like to clarify the difference between premastering and mastering. The former is done by the composer/producer right after the mix stage in order to prepare the final mixes for professional mastering or small-scale distribution. The latter is usually done by a professional mastering engineer right before going to the mass production stage, and it involves a series of steps, tools, and techniques specifically targeted at certain audio material. In other words, a premastering session usually involves tools that are directly accessible in a regular project studio, while a mastering session requires more sophisticated tools, studio equipment, and a specially trained engineer.

Nevertheless I always advise producers and composer to premaster their projects, mainly because the sound and quality of their productions will improve greatly, even with a few small premastering touches. You will be amazed by the difference a multiband compressor, a limiter, and a little bit of equalization can do to improve the overall sound of a mix. Of course premastering doesn't pretend to be a substitute for a more detailed and in-depth mastering session. Rather, it is a way to get your material as ready as possible for one. For a detailed analysis of the mastering process I recommend the excellent book by Bob Katz entitled *Mastering Audio: The Art and the Science,* published by Focal Press. Let's take a look at the steps in a premastering session and the tools involved.

6.8.1 The premastering process: to normalize or not?

To premaster the audio material you have bounced to disk, you have to reimport it to your sequencer of choice. You can choose the reimport option right before the bounce operation, or you can create a completely new project and manually reimport the audio files you want to premaster. If you choose the first option, I recommend creating a new arrangement/sequence for the premastering session in order to keep it separated from the original multitrack session. I usually choose this option if I am working with a single song or cue, since I prefer to keep the entire project in a single session file. On the other hand, if I am working with multiple cues, I prefer the second option, which allows me to create a separate project file, only for the premastering session of the entire multicue project.

Once your files are imported to your premastering project, there are a few steps to follow to get your material at the best level of audio quality possible. The first step involves making sure the beginning and end of the stereo mix contain no digital clicks or pops. These artifacts are usually generated if a waveform doesn't start with an amplitude of negative infinity (centerline). Figure 6.19a shows a waveform that would click when played back. To fix this problem you have to fade in the beginning of the file and fade out the end of it to smooth out such glitches. Figure 6.19b shows the same waveform after a fade-in was applied.

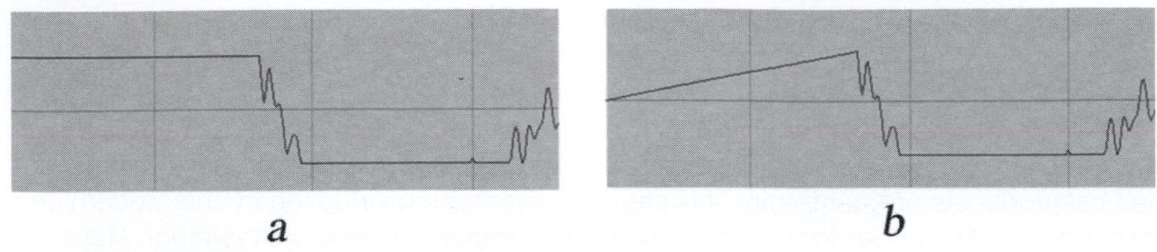

Figure 6.19 (a) A waveform with a digital click at the beginning; (b) the same waveform with a fade-in applied.

To avoid unwanted digital artifacts, always apply a very short fade-in and fade-out at the beginning and end, respectively, of the bounced audio file. Some engineers like to apply "Normalization" to the audio files before the fade-in/fade-out process. This is highly controversial among engineers. Personally I do not particularly like normalizing audio files before the final mastering process; nor, as a matter of fact, do I favor normalization at all. To normalize a file means to bring to a certain level (usually set to the maximum of 0 dB) the highest peak in amplitude of the file without changing its overall dynamic. In other words when you normalize a file you increase its amplitude just up to the level of distortion without altering the dynamic of the file. While this process might seem good since it makes the mix louder by increasing the overall volume of a file, it also increases the noise in the file, thereby, in a sense, limiting its real benefits. In addition, when we alter the original audio file even further, the signal degradation introduced by the normalization process might even be counterproductive because of the extra quantization distortion introduced during the gain-changing process. Even though controversial, the normalization function can be useful in raising the volume for material recorded at a particularly low volume. In Figure 6.20 you can compare the same waveform before normalization and after. Listen to Examples 6.13 and 6.14 to compare the same material before and after normalization. Table 6.16 sums up the procedures for applying normalization for all four sequencers.

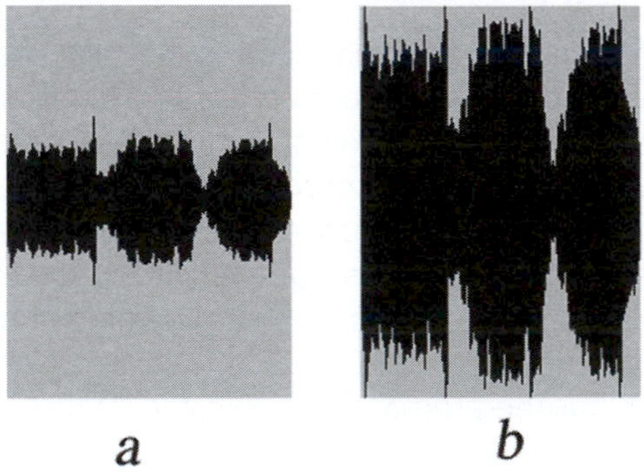

Figure 6.20 A waveform (a) before and (b) after normalization.

6.8.2 Premastering equalization

The next step in premastering is equalization. The type and amount of equalization applied for premastering can vary, depending on the audio material you are processing. Keep in

Table 6.16 The normalization process in DP, CSX, LP, and PT

Sequencer	Procedure	Parameters
DP	Open the Soundbites window and double-click on the audio file you want to normalize. This brings up the Destructive audio editor. Select the area of the audio file you want to normalize (if you want to normalize the entire file, use *command-a* to select all). Then choose the "Normalize" option from the Audio menu	There are no options in DP regarding the normalization process. Normalization is always applied at 100%
CSX	Click on the audio part you want to normalize. Select the "Normalize" function located in the Process submenu of the Audio menu	The Maximum value represents the ceiling at which the highest amplitude of the audio part will be raised. By clicking on the "More" button, you can also set two other parameters: pre- and postcrossfade. These two options allow you to control how the normalization process is applied to the file by mixing over time unprocessed and processed audio (and vice versa)
LP	Open the Sample Audio editor by double-clicking on the audio object you want to normalize. Select the area you want to process (*command-a* to select all) and select the "Normalize" option from the Functions menu	You can change the percentage of normalization under the "Settings" option found in the Functions menu (100% = 0-dB ceiling)
PT	From the Edit window select the region you want to normalize and then select the Normalize option from the AudioSuite menu	You can specify the percentage of normalization (100% = 0-dB ceiling)

mind that equalization is not always needed and so it is really up to you whether to apply it or not. In most cases you will find that a small amount of eq. at this stage can bring to life the mix, and it can also be used to fix smaller mistakes overseen at the mixing stage. At this point you don't want to apply anything extremely drastic; you shouldn't boost or cut more than 2 or 3 dB maximum per band. If you find yourself having to apply more correction, then consider going back and remixing the project.

The main idea here is to apply several small corrections in order to fine-tune the production. You can use a four- or eight-band equalizer to correct small imperfections here and there. Usually you end up boosting the high frequencies (around 12–5 kHz) a little bit to add sparkle and to correct the bass frequencies to avoid muddiness and, at the same time, increase the low end of the mix (you can start by cutting around 250 Hz and boosting a little bit around 100 Hz). I usually like to work in the middle frequencies to avoid the boxed and nasal sonority that is typical of some MIDI productions. For this reason I like to cut frequencies around 1 kHz a bit and boost frequencies between 6 and 7 kHz, paying particular attention not to affect too much the vocal range of the singer, if present. As I mentioned before, each boost or cut should be no higher than 2 or 3 dB. Listen to Examples 6.15 and 6.16 to compare a stereo mix before and after a premastering equalization session.

6.8.3 Multiband compression

When you are satisfied with the equalization, it is time to insert a multiband compressor on the audio channel you are premastering. By applying band-specific compression to the stereo material, you can effectively and precisely rebalance the dynamic relation between the main sections of your mix. As with equalization, I recommend not overcompressing your material, especially since at the final mastering stage the engineer will probably apply more compression.

The same rules learned for single-band compression apply at this stage. Start with mild settings (low ratio and high threshold) and slowly increase the ratio and lower the threshold until you can start hearing the compressor creating unwanted audio artifacts. At that point back up a little by gently raising the threshold slightly. Repeat the same process for each band until you reach a satisfying balance among the bands. The range of each band can vary extremely, depending on the material you are compressing and on the number of bands available on the compressor plug-in you are using. For a three-band compressor I recommend starting with the low range from 20 to 500 Hz, the middle range from 500 to 5000 Hz, and the high range from 5000 to 20,000 Hz. In a four-band compressor (such as the one bundled with LP that we already analyzed in the previous section) you can start with the bands divided in the following manner: 20–200 Hz, 200–1000 Hz, 1000–8000 Hz, and 8000–20,000 Hz. Once again, these are starting point settings to be adapted according to the audio material you are compressing. In general try to isolate the key elements of the music you are compressing, and target your band separation so that you have individual control over these elements. For example, if you are compressing a dance piece where the bass drum is prominent, try to focus on its signal and frequencies in order to achieve a tighter and punchier sound. While PT is not bundled with a multiband compressor (although there are plenty of third-party plug-ins that can be used), LP comes bundled with the "Multipressor," which we discussed earlier in this chapter.

DP features a three-band compressor (MasterWorks compressor) that allows you to control the usual parameters of a compressor for each of the three bands independently (Figure 6.21). In addition you can solo and/or bypass each of the three bands to monitor the compression on each band separately. The solo/bypass buttons for each band are located in the upper left corner of the MasterWorks compressor window. The band ranges can be adjusted by moving the arrows located below the band ranges area in the top left corner of the window or by punching in the values in the parameters fields.

CSX features an impressive five-band compressor called "Multiband Compressor" (Figure 6.22). The band ranges are set by moving, with the mouse, the diamond-shaped icons located to the left and right of each band range at the top of the plug-in window. You can change the gain of each band by clicking and dragging up and down the diamond-shaped icon at the top of each frequency range. To reduce the number of bands, simply drag their

Figure 6.21 The MasterWorks multiband compressor in DP.

boundaries to the extreme left or right of the frequency spectrum. To add another band (remember that you have a maximum of five available), drag one of the diamond-shaped icons from the extreme left or right toward the center of the frequency spectrum.

The highest frequency available in the compressor depends on the sampling frequency on which your project is based. If your session is set at, say, frequency *x*, then the highest frequency available will be *x*/2, according to the Nyquist theorem. Thus if you set your project at a sampling frequency of 44,100 Hz, then the highest range available on the multiband compressor will be 22,050 Hz. For each band you can control the compressor parameters by inserting breakpoints in the in/out line located in the center graphic area of the plug-in. The first point you insert will represent the threshold. If you move points above the in/out line, the effect works as an expander; if you move them below the in/out line, it works as a compressor. This approach allows you to create complex and perfectly tailored settings that can really suit each individual project. To delete a point, simply *shift-click* it.

To edit the parameters of a different band, simply click on the respective band area at the top of the window where the ranges are set. The new band will become highlighted and you can start adding breakpoints on the graphic area in the center of the window. The compressor "Mode" option allows you to choose between "Classic" and "Complex." The former sets the compressor with fixed attack and release; the latter uses an "audio signal adaptive

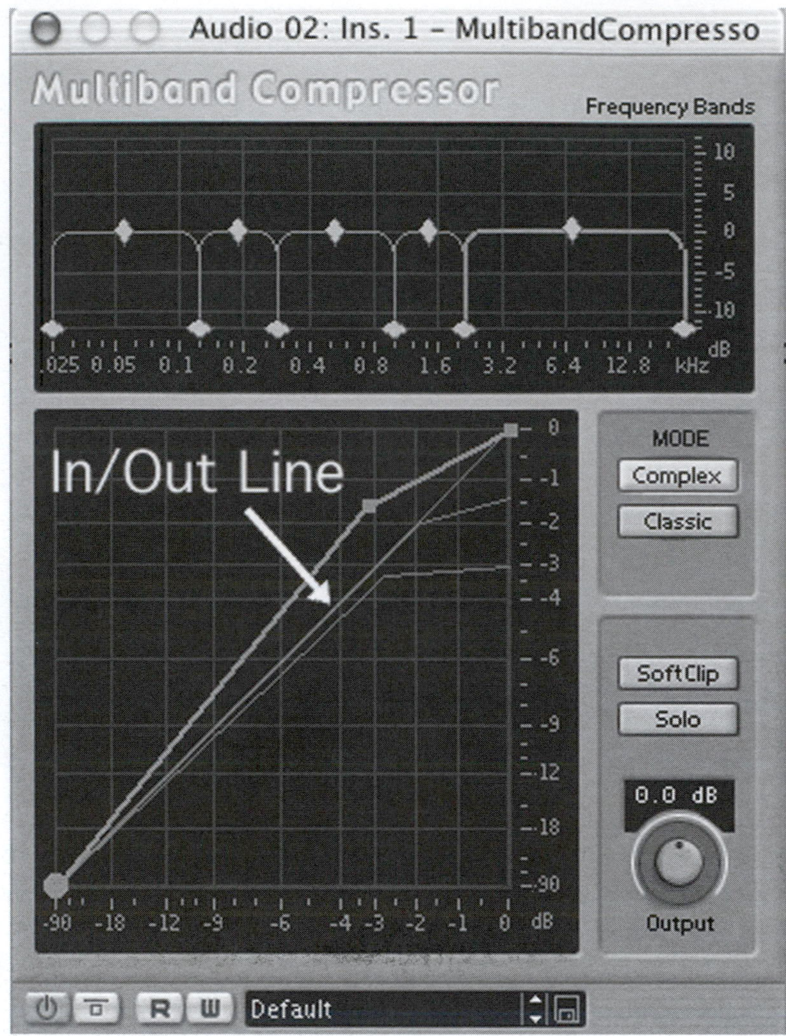

Figure 6.22 The "Multiband Compressor" in CSX.

compression algorithm" that automatically adapts itself to the incoming audio signal. If you enable the Soft Clipping option, then should the output of the compressor go over the 0-dB level, CSX will automatically limit the signal to avoid distortion, and at the same time it will add a gentle, tubelike sonority to achieve a slightly warmer sound. Listen to Examples 6.17 and 6.18 to compare the same audio file with and without the applied multiband compressor.

6.8.4 *The limiter*

The last step of a premastering session involves the limiter. By inserting this effect at the end of the chain we can accomplish two results. First we can make sure there will be no distortion generated in the final bounce, and second we will be able, if necessary, to increase the overall volume of the track without passing the distortion level of 0 dB. As explained earlier, a limiter is basically a compressor with an extremely high ratio. What a limiter accomplishes is to set a ceiling (the threshold) over which the signal cannot go. Every time the signal might go over it, thereby creating distortion, it is "smashed" right to the threshold level. This can be very effective during tracking to avoid distortion for instruments that have a very wide dynamic range (e.g., brass).

An interesting application (especially in a premastering and mastering situation) involves the use of the limiter to increase the overall volume of the material processed and reducing at the same time its dynamic range. Some of the bundled limiter plug-ins that come with your sequencer can be effectively used to increase the volume of your tracks for premastering purposes. Among them the most efficient are the MasterWorks Limiter bundled with DP and the Limiter and Adaptive Limiter (Ad-Limiter) bundled with LP. The MasterWorks Limiter (Figure 6.23) is among my favorites for a premastering session since it easily adapts to any type of audio material and is very easy to set up. The parameters of the MasterWorks Limiter can appear a bit overwhelming at the beginning, but in fact they are fairly simple to set. In Table 6.17 you can find a brief explanation of the parameters and their functions.

To increase the overall volume of the audio material you are processing, try to set a short release, set the ceiling to −0.5 dB, set the lookahead to around 10 ms, and lower the threshold until you start hearing distortion artifacts. This will give you an idea of "how far" you can push the limiter. At that point, raise the threshold again until you are satisfied with the result.

In LP there are two plug-ins that can be effectively used to increase the overall volume of a mix: the Limiter and the Adaptive Limiter (Figure 6.24). While they are both based on the same principle, the former is basically a compressor with infinite ratio, while the latter, in

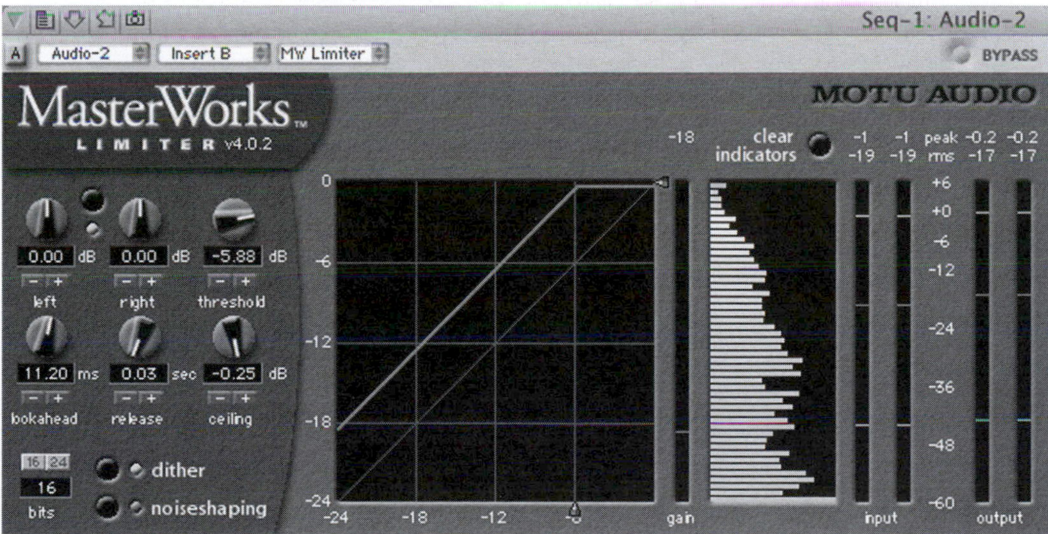

Figure 6.23 The MasterWorks Limiter in DP.

Table 6.17 The parameters of MasterWorks Limiter in DP

Parameter	Function	Comments
Gain (Left and Right)	Controls the input gain for each channel independently	The button located between the Left and Right rotary knobs is the "Link" button. If selected (green), the gains of the two channels will change together, preserving their relative values
Threshold	Sets the point at which the limiter will start operating	In practical terms, if you lower the threshold, the overall volume of the track will raise, since the average level of the audio signal increases and the limiter stops the highest peaks from distorting
Lookahead	Sets the amount of time in milliseconds that the computer uses to look ahead for incoming peaks that need to be limited	
Release	Has the same function as the Release parameter on a compressor. It sets how long the limiter will keep working after the signal goes below the threshold	
Ceiling	Sets the maximum value in decibels (output) at which the limiter will limit the signal	
Bits	Sets the output bit resolution of the material processed	If you downsize the word length of the material (e.g., you go from 24 bit to 16) it is recommended to use the Dither option
Dither and Noiseshaping	Depending on the bit resolution you selected, guarantees better downsizing of the word length of the audio material (Dither introduces a low-level white noise to reduce the quantization noise that occurs at low amplitude with a low bit resolution)	Noiseshaping can be introduced to make less noticeable the noise artificially inserted by the Dithering process
Indicators	The Peak value indicates the highest dynamic peak detected by the limiter. The RMS indicates the average level of the signal	

addition to limiting the audio signal, provides a warmer color tone that tends to resemble the one generated by analog clipping.

The parameters of the Limiter are very simple to set up. The Gain determines the amount of signal increase you want the audio material to acquire. The Lookahead sets the amount

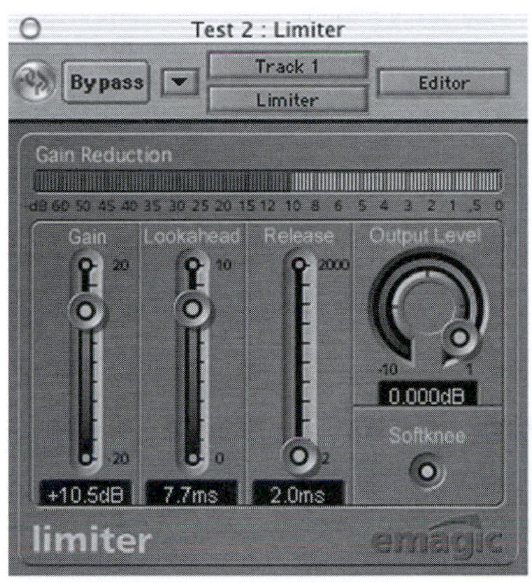

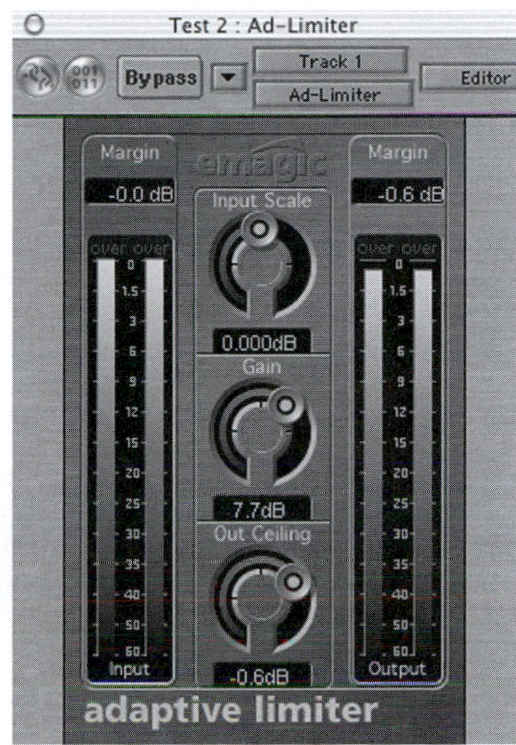

Figure 6.24 The Limiter and Adaptive Limiter in LP.

of time the plug-in foresees before the material is limited. This allows for an increased level of accuracy when high peaks reach the limiter. The Release works, as in a regular compressor, by setting the amount of time it takes to turn the effect off. The Output Level sets the ceiling of the limiter. Usually you should set it to 0 dB or slightly lower (–0.5 dB). The Softknee allows you to obtain a smooth transaction between unlimited and limited signal. If it is disabled, then the transition at the threshold point will be fairly abrupt. The rules stated in the case of the MasterWorks Limiter also apply to the limiter in LP. Set the ceiling, Lookahead, and Release parameters and then start raising the Gain until you are satisfied with both the volume and the overall sonority.

As I mentioned before, the Adaptive Limiter is a very good alternative to LP's regular limiter. It has the ability to color the sound slightly with a more "analog" texture. Its parameters are very simple. The Input Scale works in a similar way to a "recording" volume on a digital board. Set it so that the audio material never goes over the 0-dB ceiling. Then adjust the Gain to increase the overall volume of the audio material according to your taste and to the type of audio material you are working on. The Out Ceiling sets the ceiling of the limiter. I find this particular effect can add personality and warmth to a digital recording, and it can really improve the overall texture of your projects.

The Limiter bundled with PT provides basic settings, and its overall sound quality does not make it my favorite choice for a premastering session. Its settings are basically those of a compressor with a fixed ratio set to 100:1. I find this limiter particularly helpful for light uses on tracks that do not require a hard limiting process. The limiter bundled with CSX also provides fairly basic controls. It is part of the dynamic effects, along with compressor and gate. You can only change the threshold and the release, making it less than an ideal choice for premastering. If you want to increase the overall level of a track without distortion, you can, though, activate both the compressor and the limiter sections of the dynamic plug-in of CSX and use the makeup gain of the compressor section to increase the overall volume of the track. Make sure the limiter is also turned on and set to the minimum release possible, with the threshold set around −0.5 dB.

Once you have applied the limiter and double-checked that all the previous effects inserted on the channel are perfectly balanced with one another, it is time to bounce the premastered file to disk in order to make the changes permanent. The process is exactly the same as described for a multitrack session. Make sure to select the entire part/section you want to bounce and that every parameter of every inserted premastering effect is set as you wish.

In general, do not overuse any of these premastering tools. You want to be able to increase the overall quality of the track without limiting the final mastering engineer in doing his/her job. If you apply too much compression or too much limiting it is going to be impossible for the mastering engineer to take it out. So always remember that less is more in this case. I also recommend bringing to the mastering session two versions of the final mix, one premastered and one without any postmix effects. This will allow full control over the final product in the mastering session.

Listen to Examples 6.19 and 6.20 to compare an audio file before and after a full premastering session.

6.9 Summary and conclusion

Sequencing techniques and MIDI orchestration principles constitute an important part of contemporary music production. As a modern composer/producer, though, it is crucial to strive for high-level production during the entire production process. This is why it is extremely important also to be familiar with the two steps that follow the composition and the sequencing session: mixing and premastering. The mixing stage involves not only the correct balancing of the levels among tracks but also panning, equalizing, and effects handling. Before moving to the mixing stage it is important to decide how to incorporate MIDI and audio tracks together to reach a cohesive sonority. To mix your project you have two options: Mix the MIDI tracks externally from the sequencer using the effects and equalization available through your mixing board along with the audio tracks recorded in your computer, and opt for transferring all the MIDI tracks as audio tracks inside your sequencer and mix everything taking advantage of your computer's effect plug-ins. The second option offers several benefits, such as a higher integration in terms of balance between tracks (since all the tracks will be processed with a similar effect processor and eq.) and a higher flexibility in terms of automation and mixing possibilities. Depending on the bus setup of your mixing board, you can use the main output or the busses to record the MIDI tracks as audio. When recording the MIDI tracks as audio you might end up having a fairly large number of audio tracks to mix. If your sequencer limits you to a maximum number of audio tracks, you might need to create submixes to reduce the overall number of tracks. Refer to Table 6.1 for a list of suggested submixes. Once all the audio tracks are ready for the mixing session, I suggest ordering them according to a standard template that you would apply to every project to speed up the mixing process and be able to locate each track as quickly as possible.

The real mixing process begins with a rough mix that has the main goal of giving you an overall idea of how the sequence should sound. Start from the ground up with the orchestra. Begin with bass drums, then bass, the rest of the drums, followed by the piano and keys, guitars, strings, brass, woodwinds, and then leads (which could include synthesizers, acoustic instruments, or vocals). Keep checking the mix after every addition, and go back to make corrections to the volume of the tracks already mixed every time you feel it's necessary. Panning allows you to accurately place instruments in the stereo image (or in surround if you are doing a surround mix). Try to achieve two main goals when working on the panning: balance between channels and respect of frequency placement. Balance is achieved by correctly placing instruments across the stereo image without favoring the left or right channel. At the end of the mix the two should be equally featured. Frequency placement also has an impact on the way you deal with panning. Usually instruments that feature low frequencies are more naturally placed in the center of the stereo image, while instruments that feature high frequencies can be panned with more extreme settings.

While panning allows you to place instruments left and right, reverberation, along with volume, allows you to place them in a bidimensional setting. By adding more reverb (wetter signal) and by lowering the volume of the dry signal, you can position an instrument on the background level of a mix. If you lower the reverb level and slightly raise the dry volume, you can bring the same instrument closer to the listener and therefore inside the middleground area. A louder and dryer signal will have the yield sound very much in the front of the mix (foreground). Typical parameters of a reverb include: Length, Room Size, Early Reflections, Pre-Delay, Delay, High Filter, and Mix. As a general rule, a reverb time between 1.5 and 2.5 seconds is a good starting point for a wide range of instruments. Instruments that cover mainly the low end of the frequency range, such as bass and bass drums, usually require much less reverberation (or none), while on instruments that cover the mid- and high-frequency range, such as guitars, HH, and even snare drum, you can apply more reverb. All four sequencers come bundled with reverb plug-ins: Reverb A in CSX, D-Verb in PT, eVerb in DP and PlatinumVerb, and Space Designer in LP.

An equalizer allows you to boost or cut the volume of specified frequencies. In the mixing session an equalizer can be used either to correct problems that were created during the recording session or that arise due to incompatibility among instruments or simply in a creative way to produce original effects. Use equalizers as insert and not as aux sends. The several types of equalizer available these days fall into five main categories that have proven to be the most useful in a mixing situation: peak, high shelf, low shelf, high pass, and low pass. One of the most important concepts to keep in mind when equalizing is to be able to emphasize the characteristic frequencies of the track you are working on and eliminate frequencies that do not enhance its sonic features in any way. Try to listen to your mix, understand the problems, and find solutions using the tools available. For generic equalization tips, consult Table 6.9. In general, remember that it is better to cut than to boost when using an equalizer.

Dynamic effects are designed to alter, in one way or another, the dynamic range of an audio signal by controlling over time the ratios between high and low amplitude peaks of an audio track. While used mainly on acoustic tracks, they can also be effective, with parsimony, on MIDI tracks. The dynamic effects include compressor, limiter, expander, and gate. A compressor allows you to reduce the dynamic range of an audio signal. A limiter allows you to set a ceiling that the signal cannot pass, to avoid distortion. An expander is the exact opposite of a compressor: It allows you to increase the dynamic range of an audio signal. A gate is an extreme application of an expander, where the signal is muted after it goes below the set threshold. In general, I recommend using a compressor sparsely and only if really needed. In fact, synthesized tracks in general can benefit from a higher dynamic range and not a reduced one. A limiter can be effectively used to increase the overall volume of a signal without getting distortion. A noise gate can be effectively used to reduce the noise in between "quiet" passages that if note-muted would contribute to the overall noise level of the mix.

Once the mix is ready, it is time to bounce it to disk, meaning to mix down digitally the multitrack session to a stereo (or surround) file. The mix-down that will be created reflects exactly the mix you are hearing from your sequencer, so make sure to check carefully that the mix is exactly how you want it before creating the bounce. Each sequencer handles the bounce-to-disk operation in very similar ways. Common parameters that can be set during the bounce procedure include the file format of the bounce, the output channels used, resolution, sample rate, and the option to automatically reimport the bounce in the current project. The audio file formats in widest use for bounced mixes include AIFF, WAV, SD2, and BWF. Other formats that, though less popular, are part of the options offered by most sequencers are Wave 64, MP3, QuickTime, Ogg Vorbis, Sound Resource, Windows Media Audio format, Windows Media Audio Pro format, and RealAudio.

The premastering session is designed to increase the overall quality of the bounced file before sending it to the final mastering session. Generally a premastering session is conducted by the composer/producer right after the mix stage to prepare the final mixes for professional mastering or for small-scale distribution. A mastering session, on the other hand, is usually done by a professional mastering engineer right before going to the mass production stage, and it involves a series of steps, tools, and techniques that are specifically targeted to a certain audio material. I always advise producers and composers to premaster their projects, mainly because the sound and quality of their projects will improve greatly even with a few small premastering touches.

A premastering session can be conducted inside the main sequence (in this case it is preferable to create a separate arrangement/sequence in order to manage the session in an easier way). Another option is to create a separate project dedicated to the premastering session. Once you have your files imported to your premastering project, make sure the beginning and end of the stereo mix contain no digital clicks or pops. To fix this problem you have to fade in and out the beginning and end of the file. Normalization is sometimes necessary to increase the overall volume of the file without changing its dynamics. A multiband equalizer is then inserted to help bring the mix to life and to fix smaller mistakes overlooked at the mixing stage. Use the eq. sparsely without boosting or cutting more than 2 or 3 dB maximum per band. The main idea here is to apply several small corrections in order to fine-tune the overall sonority of the mix.

After equalization, insert a multiband compressor. By applying a band-specific compression to the stereo material you can effectively and precisely rebalance the dynamic relation between the main sections of your mix. Do not overcompress your material, especially since at the final mastering stage the engineer will probably apply more compression. At the end of the premastering process, use a limiter to avoid distortion and to increase the overall volume of the stereo mix. Try not to overlimit the material before sending it to the final mastering facility. Always bring to the mastering session two versions of the final mix, one premastered and one without any postmix effects. This will allow the mastering engineer maximum flexibility. Once the premastering effects are set, bounce your mix to disk.

6.10 Exercises

Exercise 6.1

- Set up and record a sequence with the following features:
  - Twelve MIDI tracks
  - Two audio tracks with loops of live instruments
  - Free instrumentation and tempo

Exercise 6.2

- Using the sequence created in Exercise 6.1, record each MIDI track as a separate audio track using the bus system of your mixing board (if you don't have a mixing board with a bus system, use the main outputs, as explained in the chapter).

Exercise 6.3

- Mix the previous project using at least two reverbs as aux sends, equalization on each channel, and four dynamic effects. When done, bounce the final mix to disk using at least two different file formats.

Exercise 6.4

- Start a new project and reimport the bounce created in the previous exercise. Following the steps and techniques indicated in this chapter, create a premastered file.

Appendix: List of Examples Contained on the CD-ROM

Chapter	Example		CD track
1	**1.1**	Audio File 44.1 kHz, 16 bit	1
	1.2	Audio File 22.05 kHz, 16 bit	2
	1.3	Audio File 11 kHz, 16 bit	3
	1.4	Audio File 44.1 kHz, 16 bit	4
	1.5	Audio File 44.1 kHz, 10 bit	5
	1.6	Audio File 44.1 kHz, 8 bit	6
2	**2.1**	MIDI drum part nonquantized	7
	2.2	MIDI drum part quantized	8
	2.3	MIDI drum part nonquantized	9
	2.4	MIDI drum part quantized	10
	2.5	MIDI drum part quantized with quantization value of quarter note	11
	2.6	Audio drum loop at 150 BPM	12
	2.7	Audio drum loop adapted to 140 BPM	13
	2.8	Audio drum loop adapted to 160 BPM	14
	2.9	Audio drum loop adapted to 170 BPM	15
	2.10	MIDI string parts sequenced without volume automation	16
	2.11	MIDI string parts sequenced with volume automation	17
	2.12	MIDI oboe part sequenced without volume automation	18
	2.13	MIDI oboe part sequenced with volume automation	19
	2.14	Guitar track without volume automation	20
	2.15	Guitar track with volume automation	21
	2.16	MIDI synth arpeggio without panning automation	22
	2.17	MIDI synth arpeggio with panning automation	23
	2.18	MIDI synth pad without panning automation	24
	2.19	MIDI synth pad with panning automation	25
	2.20	MIDI mallet crescendo on cymbals with panning automation	26
	2.21	MIDI mark three with panning automation	27
	2.22	Harp glissando with panning automation	28
	2.23	Audio guitar track without panning automation	29
	2.24	Audio guitar track with slow panning automation	30
	2.25	MIDI synth pad with random mute automation	31
	2.26	MIDI synth pad with 16th notes mute automation	32
	2.27	Combination of pan, volume and mute automation	33

(*Continued*)

Chapter	Example		CD track
3	**3.1**	MIDI drum part quantized with Sensitivity = 100%	34
	3.2	MIDI drum part quantized with Sensitivity = 35%	35
	3.3	MIDI drum part quantized with Sensitivity = −80%	36
	3.4	MIDI drum part quantized with Strength = 100%	37
	3.5	MIDI drum part quantized with Strength = 70%	38
	3.6	MIDI drum part quantized with Strength = 50%	39
	3.7	MIDI drum part quantized with Strength = 30%	40
	3.8	Straight 8th notes versus swing 8th notes	41
	3.9	MIDI HH part quantized with Swing = 100%	42
	3.10	MIDI HH part quantized with Swing = 120%	43
	3.11	MIDI HH part quantized with Swing = 70%	44
	3.12	MIDI drum part quantized entirely with Swing = 60%	45
	3.13	MIDI drum part quantized with different Swing values	46
	3.14	MIDI drum part quantized without Groove Quantization	47
	3.15	MIDI drum part quantized with Groove Quantization	48
	3.16	MIDI drum part quantized without Groove Quantization	49
	3.17	MIDI drum part quantized with Groove Quantization	50
	3.18	MIDI strings part	51
	3.19	MIDI strings part with addition of acoustic strings	52
	3.20	MIDI drum part with all velocities set to 95	53
	3.21	MIDI drum part with varied velocities	54
	3.22	MIDI drum part with varied velocities and HH 32nd note variations	55
	3.23	MIDI synth arpeggio programmed using the drum editor	56
	3.24	MIDI guitar part sequenced using a keyboard controller	57
	3.25	MIDI guitar part sequenced using a guitar MIDI controller	58
	3.26	MIDI drum part sequenced using a keyboard controller	59
	3.27	MIDI guitar part sequenced using a MIDI pad controller	60
	3.28	Sequence with a steady tempo track	61
	3.29	Sequence with subtle increasing tempo changes	62
	3.30	MIDI piano part sequenced using "Tap Tempo" for tempo changes	63
4	**4.1**	SMPTE signal example	64
5	**5.1**	MIDI guitar part sequenced from keyboard controller, no amplifier	65
	5.2	MIDI guitar part sequenced from MIDI guitar controller, with amplifier	66
	5.3	MIDI drum part	67
	5.4	MIDI drum part with acoustic percussions	68
	5.5	MIDI strings ensemble with synthesized patches only	69
	5.6	MIDI strings ensemble with sampled and synthesized patches layer	70
	5.7	MIDI strings ensemble with sampled and acoustic strings layer	71
	5.8	MIDI woodwinds ensemble without volume automation	72
	5.9	MIDI woodwinds ensemble with volume automation	73

(Continued)

Chapter	Example	CD track
	5.10 Subtractive synthesis example 1	74
	5.11 Subtractive synthesis example 2	
	5.12 Subtractive synthesis example 3	
	5.13 Subtractive synthesis example 4	
	5.14 FM synthesis example 1	75
	5.15 FM synthesis example 2	
	5.16 FM synthesis example 3	
	5.17 FM synthesis example 4	
	5.18 Granular synthesis example 1	76
	5.19 Granular synthesis example 2	
6	**6.1** Snare drum near center	77
	6.2 Snare drum far left	78
	6.3 Snare drum very far center	79
	6.4 Snare drum far right	80
	6.5 Reverb time set at 1 second	81
	6.6 Reverb time set at 3 seconds	82
	6.7 Reverb pre-delay set at 11 ms	83
	6.8 Reverb pre-delay set at 90 ms	84
	6.9 Reverb high cut filter set at 8500 Hz	85
	6.10 Reverb high cut filter set at 2000 Hz	86
	6.11 Reverb diffusion set at 40%	87
	6.12 Reverb diffusion set at 100%	88
	6.13 Audio file not normalized	89
	6.14 Audio file normalized	90
	6.15 Audio file without pre-mastering equalization	91
	6.16 Audio file with pre-mastering equalization	92
	6.17 Audio file without pre-mastering multi-band compressor	93
	6.18 Audio file with pre-mastering multi-band compressor	94
	6.19 Audio file without pre-mastering session	95
	6.20 Audio file with pre-mastering session	96

All tracks composed, programmed, arranged and performed by Andrea Pejrolo. Melissa Bull plays Violin on tracks 52 and 71, Stephanie Fong plays Viola on tracks 52 and 71, Michal Palzewicz plays Cello on tracks 52 and 71.

Special thanks to: Vienna Symphonic Library, FXpansion-BFU, Tascam GigaStudio, Motu, Apple Computer, Steinberg-Pinnacle, DigiDesign, Roland U.S., Anom-Audio.

Index